T0134489

Advances in Intelligent Systems and Computing

Volume 804

Series editor

Janusz Kacprzyk, Polish Academy of Sciences, Warsaw, Poland
e-mail: kacprzyk@ibspan.waw.pl

The series "Advances in Intelligent Systems and Computing" contains publications on theory, applications, and design methods of Intelligent Systems and Intelligent Computing. Virtually all disciplines such as engineering, natural sciences, computer and information science, ICT, economics, business, e-commerce, environment, healthcare, life science are covered. The list of topics spans all the areas of modern intelligent systems and computing such as: computational intelligence, soft computing including neural networks, fuzzy systems, evolutionary computing and the fusion of these paradigms, social intelligence, ambient intelligence, computational neuroscience, artificial life, virtual worlds and society, cognitive science and systems, Perception and Vision, DNA and immune based systems, self-organizing and adaptive systems, e-Learning and teaching, human-centered and human-centric computing, recommender systems, intelligent control, robotics and mechatronics including human-machine teaming, knowledge-based paradigms, learning paradigms, machine ethics, intelligent data analysis, knowledge management, intelligent agents, intelligent decision making and support, intelligent network security, trust management, interactive entertainment, Web intelligence and multimedia.

The publications within "Advances in Intelligent Systems and Computing" are primarily proceedings of important conferences, symposia and congresses. They cover significant recent developments in the field, both of a foundational and applicable character. An important characteristic feature of the series is the short publication time and world-wide distribution. This permits a rapid and broad dissemination of research results.

More information about this series at http://www.springer.com/series/11156

Tania Di Mascio · Pierpaolo Vittorini
Rosella Gennari · Fernando De la Prieta
Sara Rodríguez · Marco Temperini
Ricardo Azambuja Silveira
Elvira Popescu · Loreto Lancia
Editors

Methodologies and Intelligent Systems for Technology Enhanced Learning, 8th International Conference

Editors
Tania Di Mascio
Department of Information Engineering, Computer Science and Mathematics
University of L'Aquila
L'Aquila, Italy

Pierpaolo Vittorini
Department of Life, Health and Environmental Sciences
University of L'Aquila
L'Aquila, Italy

Rosella Gennari
Computer Science Faculty
Free University of Bozen-Bolzano
Bolzano, Italy

Fernando De la Prieta
BISITE Digital Innovation Hub
University of Salamanca
Salamanca, Spain

Sara Rodríguez
BISITE Digital Innovation Hub
University of Salamanca
Salamanca, Spain

Marco Temperini
Dipartimento di Ingegneria Informatica
Sapienza Università di Roma
Rome, Italy

Ricardo Azambuja Silveira
Department of Computer Science and Statistics
Federal University of Santa Catarina
Florianópolis, Brazil

Elvira Popescu
Faculty of Automation Computers and Electronics Computers and Information Technology Department
University of Craiova
Craiova, Romania

Loreto Lancia
Department of Life, Health and Environmental Sciences
University of L'Aquila
Coppito, Holy See (Vatican City State)

ISSN 2194-5357 ISSN 2194-5365 (electronic)
Advances in Intelligent Systems and Computing
ISBN 978-3-319-98871-9 ISBN 978-3-319-98872-6 (eBook)
https://doi.org/10.1007/978-3-319-98872-6

Library of Congress Control Number: 2018954639

This Springer imprint is published by the registered company Springer Nature Switzerland AG
The registered company address is: Gewerbestrasse 11, 6330 Cham, Switzerland

Preface

Education is one of the pillars of our societies, as it shapes many of their social values and characteristics. Knowledge-based societies offer significant opportunities for novel ICT-based solutions in the area of Technology Enhanced Learning (TEL). Intelligent or smart systems, rooted in Artificial Intelligence (AI), have become increasingly relevant for education and a pillar of the TEL field. New smart solutions can be stand-alone or interconnected to others. They target not only cognitive processes but also social, motivational, and emotional factors. They can cater for different users and be personalized for them, e.g., "fragile users," as for example children, elderly people, and people with special needs.

Nowadays, it is crucial noting that learning takes place outside as well as inside of classrooms (ecological environments), and new smart learning ecosystems have developed in diverse contexts. FabLabs, makerspaces, and other fabrication spaces have in particular emerged as novel smart environments for learning, thanks to educators striving to place learners at the center of an experience- and interaction-based educational process, in the tradition of pioneers such as Montessori and Papert, and more recent experiences about the ecological learning contexts.

The 8th edition of this conference expands the topics of the previous editions in order to provide its participants with an open forum for discussing intelligent systems and smart environments for TEL, stand-alone solutions, or interconnected ones, as well as their roots in novel learning theories, methodologies for their design or evaluation, also fostering entrepreneurship and increasing business start-up ideas. The conference intends to bring together researchers, educators, entrepreneurs, and developers from industry to discuss the latest scientific research, technical advances, and methodologies.

This volume presents all papers that were accepted at MIS4TEL 2018. All underwent a peer review selection: Each paper was assessed by three different reviewers, from an international panel composed of about 100 members of 20 countries. The program of MIS4TEL counted 24 contributions in the main track. This edition of MIS4TEL also included two dedicated workshops: the first focusing on *Social and Personal Computing for Web-Supported Learning Communities* with four accepted papers and the second regarding *TEL in nursing education program* with six accepted contributions. All authors come from diverse countries, such as Austria, Brazil, Colombia, Cyprus, Czech Republic, Denmark, Germany, Italy, Mexico, Norway, Portugal, Romania, Serbia, Spain, Switzerland, and USA. The quality of papers was on average good, with an acceptance rate of approximately 70%, and a total number of submissions that consistently increased with respect to the previous editions.

The MIS4TEL series has grown across years in quality and visibility at international level. As we are keen on re-stating every year, that would not have been possible without the interest of MIS4TEL authors in the conference as well as the help of the Program Committee who assisted the editors in the review process for giving constructive feedback to all authors. Therefore, we would like to thank, once more, all the contributing authors, reviewers, and sponsors (IBM, Indra, and IEEE SMC Spain), as well as the Organizing Committee for their hard and highly valuable work. The work of all such people crucially contributed to the success of MIS4TEL 2018 and to shape the future and practice of TEL research.

Tania Di Mascio
Pierpaolo Vittorini
Rosella Gennari
Fernando De la Prieta
Sara Rodríguez
Ricardo Azambuja Silveira
Marco Temperini
Elvira Popescu
Loreto Lancia

Organization of MIS4TEL 2018

http://www.mis4tel-conference.net/

General Chair

Tania Di Mascio	University of L'Aquila, Italy

Technical Chair

Pierpaolo Vittorini	University of L'Aquila, Italy

Paper Co-chairs

Rosella Gennari	Free University of Bozen-Bolzano, Italy
Fernando De la Prieta	University of Salamanca, Spain
Ricardo Azambuja Silveira	Universidade Federal de Santa Catarina, Brazil
Marco Temperini	Sapienza University of Rome, Italy

Proceedings Chairs

Fernando De la Prieta	University of Salamanca, Spain
Rosella Gennari	Free University of Bozen-Bolzano, Italy

Workshop Chair

Elvira Popescu	University of Craiova, Romania

Steering Committee

Pierpaolo Vittorini	University of L'Aquila, Italy
Rosella Gennari	Free University of Bozen-Bolzano, Italy
Tania Di Mascio	University of L'Aquila, Italy
Fernando De la Prieta	University of Salamanca, Spain

Local Organization

Antonio Fernández	University of Castilla-La Mancha, Spain
Elena Navarro	University of Castilla-La Mancha, Spain
Pascual González	University of Castilla-La Mancha, Spain

Program Committee

Vicente Julian	Universitat Politècnica de València, Spain
Jorge Gomez-Sanz	Universidad Complutense de Madrid, Spain
Paulo Novais	University of Minho, Portugal
Florentino Fdez-Riverola	University of Vigo, Spain
Margarida Figueiredo	Universidade de Évora, Portugal
Antonio J. Sierra	University of Seville, Spain
Orazio Miglino	NAC Lab, University of Naples "Federico II" and LARAL, Institute of Cognitive Sciences and Technologies, CNR, Italy
Ana Belén Gil González	University of Salamanca, Spain
Alessandra Melonio	Free University of Bolzano/Bozen, Italy
Pablo Chamoso	University of Salamanca, Spain
Angélica González Arrieta	Universidad de Salamanca, Spain
Javier Bajo	Universidad Politécnica de Madrid, Spain
Besim Mustafa	Edge Hill University, UK
Marcelo Milrad	Linnaeus University, Sweden
Michela Ponticorvo	NAC Lab, University of Naples "Federico II", Italy
Vincenza Cofini	University of L'Aquila, Italy
Mauro Caporuscio	Linnaeus University, Sweden
Victor Sanchez-Anguix	Coventry University, UK
Elvira Popescu	University of Craiova, Romania
Margherita Pasini	Università degli Studi di Verona, Italy
Margherita Brondino	University of Verona, Italy
Jose Neves	University of Minho, Portugal
Ana Almeida	ISEP-IPP, Portugal
Constantino Martins	Knowledge Engineering and Decison Support Research (GECAD)- Institute of Engineering - Polytechnic of Porto, Porto, Portugal

Organization of Workshop on Social and Personal Computing for Web-Supported Learning Communities (SPeL)

The workshop follows the previous SPeL 2008, SPeL 2009, SPeL 2010, SPeL 2011, DULP & SPeL 2012, SPeL 2013, SPeL 2014, SPeL 2015, SPeL 2016, and SPeL 2017 workshops, held in conjunction with the SAINT 2008 conference, WI/IAT 2009 conference, DEXA 2010 conference, ICWL 2011 conference, ICALT 2012 conference, ICSTCC 2013 conference, ICWL 2014 conference, ICSLE 2015 conference, ICWL & SETE 2016 conferences, and ICWL & SETE 2017 conferences, respectively. The general topic of the workshop is the social and personal computing for Web-supported learning communities.

Web-based learning is moving from centralized, institution-based systems to a decentralized and informal creation and sharing of knowledge. Social software (e.g., blogs, Wikis, social bookmarking systems, media sharing services) is increasingly being used for e-learning purposes, helping to create novel learning experiences and knowledge. In the world of the pervasive Internet, learners are also evolving: The so-called digital natives want to be in constant communication with their peers; they expect an individualized instruction and a personalized learning environment, which automatically adapt to their individual needs. The challenge in this context is to provide intelligent and adaptive support for collaborative learning, taking into consideration the individual differences between learners.

This workshop deals with current research on the interplay between collaboration and personalization issues for supporting intelligent learning environments. Its aim is to provide a forum for discussing new trends and initiatives in this area, including research about the planning, development, application, and evaluation of intelligent learning environments, where people can learn together in a personalized way through social interaction with other learners.

The workshop is targeted at academic researchers, developers, educationists, and practitioners interested in innovative uses of social media and adaptation techniques for the advancement of intelligent learning environments. The proposed field is interdisciplinary and very dynamic, taking into account the recent advent of Web 2.0 and ubiquitous personalization, and it is hoped to attract a large audience.

Organizing Committee

Elvira Popescu	University of Craiova, Romania
Sabine Graf	Athabasca University, Canada

Program Committee

Marie-Hélène Abel	Université de Technologie de Compiègne, France
Yacine Atif	Skövde University, Sweden
Silvia Margarita Baldiris Navarro	Universitat de Girona, Spain
Tharrenos Bratitsis	University of Western Macedonia, Greece
Ting-Wen Chang	Beijing Normal University, China
Maria-Iuliana Dascalu	Politehnica University of Bucharest, Romania
Mihai Dascalu	Politehnica University of Bucharest, Romania
Stavros Demetriadis	Aristotle University of Thessaloniki, Greece
Giuliana Dettori	Institute for Educational Technology (ITD-CNR), Italy
Gabriela Grosseck	West University of Timisoara, Romania
Hazra Imran	Athabasca University, Canada
Malinka Ivanova	TU Sofia, Bulgaria
Mirjana Ivanovic	University of Novi Sad, Serbia
Jelena Jovanovic	University of Belgrade, Serbia
Ioannis Kazanidis	Technological Educational Institute of Kavala, Greece
Milos Kravcik	German Research Center for Artificial Intelligence, Germany
Zuzana Kubincova	Comenius University Bratislava, Slovakia
Amruth Kumar	Ramapo College of New Jersey, USA
Frederick Li	University of Durham, UK
Anna Mavroudi	Norwegian University of Science and Technology, Norway
Wolfgang Mueller	University of Education Weingarten, Germany
Kyparisia Papanikolaou	School of Pedagogical and Technological Education, Greece
Alexandros Paramythis	Contexity AG, Switzerland
Ricardo Queirós	Polytechnic Institute of Porto, Portugal
Demetrios Sampson	Curtin University, Australia

Olga Santos	Spanish National University for Distance Education, Spain
Marco Temperini	Sapienza University of Rome, Italy
Stefan Trausan-Matu	Politehnica University of Bucharest, Romania
Rémi Venant	University of Toulouse, France
Riina Vuorikari	Institute for Prospective Technological Studies (IPTS), European Commission

Organization of Workshop on TEL in Nursing Education

In the field of nursing, learning outcomes involve nurses both as learners and as educators.

As learners, they are involved in basic and post-basic academic programs, whereas they act as educators when they are engaged in health educational programs aiming to enhance community health-literacy levels.

According to some evidence, the quality of learning outcomes in basic and post-basic nursing academic programs could be potentially improved by technology-based systems like simulation and blended learning models. However, little is known about the use of technology to enhance community health-literacy levels.

This workshop aims to share the best available knowledge about the application of technology-based systems into basic and post-basic nursing academic programs and into health educational programs aiming to enhance community health-literacy levels.

In order to pursue this intent, workshop topics have been grouped into the following three main discussion aims.

First, topics on education in nursing academic programs aim to discuss the effects of simulation and other technology-based systems on learning quality, including ethical and legal aspects.

Secondly, topics on community health educational programs aim to discuss the impact of technology in improving community health-literacy levels.

Finally, the workshop intends to provide a complete overview of technology-based methods as useful tools to improve the learning of the nursing process in clinical settings.

Organizing Committee

Loreto Lancia	University of L'Aquila, Italy
Rosaria Alvaro	University of Tor Vergata, Italy

Program Committee

Antonello Cocchieri	University Policlinico Agostino Gemelli, Italy
Fabio D'Agostino	University Tor Vergata, Italy
Angelo Dante	University of L'Aquila, Italy
Cristina Petrucci	University of L'Aquila, Italy
Gianluca Pucciarelli	University Tor Vergata, Italy
Ercole Vellone	University Tor Vergata, Italy
Pierpaolo Vittorini	University of L'Aquila, Italy

Contents

Main Track

Blending Classroom, Collaborative, and Individual Learning Using Backstage 2

Sebastian Mader(✉) and François Bry

Institute for Informatics, Ludwig-Maximilian University of Munich, Munich, Germany
sebastian.mader@ifi.lmu.de

Abstract. Seminars are difficult and therefore often neglected classes in STEM education even though they greatly contribute to the students' scientific maturity. Seminars are a traditional educational format blending classroom, collaborative, and individual learning: Seminar participants are tasked to discover, understand, and convey a scientific or technical issue to the other seminar attendees in an essay and in an oral presentation engaging them into a fruitful discussion. Seminars are rightly considered a cornerstone of STEM education, yet they are often frustrating experiences for both learners and teachers due to insufficient supervision and practice. This article reports on using Backstage 2, a web platform that, by offering a virtual space and tools for a fruitful communication, bridges classroom, collaborative, and individual learning activities. The contribution of this article is threefold: First, a class format aimed at boosting collaboration in seminars, second, technological tools supporting collaboration among seminar attendees, and third, an evaluation of the approach demonstrating its effectiveness.

Keywords: Online learning environments · Blended learning
Peer review · Computer-supported collaborative learning

1 Introduction

Seminars are essential components of the training in Science, Technology, Engineering, and Mathematics (STEM). During STEM seminars, students learn to discover, understand, and present scientific or technical issues beyond what is taught in other classes. A STEM seminar typically requires from an attendee the following: Selecting relevant literature on an issue, reading and understanding it, summarizing the essential aspects of the issue in a written overview, presenting the overview in a talk, and finally answering questions posed by seminar attendees and the teacher. Thus, seminars are a traditional educational format blending classroom, collaborative, and individual learning long before "blended learning" became fashionable. Collaboration is essential among other because answers can only be given if questions are posed.

T. Di Mascio et al. (Eds.): MIS4TEL 2018, AISC 804, pp. 3–11, 2019.
https://doi.org/10.1007/978-3-319-98872-6_1

Ideally and traditionally, the work for a seminar is both, closely supervised by teachers and collaborative in the sense that seminars attendees assist each other in most of the afore-mentioned phases. In practice, however, STEM teachers hardly have the time to sufficiently supervise the attendees of their seminars and collaboration among seminar attendees is very limited if existing at all. The reasons for this unsatisfactory state of affairs are large numbers of seminar attendees, lack of locations where seminar attendees can work together on their own initiative, and disinterest among students and teachers alike for a type of class often considered to be "worth only a few credits".

STEM seminars also face a considerable educational obstacle: STEM education does not sufficiently prepare for the self-regulated learning seminars require. Since the concepts and techniques conveyed in most STEM classes are highly abstract and rather complex, STEM education favours lectures and tutorials giving little opportunity for self-regulated work.

STEM seminars nowadays face an additional obstacle: Even though they are focused at training communication skills, they still are mostly run without recent, if at all, communication technology. As a consequence, seminars are often perceived by STEM students as obligatory steps in their courses imposed by tradition and without much relation to real life and their future careers.

This state of affairs and the conviction that well-run seminars can greatly contribute to a good training of STEM students led us first to reconsider the formats of the seminars we give, second to reflect on what communication tools could help restore a good seminar practice among students, third conceive, implement, and deploy a tool, Backstage 2[1], and fourth to evaluate Backstage's impact on seminar attendees. This article reports on that endeavour and on the findings of an evaluation demonstrating the effectiveness of the technology-enhanced novel seminar format.

This article is structured as follows. Section 1 is this introduction. Section 2 is devoted to related work. Section 3 presents a teaching format for seminars fostering collaboration. Section 4 presents those functions of the platform Backstage 2 supporting the proposed teaching format for seminars. Section 5 describes an evaluation of the technology-enhanced seminar format and its findings. Section 6 concludes the article and gives perspectives for future work.

2 Related Work

The technology-enabled blending of classroom, collaborative, and individual learning in STEM seminars reported about in this article is a contribution to blended learning and relates to audience response systems, backchannels, collaborative annotation systems, and peer review in learning.

Blended Learning. Even though dated, the book [2] still is a good introduction to blended learning. Among other, the book discusses different definitions of blended learning with their main differentiator being what is being blended – for

[1] https://backstage.pms.ifi.lmu.de:8080.

this article, both classroom learning and individual learning as well as different instructional methods are blended. The research literature on blended learning is too vast for being reviewed here. The article [9] is a recent overview of that research. The following publications have inspired or are related to the present article: The articles [8,13,15,18] hint at the potential of various forms of blended learning and the articles [1,16] address the design of blended learning educational formats.

Backchannels and Audience Response Systems. A Backchannel is "a secondary or background complement to an existing frontchannel" [22, p. 852]. Backchannels can be deployed during lectures, allowing students to "exchange questions, comments and thoughts on the subject matter synchronously to the lecturer's presentation" [14, p. 6]. Most backchannels designed for lectures [3,7] provide audience response systems, which allow students to answer lecturers' questions during lectures and give immediate feedback about the classes' performance. The article [11] provides an overview of the research on ARSs.

Collaborative Annotation Systems. Using a collaborative annotation system (CAS), groups can collaborate on annotations for documents and their editing. The CAS PAMS 2.0 is discussed in [17]. That article stresses the positive response of students to the CAS and the positive impact on students' learning. The CAS MyNote is discussed in [4] where a positive response of its users is mentioned. CAS are enablers of peer review in learning.

Peer Review in Learning. Peer Review and peer assessment refer to the enrolment of students for providing feedback, in place of or in addition to the teacher's feedback, to their fellow students, their peers. A positive impact of peer review on both reviewers and reviewees has been demonstrated in various studies: Peer review has been for example shown to significantly improve the peer reviewers' own writing abilities [12]; an effect this article attributes to the reflection about one's own work triggered by doing peer review. This hypothesis is also made in [19] where a further explanation is hypothesised: The better learning of peer reviewers might result from the increased time they have to spend on the issue they learn.

The article [10] reports on students positively appreciating peer review because it makes them discover alternative solutions or answers and because they get more feedback and negatively because of their difficulties in discriminating between good and bad solutions or answers. According to the article [21], students see three main benefits in peer reviews: the chances "to compare different approaches", to "compare standard of work", and the "exchange of information and ideas" [21, p. 52].

3 A Novel STEM Seminar Format, Its Promotion, and Its Limitations

A participant in a STEM seminar is expected to select relevant literature on an issue, to read and understand it, to write an overview of the issue's core aspects, to present that overview in a talk, and to answer the subsequent questions of the seminar attendees and of the teacher. In practice, however, many STEM seminars are far from this. Instead, they often consist of insufficiently prepared and poorly written overviews, talks hard to follow with low attendance. Collaboration among seminar attendees is minimal, if at all existing, and often limited to half-hearted questions that are given half-hearted answers. This often leads teachers to step in and to lecture on the presentation's issue, what in turn reduces the students' participation.

The didactic benefits of STEM seminars, as they are often run, are questionable. This unsatisfying state of affairs led us to reflect on how to revive the seminars we run for bachelor students in informatics.

First, peer review was introduced: In addition to the afore-mentioned activities, each participant has to deliver written reviews of the written overviews of two other seminar participants. A schedule is devised with deadlines for delivering a first draft of an overview, for delivering reviews, for discussing overviews and peer reviews in classroom sessions, and for finalizing overviews taking into account the reviewers' feedback. Thus, the novel seminar format blends synchronous classroom sessions and asynchronous homework. The grading of seminar attendees' work is based not only on their written overviews and talks but also on their reviews, what makes peer review an integral component of the work expected from seminar participants.

Second, we chose to promote the seminar format described above by introducing it with a subject that first, had never been offered before, second, makes peer review necessary, and third, appeals to a large number of students. Indeed, students are unlikely to accept additional work if they do not see the point of it. We chose job applications as subject, an issue of high interest to most students, often subject to different viewpoints and therefore good at triggering reflections and debates. Indeed, there is more than one way to write a good job application and "[p]eer assessment may prove especially valuable in cases in which structured and formal education is neither preferred nor even available" [5, p. 89].

In a Job Application Seminar, the literature to select is about writing applications (letters and curricula vitae) stressing the usages in different countries, the overviews are replaced by applications, and the talks by simulated job interviews. The peer review is extended to the interviews, that is, every seminar attendee has to play two roles at different times, that of a job applicant and that of a member in the selection committee.

The Job Application Seminar, in German "Bachelorseminar Bewerbungen", has been offered once a year since 2014. Since its first realisation, it has been very well received by students. At each of its realisations, all students engaged heavily in researching the issue and in discussing it with fellow attendees both outside and in the classroom. The role playing, both as applicant and as member in a selection committee, has always been popular among the seminar attendees who all performed it earnestly.

In spite of its success, the students' participation in the Job Application Seminar was not fully satisfying and the organisation load it imposed on teachers was high. Students read only those applications they had to review and the reviews were very heterogeneous what made it difficult to work with them. The timely collecting and dispatching of the written material of the seminar attendees turned out to be a time-consuming and dull task for the teachers. We realized that technology could overcome both problems what led us to reflect on, and develop an appropriate technological support.

4 Towards Overcoming the Seminar Format's Limitations

In the seminar realisation of the summer term of 2017, we provided the seminar participants and the teachers a tool, Backstage 2, for collaborative content collection and creation to be used in collecting and discussing literature and for peer review. The relevant functionalities of Backstage 2, an education software developed by the authors built upon the foundation of the first version of Backstage [14], are briefly presented in the following.

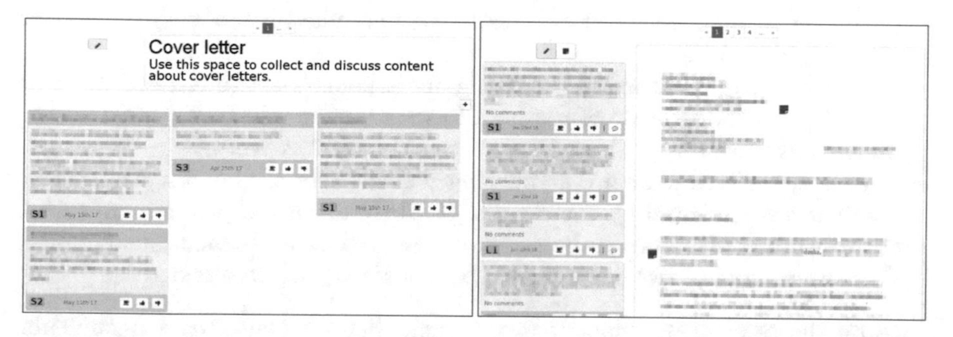

Fig. 1. Left: vertical stream of a unit (top) with four student contributions (bottom). Right: annotations (left) created for a peer review of an application page (right). The text was enlarged and translated from german

With Backstage 2, the documents of a course (like a seminar or a lecture) are stored in a folder called "course" that contains one or several "units". A unit consists of one or more pages of arbitrary media type. For the seminar, pages for the different parts of an application, such as *cover letter*, were created. Each set of pages is called a *horizontal stream.*

Every page can be annotated by every seminar participant as follows: A region is selected on the page and a textual comment referring to that region is created. Annotations can be seen by every seminar participant who can rate and comment it – see Fig. 1 for an example.

Every page has a *vertical stream* located below the page using which participants can attach additional documents referring to that page – see Fig. 1. Vertical stream documents can be seen and annotated by all attendees.

A couple of features of Backstage 2 are worth stressing. First, annotations are given a context in the form of the region of a page they refer. Second, the structure of a "course" including those of its "pages" that are given a vertical stream are specified by the teacher. These specifications act as scripts [6,20] and care for homogeneous students' contributions what, in turn, helps their sharing. Collaborative content collection is enabled by the vertical stream, and peer review is done using both the vertical stream and annotations.

5 Evaluation and Findings

During the summer term of 2017, the Job Application Seminar has been held in its novel form using Backstage 2. 20 participants have attended the seminar. During the peer review, 317 annotations (average: 15.85) and 26 vertical stream documents (average: 1.3) were created. During the collaborative content collection, a total of 12 documents (average: 0.6) were created.

Method. A survey consisting of four parts was conducted during the final session of the seminar:

- Part 1 was a self-assessment of the student's activity on Backstage consisting of yes/no-questions.
- Part 2 was a questionnaire measuring the student's attitude towards using Backstage in the seminar consisting of five statements to be rated on a four-point Likert scale from *strongly agree* to *strongly disagree.*
- Part 3 was a questionnaire estimating which parts of the Job Application Seminar are perceived by the student as likely to have a positive impact on their future applications to be rated on the same scale as mentioned above.
- Part 4 consisted of two questions to be answered with free text.

Data on the user-created contributions, including vertical stream documents, annotations, and comments, was collected from Backstage.

Results. 17 (out of 20) participants attended the final session each of whom participated in the survey; 12 were male, 5 female. The answers of 3 participants, except for their free text answers, were discarded because they included contradictory answers. Part 1 shows that the majority of the participants used all of the functionality offered by Backstage. A detailed discussion of these results is out of the scope of this article.

The left image in Fig. 2 shows the results for Part 2; the right image the results for Part 3. Part 2 clearly shows a positive attitude towards to usage of Backstage 2. Part 3 shows that all parts of the course were perceived as positive even though at varying degrees.

All participants gave free text answers in Part 4 yielding 67 statements. A content analysis and categorization of these statements was performed by three human judges resulting in a Fleiss' kappa of 0.848. Most liked in the seminar and its use of Backstage were: access to other attendees' applications (6 statements), peer review (4 statements), and user-friendliness and usability (7 statements). Negatively perceived were: information overload and a lack of overview for annotations (4 statements).

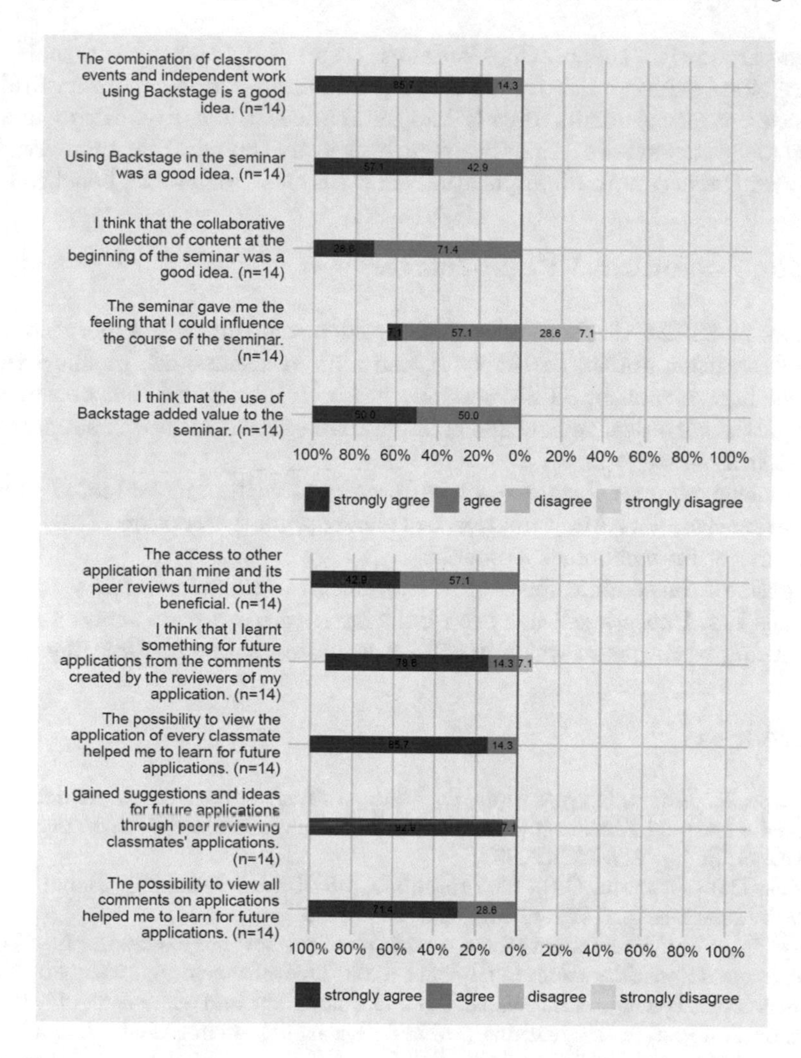

Fig. 2. Results of the Likert questionnaires on *Attitude towards the use of Backstage in the course* (top) and *Perceived value of parts of the course design* (bottom)

Discussion. The results indicate that Backstage 2 well supports the novel seminar format. The answers of Fig. 2 suggest that Backstage 2 had a significant impact on the acceptance of the novel seminar format – and of its increased workload for students. An evaluation comparing the reception of the novel seminar format with and without the technological support of Backstage is none the less outstanding and will be performed in the forthcoming months.

Only 6 students participated actively in collaboratively collecting 12 documents. Since the majority of the participants expressed positive appreciations of collaborative content collection, most of the students probably were passive consumers of their peers' submissions.

Compared with the previous summer term, the teachers felt that classroom sessions gained much from Backstage 2: Discussion felt more natural and advanced because students already had familiarized themselves with the applications and peer reviews. An assessment backed by the fact that many students mentioned the access to all applications and the peer reviews as beneficial.

6 Conclusion and Perspectives

Seminars in STEM teaching often face problems: insufficient supervision, limited collaboration among participants, and limited discussions. In this article, a novel seminar format for STEM teaching has been introduced that boosts active participation with peer review and collaborative content collection supported by the system Backstage 2.

The novel seminar format has been promoted with, and evaluated in a Job Application Seminar. An evaluation has shown a good reception of the seminar format and of its technological support.

Aspects of the seminar format are relevant for other courses like lectures and practicals too. Backstage 2 has been built for covering a wide array of teaching and learning scenarios in which it will be evaluated in the forthcoming months.

References

1. Alonso, F., López, G., Manrique, D., Viñes, J.M.: An instructional model for web-based e-learning education with a blended learning process approach. Br. J. Educ. Technol. **36**(2), 217–235 (2005)
2. Bonk, C.J., Graham, C.R.: The Handbook of Blended Learning: Global Perspectives, Local Designs. Wiley, San Francisco (2006)
3. Bry, F., Pohl, A.: Backstage: a social medium for large classes. In: Campus Transformation-Education, Qualification and Digitalization, pp. 255–280 (2014)
4. Chen, Y.C., Hwang, R.H., Wang, C.Y.: Development and evaluation of a Web 2.0 annotation system as a learning tool in an e-learning environment. Comput. Educ. **58**(4), 1094–1105 (2012)
5. Dunning, D., Heath, C., Suls, J.M.: Flawed self-assessment: implications for health, education, and the workplace. Psychol. Sci. Public Interest **5**(3), 69–106 (2004)
6. Fischer, F., Bruhn, J., Gräsel, C., Mandl, H.: Fostering collaborative knowledge construction with visualization tool. Learn. Instr. **12**(2), 213–232 (2002)
7. Flint, J.: Feedback-proxys zur digitalisierung von classroom response systemen. Doctoral thesis, Faculty of Computer Science and Electrical Engineering, University of Rostock, Germany (2017)
8. Garrison, D.R., Kanuka, H.: Blended learning: uncovering its transformative potential in higher education. Internet High. Educ. **7**(2), 95–105 (2004)
9. Halverson, L.R., Spring, K.J., Huyett, S., Henrie, C.R., Graham, C.R.: Blended learning research in higher education and k-12 settings. In: Learning, Design, and Technology, pp. 1–30 (2017)
10. Hanrahan, S.J., Isaacs, G.: Assessing self- and peer-assessment: the students' views. High. Educ. Res. Dev. **20**(1), 53–70 (2001)

11. Kay, R.H., LeSage, A.: Examining the benefits and challenges of using audience response systems: a review of the literature. Comput. Educ. **53**(3), 819–827 (2009)
12. Lundstrom, K., Baker, W.: To give is better than to receive: the benefits of peer review to the reviewer's own writing. J. Second Lang. Writ. **18**(1), 30–43 (2009)
13. McGee, P., Reis, A.: Blended course design: a synthesis of best practices. J. Asynchronous Learn. Netw. **16**(4), 7–22 (2012)
14. Pohl, A.: Fostering awareness and collaboration in large-class lectures. Doctoral thesis, Institute of Iformatics, Ludwig-Maximilian University of Munich, Germany (2015)
15. Sheaa, P., Bidjeranob, T.: Learning presence: towards a theory of self-efficacy, self-regulation, and the development of a communities of inquiry in online and blended learning environments. Comput. Educ. **55**(4), 1721–1731 (2010)
16. Singh, H.: Building effective blended learning programs. Educ. Technol. **43**(6), 51–54 (2003)
17. Su, A.Y., Yang, S.J., Hwang, W.Y., Zhang, J.: A Web 2.0-based collaborative annotation system for enhancing knowledge sharing in collaborative learning environments. Comput. Educ. **55**(2), 752–766 (2010)
18. Sun, Z., Liu, R., Luo, L., Wu, M., Shi, C.: Exploring collaborative learning effect in blended learning environments. J. Comput. Assist. Learn. **33**(6), 575–587 (2017)
19. Topping, K.: Peer assessment between students in colleges and universities. Rev. Educ. Res. **68**(3), 249–276 (1998)
20. Weinberger, A., Ertl, B., Fischer, F., Mandl, H.: Epistemic and social scripts in computer-supported collaborative learning. Instr. Sci. **33**(1), 1–30 (2005)
21. Williams, E.: Student attitudes towards approaches to learning and assessment. Assessm. Eval. High. Educ. **17**(1), 45–58 (1992)
22. Yardi, S.: The role of the backchannel in collaborative learning environments. In: Proceedings of the 7th International Conference on Learning Sciences, pp. 852–858. International Society of the Learning Sciences (2006)

Predicting Learners' Behaviours to Get It Wrong

Niels Heller(✉) and François Bry

Institute for Informatics, Ludwig-Maximilian University of Munich, Munich, Germany
niels.heller@ifi.lmu.de

Abstract. One of the most vexing aspects of tertiary education is the learning behaviour of many beginners: Late drop-outs after much time has already been invested in attending a course, incomplete homework even though completed homework is a sufficient condition for success at examinations, and misconceptions that are not overcome early enough. This article presents three predictors related to these learning-impairing behaviours that have been built from data collected with a learning platform and by examining homework assignments, and developed as Hidden Markov Model, by relying on Collaborative Filtering, and by using Multiple Linear Regression. The sensitivities and specificities of the first two predictors are above 70% and the R^2-error of the third predictor is about 20%. Considering the large numbers of unknown parameters like course-independent learning, this quality is satisfying. The predictors have been developed for fostering a better learning by raising the learners' consciousness of the deficiencies of their learning. In other words, the predictors aim at "getting it wrong". The article reports on the predictors and their evaluation.

Keywords: Learning analytics · Behaviour prediction
User modelling and adaptation in TEL · Evidence-based studies

1 Introduction

One of the most vexing aspects of tertiary education, especially in Science, Technology, Engineering, and Mathematics (STEM), is the learning behaviour of many beginners: Late drop-outs after much time has already been invested in attending a course, incomplete homework even though completing homework is known to be a sufficient condition for success at STEM examinations, and misconceptions or common fallacies related to a misunderstanding of mathematical and other abstractions that are not overcome early enough.

Late drop-outs, incomplete homework and misconceptions are arguably among the main reasons for failing at examinations. Reasons for these symptoms include the difficulty for beginners to adjust to a teaching style significantly different from that of secondary schools and the very limited individual feedback

T. Di Mascio et al. (Eds.): MIS4TEL 2018, AISC 804, pp. 12–19, 2019.
https://doi.org/10.1007/978-3-319-98872-6_2

from experienced teachers resulting from high numbers of students per teacher (commonly 70 or more in computer science at German universities).

The research reported about in this article aims at compensating for the insufficient human coaching of STEM students for their better learning with algorithmically-generated individual feedback. Using predictors referring to the afore-mentioned three learning obstacles –late drop-outs, incomplete homework, and misconceptions– learners are made conscious of currently sub-optimal aspects of their learning and are motivated to improve it. This article reports on the first stage of this endeavour, the development and evaluation of three predictors of the afore-mentioned three learning obstacles. More precisely, the three predictors respectively forecast:

1. Skipping, that is, not fully participating or not participating at all in learning activities such as a weekly homework or a lecture
2. Examination fitness measured as a mark, that is, a percentage of the total score obtainable at an examination
3. Misconceptions related to a course's content

All three predictors have been developed using data on learners' behaviours collected from computer science courses given at Ludwig-Maximilian University of Munich. Some data has been collected through the teaching and learning platform Backstage [3, 15], other by examining homework assignments and examinations. The Skipping Predictor is a Hidden Markov Model [17], the Examination Fitness Predictor is a Multiple Linear Regression Model [4, chapter 6] and the Misconception Predictor relies on Collaborative Filtering [12]. The predictors' qualities are as follows:

1. Skipping is predicted with a sensitivity of 72.9% and a specificity of 84.7%.
2. Examination fitness is predicted with a R^2-error of 20.0% in one dataset and 19.8% in another.
3. Misconceptions are predicted with a sensitivity of 71.8% and a specificity of 80.7%

This article consists of 7 sections. Section 1 is this introduction. Section 2 presents related work. Each of Sects. 3, 4, and 5 describes one of the three predictors, reports on its quality, and on the individual feedback it gives to learners. Section 6 reports on perspectives for future work. Section 7 is the conclusion.

2 Related Work

This article is a contribution to "learning analytics", that is, "the measurement, collection, analysis and reporting of data about learners and their contexts, for purposes of understanding and optimizing learning and the environments in which it occurs" [20]. Most learning analytics refer to Massive Open Online Courses (MOOCs). In contrast, the learning analytics presented in this article refer to formal education and presence courses.

Drop-out Prediction. There is no widely accepted definition of drop-out. Some authors define it as a discontinued participation in a formal course of study [13]. Other authors define it in terms of periods of inactivity during a course that can span from several weeks [11] to several years [21]. The Skipping Predictor presented below in Sect. 3 refers to skipping defined as not taking part in the next learning activity like a lecture or a homework assignment. Skipping is thus a narrow form of drop-out. Drop-out predictions often rely on measures of engagement in, or satisfaction with, a course [7,9]. In contrast, the Skipping Predictor presented below is based on learners' gaps in knowledge. The drop-out predictor presented in [13] uses both, time invariant data gathered from registration forms and time dependent data gathered from multiple-choice tests and aims at good predictions already at the beginning of a course. Different methods have been used for drop-out prediction: Support Vector Machines [13], Neural Networks [10,13], Decision Trees and Bayesian Classifiers [8].

Examination Performance Prediction. Both, time invariant data (such as grades in previously attended courses or demographic data) and time dependent data (such as engagement measures) have been used in predicting examination performances [1,6]. The predictors described in that articles use Neural Networks and Multiple Linear Regression respectively. The Examination Fitness Predictor presented below in Sect. 4 is based on learners' participation in learning activities like lectures and homework assignments. Indeed, that predictor's aim is to suggest remedies in case insufficient examination performances are predicted. Basing predictions on parameters (like demographic data) learners cannot influence would be counterproductive and unethical. A correlation between emotional affects and examination performances has been observed by Pardos et al. [14]. Relying on experts for estimating learners' affects, these authors built a dataset and used it in training a machine learning method.

Learners' Misconceptions. The predictor of learners' misconceptions presented below in Sect. 5 is to the authors' knowledge the first of its kind. It is in the constructivist tradition of Posner et al. [16] who proposed the "conceptual change model" stating that learners have some, possibly erroneous, conceptions when they engage in a learning activity and that these conceptions change when certain conditions are met. That model is widely applied in science education [19] and is the basis of the Misconception Predictor presented below. That predictor also relates to research on provoking conceptual changes among learners [22] and on Radatz' cognitivist investigations of procedural errors in mathematical problem solving [18]. Procedural errors relate to misconceptions because "they result from non random applications of rules based on certain beliefs" [5, p. 33].

3 Skipping Predictor

The Skipping Predictor is based on a human labelling of homework submissions according to the following scheme:

- **SKIP**, for skipped, when a homework assignment is not delivered.

- **IK**, for insufficient knowledge, reflected by an incorrect use of symbols, statements like "I don't know how to solve this", or an answer not fitting a question, and requiring learning again part of the course material.
- **OE**, for other error, that is, errors not due to an insufficient knowledge.
- **NE**, for no errors, otherwise.

The homework submissions of 80 randomly selected students enrolled in an introductory course on computer science theory have been used in building the Skipping Predictor. The course had 11 weekly homework assignments. 80 label sequences of length 11 were thus generated. The labelling has been performed by the course's teaching team. For testing purposes, four members of the teaching team categorized independently from each other 30 of the $80 \times 11 = 880$ submissions yielding an inter rater reliability Fleiss-κ of 0.78. For evaluation purposes, one member of the teaching team categorized once again all 880 submissions.

The predictor is a Hidden Markov Model, that is, a dynamic state system consisting of "hidden states" of which one is currently active [17]. The active state changes at each step, that is, with each weekly homework, according to predetermined state transition probabilities. A state depends on all assignment submissions of each student so far. An "emission" (or "observation"), taken from a fixed set, is observed in each step. The probability of observing an emission is defined by the active state. The afore-mentioned set of labels –SKIP, IK, OE, and NE– was used as emission set. Two hidden states were used in the model. The model was trained using the Baum-Welch Algorithm [2]. The evaluation was a 10-fold cross validation.

The predictor's quality is, as usual, estimated by its sensitivity (or recall), that is, the proportion of predicted skips that have taken place, and by its specificity" (or true negative rate), that is, is the proportion of non-skips that have been predicted. The Skipping Predictor has a sensitivity of 72.9% and a specificity of 84.7%.

These results are satisfying, yet higher inter rater reliability of the labels and specificity would be preferable. Indeed, it is preferable to nudge against skipping as few learners as possible that are not at risk of skipping. We expect the labellers' competence to improve over time what should result in a higher inter rater reliability and a higher specificity.

4 Examination Fitness Predictor

The Examination Fitness Predictor is based on the numbers of SKIP, IK, OE, and NE labels described in the previous Sect. 3 assigned to the homework submissions of each student. The predictor applies a Multiple Linear Regression Model, that is, it expresses a dependent numerical variable as a linear combination of independent numerical variables. The predictor's dependent variable is the examination fitness expressed as a mark, that is, a percentage of the total score obtainable at the examination. Its independent variables are the numbers of each label SKIP, IK, OE, and NE assigned to the homework submissions of each student.

Two Examination Fitness Predictors have been built from two datasets. The first dataset, the "homework dataset", consists in the label numbers of the dataset described above in Sect. 3. The second dataset, the "weekly learning assessment dataset", consists in the label numbers of quiz answers collected by the system Backstage [3,15] during an introductory course on functional programming. Using Backstage, a multiple-choice quiz, each possible answer of which had been labelled SKIP, IK, OE, or NE, was run at the beginning of each weekly lecture so as to assess how well the content of the last week's lecture had been learned. The second dataset is needed for the following reasons: It allows an evaluation on data that do not require any human labelling and on data referring to a different educational setting: Answering quizzes on last week lecture and homework assignment on the current week lecture, respectively. Interestingly, with these datasets different independent variables, including in both cases the variable IK (insufficient knowledge), have been identified as impacting on examination fitness:

- With the homework dataset, SKIP (missed learning activity) and IK (insufficient knowledge) impact on examination fitness with a significance level of 0.05%.
- With the weekly learning assessment dataset, OE (other error) and IK (insufficient knowledge) impact on examination fitness with a significance level of 0.05%.

The Examination Fitness Predictor's quality, expressed as usual for a Multiple Linear Regression Model by the coefficient of determination (R^2), is as follows:

- Homework dataset with the independent variables SKIP and IK: $R^2 = 20.0\%$
- Weekly learning assessment with the independent variables OE and IK: $R^2 = 19.8\%$

Examination Fitness Predictors could yield better predictions if, in addition to the afore-mentioned variables, they would refer as well to measures of activity and to demographics. We rejected such an improvement for two reasons. Firstly, the resulting predictions could wrongly suggest to students that more activity, whatever its nature, could positively impact on examination fitness. Secondly, demographics cannot be influenced by students. Since the goal of our predictors is to nudge students to a better learning, the predictors' quality must be subordinated to the impact of their predictions on the students' behaviours.

The forecasts of the Skipping and Evaluation Fitness Predictors allows for the following intervention. The Skipping Predictor recognizes students at risk of skipping. The Evaluation Fitness Predictor forecasts that skipping a homework assignment reduces examination fitness on average by 2,5%, as the evaluation has shown. Such an automatically generated feedback keeps many, if not all, students "on tracks", by nudging them not to skip their next homework assignment. A description of this nudging and its evaluation are out of the scope of this article.

5 Misconception Predictor

The Misconception Predictor relies, like the Skipping Predictor, on a human labelling of the students' weekly homework submissions. In contrast to the Skipping Predictor, however, no pre-defined labels were used. Instead, the labellers introduced new labels as they needed them. The labelling was performed on the project platform of our Backstage 2 system what ensured the sharing of misconception labels between the labellers. The homework submissions of 80 students enrolled in an introductory course on the theory of computer science have been labelled by the teaching team in two steps: Firstly, a labeller read a large number of submissions for a specific assignment so as to identify and label common misconceptions. Secondly, all submissions were labelled according to the identified misconceptions. This approach is the standard correction routine of teaching teams. In order to test whether the generated labels were reliable, two measures were computed. Firstly, two labellers independently examined different sets of submissions for 4 assignments. Both labellers recorded those misconceptions they frequently encountered. Even though they were not told how many frequent misconceptions they should record, they both recorded as frequent 2 to 3 misconceptions. For 3 of the 4 examined assignments, the labellers fully agreed. For the remaining assignment, the labellers agreed on two misconceptions but they reported a different third one. Secondly, 4 further labellers where given descriptions of the frequent misconceptions formerly recorded and independently labelled 20 other submissions. On average and for all misconceptions, the inter rater reliability was Fleiss-$\kappa = 0.70$.

The Misconception Predictor relies on collaborative filtering [12], following a simple intuition: If two learners had similar (specific) misconceptions in the past, they are likely to have similar misconceptions in the future.

10-fold cross validation yielded the following quality measures:

- Sensitivity 71.8%
- Specificity 80.7%

First investigations with data collected from different lectures point to the effectiveness of the approach to build Misconception Predictors. Predictions on misconceptions are highly beneficial for keeping learners "on tracks" by warning them against "shallow learning". This nudging is out of the scope of this article.

6 Perspectives for Future Work

The predictors presented above are based on automatically gathered data using multiple-choice quizzes, and data collected by a human labelling that exhibits a high inter rater reliability. We expect the teaching teams' competence to improve over time what should result in a higher inter rater reliability and a higher specificity. Peer review, or peer assessment, that is, the enrolment of students to providing feedback, in place of or in addition to the teacher's feedback, to their fellow students, could also explicitly or implicitly contribute to a better labelling, especially of misconceptions. Adopting peer review in this manner is likely to be well accepted since students see three main benefits in peer reviews: the chances "to compare different approaches", to "compare standard of work," and the "exchange of information and ideas" [23, p. 52].

An automatic labelling of student behaviour can be easily realized for tasks such as multiple-choice quizzes or programming (unit tests can be used to asses programs). The labelling of student behaviour on other tasks might require advanced methods such as machine.

The approach described above is a proof-of-concept. For deploying its full potential, it has to be applied with datasets referring to many more courses and many more course venues. We expect that a clustering of courses will result in improved predictors for the course clusters. Indeed, the educational context is likely to matter: Is for example the students' skipping behaviours in a Calculus and Linear Algebra courses similar? In programming courses?

7 Conclusion

This article has presented three predictors, the Skipping, Examination Fitness, and Misconception Predictor, related to learning-impairing behaviours, skipping (a form of drop-out), improper homework, and misconceptions. The predictors have been built from data collected from courses in computer science using the system Backstage [3,15] and by examining homework assignments. One predictor is a Hidden Markov Model, a second predictor relies on Collaborative Filtering, and the third predictor uses Multiple Linear Regression. The sensitivities and specificities of the first two predictors are above 70% and the R^2-error of the third predictor is about 20%. Considering the large numbers of unknown parameters like course-independent learning, this quality is satisfying.

The predictors have been developed for giving an algorithmically generated feedback to students, so as to make them conscious of deficiencies in their learning and to nudge them to improve it. Thus, the predictors have been developed for "getting it wrong" because warned students better their learning and, in doing so, invalidate the predictors' forecasts. Hints at this use of the predictors, an issue out of the scope of this article, have been given. The authors expect that extensive evaluations will point to the effectiveness of nudging students towards better learning with the predictions described in this article.

References

1. Abdous, M., Wu, H., Yen, C.J.: Using data mining for predicting relationships between online question theme and final grade. J. Educ. Technol. Soc. **15**(3), 77 (2012)
2. Baum, L.E., Petrie, T., Soules, G., Weiss, N.: A maximization technique occurring in the statistical analysis of probabilistic functions of Markov chains. Ann. Math. Stat. **41**(1), 164–171 (1970)
3. Bry, F., Pohl, A.Y.S.: Large class teaching with Backstage. J. Appl. Res. High. Educ. **9**(1), 105–128 (2017)

4. Christensen, R.: Plane Answers to Complex Questions: The Theory of Linear Models. Springer Science & Business Media, Berlin (2011)
5. Confrey, J.: Chapter 1: a review of the research on student conceptions in mathematics, science, and programming. Rev. Res. Educ. **16**(1), 3–56 (1990)
6. Cripps, A.: Using artificial neural nets to predict academic performance. In: Proceedings of the 1996 ACM Symposium on Applied Computing, pp. 33–37. ACM (1996)
7. Dejaeger, K., Goethals, F., Giangreco, A., Mola, L., Baesens, B.: Gaining insight into student satisfaction using comprehensible data mining techniques. Eur. J. Oper. Res. **218**(2), 548–562 (2012)
8. Dekker, G., Pechenizkiy, M., Vleeshouwers, J.: Predicting students drop out: A case study. In: Educational Data Mining 2009 (2009)
9. Giesbers, B., Rienties, B., Tempelaar, D., Gijselaers, W.: Investigating the relations between motivation, tool use, participation, and performance in an e-learning course using web-videoconferencing. Comput. Hum. Behav. **29**(1), 285–292 (2013)
10. Guo, W.W.: Incorporating statistical and neural network approaches for student course satisfaction analysis and prediction. Expert Syst. Appl. **37**(4), 3358–3365 (2010)
11. Halawa, S., Greene, D., Mitchell, J.: Dropout prediction in MOOCs using learner activity features. eLearning Papers **37** (2014)
12. Linden, G., Smith, B., York, J.: Amazon.com recommendations: item-to-item collaborative filtering. IEEE Internet Comput. **7**(1), 76–80 (2003)
13. Lykourentzou, I., Giannoukos, I., Nikolopoulos, V., Mpardis, G., Loumos, V.: Dropout prediction in e-learning courses through the combination of machine learning techniques. Comput. Educ. **53**(3), 950–965 (2009)
14. Pardos, Z.A., Baker, R.S., San Pedro, M.O., Gowda, S.M., Gowda, S.M.: Affective states and state tests: investigating how affect throughout the school year predicts end of year learning outcomes. In: Proceedings of the Third International Conference on Learning Analytics and Knowledge, pp. 117–124. ACM (2013)
15. Pohl, A.: Fostering awareness and collaboration in large-class lectures. Doctoral thesis, Ludwig-Maximilian University of Munich, Germany (2015)
16. Posner, G.J., Strike, K.A., Hewson, P.W., Gertzog, W.A.: Accommodation of a scientific conception: toward a theory of conceptual change. Sci. Educ. **66**(2), 211–227 (1982)
17. Rabiner, L.R.: A tutorial on hidden Markov models and selected applications in speech recognition. Proc. IEEE **77**(2), 257–286 (1989)
18. Radatz, H.: Error analysis in mathematics education. J. Res. Math. Educ. **10**(3), 163–172 (1979)
19. Read, J.R.: Children's misconceptions and conceptual change in science education (2004). http://asell.org/global/docs/conceptual_change_paper.pdf
20. Siemens, G. (ed.): 1st International Conference on Learning Analytics and Knowledge 2011 (2010)
21. Tan, M., Shao, P.: Prediction of student dropout in e-learning program through the use of machine learning method. iJet **10**(1), 11–17 (2015)
22. Tippett, C.D.: Refutation text in science education: a review of two decades of research. Int. J. Sci. Math. Educ. **8**(6), 951–970 (2010)
23. Williams, E.: Student attitudes towards approaches to learning and assessment. Assess. Eval. High. Educ. **17**(1), 45–58 (1992)

A Run-Time Detector of Hardworking E-Learners with Underperformance

Diego García-Saiz(✉), Marta Zorrilla, Alfonso de la Vega, and Pablo Sánchez

University of Cantabria, Av. de los Castros SN, Santander, Spain
{diego.garcia,marta.zorrilla,alfonso.delavega,p.sanchez}@unican.es

Abstract. Due to the lack of a face-to-face interaction between teachers and students in virtual courses, the identification of at-risk learners among those who appear to show normal activity is a challenge. Particularly, we refer to those who are very active in the Learning Management System, but their performance is low in comparison with their peers. To fix this issue, we describe a method aimed to discover learners with an inconsistent performance with respect to their activity, by using an ensemble of classifiers. Its effectiveness will be shown by its application on data from virtual courses and its comparison with the results achieved by two well-known outlier detection techniques.

Keywords: At-risk students · Warning system
Educational data mining

1 Introduction

As stated by Hara and Kling [1], "Lack of timely feedback can result in learners ambiguity about their performance in the Web-based course and can contribute to their frustration". This asseveration denotes the importance of detecting and identifying those students in virtual courses who appear to show normal activity, but are at-risk of having low performance. Regrettably, regular Learning Management System (LMS) do not suffice to provide just-in-time feedback that would allow to accommodate learners at risk. This fact makes difficult for teachers to intervene and give advice to these learners as soon as some evidence appears.

Therefore, we propose an method, based on an ensemble of classifiers, that deals with this challenge and can be, later, easily integrated in LMS systems, providing teachers with a useful tool to detect active learners with underperformance at run-time. Our method aims at analysing the behaviour of those students who perform a remarkable activity in the LMS before sending a delivery but they do not pass it. If their effort is similar or even higher than the one of the latter, only a quick remedial message sent by the teacher could avoid their demotivation and possible dropout [2]. Our method thus allows teachers to both classify learners according to their effort-performance ratio (success, fail or irregular behaviour). In such a way, teachers could quickly write specific and targeted oriented messages to each learner, for instance, motivating messages with a piece of advice for our risky students.

T. Di Mascio et al. (Eds.): MIS4TEL 2018, AISC 804, pp. 20–28, 2019.
https://doi.org/10.1007/978-3-319-98872-6_3

2 Related Work

Supervised learning techniques are widely used in the educational data mining (EDM) arena for different purposes, being the performance and dropout prediction by using the activity performed by the students in LMS two hot topics [3]. The majority of available literature is reflective in nature, meaning the reported studies address the prediction problem using data sets that collect data corresponding to courses already finished, with the aim of gathering maximum information in order to achieve the highest accuracy [5]. Concurrently, some authors are focused on predicting progress through predicting performance at specific intervals within a course [6].

The detection of undesirable student behaviour has been also studied in the EDM literature. According to Romero et al. [3], the objective is to discover those students who have some kind of problem or unusual behaviour, such as learning difficulties or irregular learning. There are some works in the literature which aim to detect at-risk students with irregular learning processes [4]. However, to the best of our knowledge, the issue treated in this work, detecting hardworking students with underperformance, has not yet been profusely studied. A strategy for dealing with this issue is using class-based anomalous behaviours detection (outliers) techniques. An interesting and complete survey about classification in presence of noise is provided by Frenay and Verleysen [7], in which the authors include a classification of suitable techniques depending on the kind of problem to be solved. According to this review, our method for detecting hardworking students with underperformance can be classified among those techniques addressed to detect misclassified instances.

3 Method Description

There are four types of situations that can be found according to the students' activity in the course, their real performance in the assignments, and the predicted performance by the classifier: students who passed and are classified as passed, students who passed and are classified as failed, students who failed and are classified as failed, and students who failed and are classified as passed. These last students are characterised by having an average to high activity in the course, most of the times similar to the one of the well-classified passed students. However, they failed the assignments. For this reason, these students are at a risky situation. Therefore, these are the students targeted by our method, since a suitable and encouraged message could avoid the their drop-out.

Our method relies on classification techniques. In particular, we propose to use an ensemble of classifiers, which combines a set of classifiers to construct a new classifier that is often more accurate than any of its constituent classifiers [8]. In this study, we use ensemble methods to identify mislabelled instances which will lead to discover our targeted learners. These instances will be those marked as false positives by the majority of classifiers.

Next, we explain how our method performs step by step:

1. A data set that gathers the activity carried out by each learner during the period under study, that means, the period established for doing an assessable assignment, and the mark achieved (pass/fail), is loaded.
2. Then, an ensemble is built by means of a set of classifiers applied to this data set and evaluated with the leave-one-out method. The misclassified instances of the fail class of each classifier are marked.
3. Next, the process evaluates how many times each instance of the fail class has been marked, that means, has been misclassified. Instances marked by at least the 50% of classifiers (half or more than a half) are our targeted instances since it is highly likely that these instances correspond to the students who failed having a high activity level.
4. Finally, teachers are automatically informed by our process about these failed students with irregular activity, so they can identify and help them.

4 Configuration of the Experiments

We apply our method on students' activity data from two ended e-learning courses hosted in the LMS Moodle at a Spanish University. Both are cross-curricular courses open to students from all degrees, and their students must carry out different assignments along the weeks to pass. The number of students enrolled in these courses was 43 and 119 respectively.

Two data sets were generated for the current case study, one for each course, with the activity data corresponding to the period of the first assignment, named "d1a1" and "d2a1" respectively. The attributes used were: the number of actions performed by the student ("act"), the number of visits to the .pdf content files ("v-re"), to the SCORM resources ("v-sc"), to the statistics page about the progress in the course ("v-da") and to the html pages which collect the course syllabus ("v-co"); the number of the visits to feedback messages provided by the instructor ("v-fe"), the number of messages read ("v-fo"), posted ("a-di") and answered ("p-fo") by the student in the forum and the sum of the "a-di" and "p-fo" attributes ("pa-fo"). As class attribute, we used the pass or fail grade achieved by the learner in this assignment. It must be pointed out that we used the data from the first assignment because the students' behaviour pattern remained regular after that first delivery. This could be due to the non-compulsory nature of these courses. Once students engage to the course and check that it meets their expectations, they progress adequately.

The algorithms used to build our ensemble were the ones proposed by Smith and Martínez [9], since this selection was proven to have a good performance in order to detect outliers. It comprises classifiers that belong to five different learning paradigms with the aim of compensating the bias of each method. Our experimentation utilised the Weka implementation of the following classifiers with their default parameters: RIPPER, NNge and Ridor (Rule based); MultilayerPerceptron (Neural Networks), NaïveBayes (Bayesian); C4.5 and RandomForest (Tree-based); LWL and 5-NN (Lazy). Before the application of the classifiers, the CfsSubsetEval technique provided by Weka was run to select the most relevant activity attributes to predict the students' performance.

5 Results and Discussion

In this Section we show the results obtained by applying the proposed method on the two courses. Table 1 gathers the nine students at risk detected in the "d1a1" data set. The column "id" contains the identification of each student and the four following columns correspond to the four attributes selected by the attribute selection method. These data are shown normalised. The column "%miss" indicates the percentage of times that a failed student was classified as passed. Columns "1a" and "2a" show the mark of the students, in a grading scale from 0 to 10, achieved by each learner in the first and the second assignments of the course, meaning "dp" that the student dropped out of the course before sending the assignment and "nd" that the assignment was not delivered but the student continued in the course. Finally, column "Gr." groups these students according to the percentage of classifiers that were not able to classify them rightly: in "Group#1" there are failed students misclassified by 50% to 74% of the classifiers; "Group#2" contains failed students misclassified by 75% to 99%; and "Group#3" have failed students misclassified by 100% of the classifiers.

Table 1. Avg. values of the activity attributes of detected learners at risk in d1a1

Gr.	id	act	v-re	v-co	v-da	%miss	1a	2a
1	c1e1	0.18	0.09	0.10	0.04	55.56	dp	-
	c1e2	0.20	0.16	0.16	0.15	66.67	3	dp
2	c1e3	0.23	0.30	0.18	0.24	88.89	dp	-
	c1e4	0.31	0.16	0.33	0.24	88.89	nd	6
3	c1e5	0.45	0.35	0.27	0.48	100.0	3	dp
	c1e6	0.91	1.00	0.51	0.72	100.0	4	9
	c1e7	0.50	0.51	0.38	0.20	100.0	nd	9
	c1e8	0.58	0.42	0.60	0.44	100.0	3	9
	c1e9	0.37	0.30	0.44	0.50	100.0	nd	9

Next, we show the fitness of the method by studying these instances in detail. For this end, Table 2 displays the average activity for each group independently as well as the average activity of the nine students ("Avg."). This table likewise includes a row with the average values which belong to the students who passed ("Avg. p.") and another one for those who were well-classified as failed students ("Avg. f."). In order to visualise the differences among these groups, Table 3 shows the Euclidean distance between the average of each activity attribute belonging to the nine detected failed students with respect to the average values

Table 2. Average of the activity of each group of detected learners in d1a1

Group	Act	v-re	v-co	v-da
1	0.19	0.13	0.13	0.10
2	0.27	0.23	0.26	0.24
3	0.56	0.52	0.44	0.47
Avg.	0.41	0.38	0.33	0.33
Avg. f.	0.08	0.11	0.09	0.04
Avg. p.	0.40	0.35	0.39	0.44

Table 3. Euclidean dist. of each group from d1a1 to the fail and pass class

Group	D(fail)	D(pass)	Diff.
1	0.13	0.53	−0.40
2	0.34	0.30	0.04
3	0.84	0.24	0.60
All	0.57	0.13	0.44

of those who passed (23 learners) and with respect to those who were well-classified as failed students (20 learners).

As expected, the average activity of the passed students is noticeable higher than the one of the well-classified failed students. As can be observed, the normalised average value of the total number of actions, "act", of the passed students takes a value of 0.40, meanwhile this value is 0.08 for failed learners. The same happens with the other attributes. It must also be pointed out that the average activity of all detected students (9 in total) is noticeable closer to the activity of passed students, with an Euclidean distance of 0.13, than to the well-classified failed students with an Euclidean distance of 0.57. With this analysis, it can be stated that the detected students have, in general terms, an activity more similar to the passed students.

The detected students in the Group#1 had a lower average activity than the rest of the detected students, and this activity is more similar to the one performed by failed learners than to the passed ones. In fact, one of these students, c1e1, dropped out of the course before sending the first assignment, and the other one, c1e2, failed it, and he also dropped out later. These dropouts might have been avoided if the teacher had sent them a message addressed to know the difficulties they found in the course and to show his/her support to help them.

Focusing on Group#2, it can be observed that the student c1e3 showed a similar behaviour to student c1e2 in the Group#1, and dropped out before the first assignment. The other student, c1e4, did not deliver the first assignment, however, passed the second one. This case shows it may be beneficial to send motivating and differentiated messages to those learners with a certain degree of activity, although they seem to have dropped out, so that teachers could encourage them to continue.

The remaining students who are in Group#3 show clearly that the method works properly since it is able to detect learners who carried out a great effort and achieved bad results. It must be first highlighted that the activity of these students on average is quite similar to the one carried out by the passed students, with a distance of 0.24, with respect to the well-classified failed students, with a distance of 0.84. We found two students in this group, c1e7 and c1e9, who did not deliver the first assignment, but passed the second one with a mark of nine out of ten, students c1e6 and c1e8 who failed the first assignment but passed the second one and, finally, the student c1e5 who delivered and failed the first assignment, and then dropped out.

Next, we present the results obtained with the data set "d2a1". Table 4 contains the information about the activity performed by the failed students detected by all classifiers, since it is the most reliable setting. Likewise, it includes their average activity as well as the average activity of the passed students and that one of those learners who were well-classified as failed. The two detected students had a high activity in the course, even higher than the average activity of students who passed. In the case of student c2e2, he achieved a grade of 4.5 out of 10, which is close to the pass/fail threshold of 5. However, student c2e1, who showed an even more remarkable activity level with values very close to 1 in "act", "v-re" and "v-fo", failed the assignment with a 3. Both students could have suffered from any kind of punctual problem since they achieved a good mark in the second assignment.

Table 4. Avg. values of the activity attributes of detected learners at risk in d2a1

Id	Act	v-re	v-co	a-di	v-fo	%miss	1a	2a
c2e1	0.84	0.89	0.45	0.00	0.86	100.0	3	9
c2e2	0.51	0.44	0.48	0.00	0.28	100.0	4.5	9
Avg.	0.68	0.67	0.47	0.00	0.57	-	-	-
Avg. f.	0.01	0.02	0.01	0.00	0.00	-	-	-
Avg. p.	0.28	0.27	0.18	0.09	0.16	-	-	-

In short, we have tested that our method is able to detect learners with a behaviour that is irregular in comparison with the overall population, enabling teachers to send personalised messages in order to keep students who fit in an at-risk classification active and interested in the course.

5.1 Comparison of Our Proposed Method with Other Alternatives

The process of detecting irregular behaviours can be also studied following the classical paradigm of outliers' detection. In order to highlight the advantages of our method with respect to this kind of methods, we performed an analysis by using two of the most popular outlier detection techniques, LOF [10] and ECODB [11], and compared their results with the results of our proposed method presented in the previous section.

LOF, acronym of Local Outlier Factor, uses k-Nearest Neighbours to establish the density of local groups. Its main disadvantage is that it is not a class-outlier technique, meaning it does not take into account the class attribute in its execution. The consequences of this fact can be observed in Table 5, which displays the first 8 students detected as outliers by this technique in the first course. C1e6 is the only student who failed having performed a high activity in the course. The remaining students detected are learners who carried out the lowest activity in the course and who should not have been detected as outliers. But, due to the fact that LOF is not class-outlier, it detected them as extreme values. This states that LOF is not suitable for our purpose since it is not able to detect irregular learners.

Unlike LOF, ECODB technique takes the class value into account in order to detect outliers. By using k-Nearest Neighbours, it calculates a value named Class Outlier Factor (COF), which indicates the membership of an instance to its class. ECODB needs a previous configuration of two parameters before being executed: the "k" value for the "Nearest Neighbours" search and the number of outliers to be detected. Table 6 shows the students detected by ECODB with k = 3 and number of outliers = 12. As can be observed, there are two students who were previously detected with the ensemble and were not detected by ECODB. One of them is c1e1, who, indeed, performed the lowest activity among all the students who failed detected by the ensemble method. The other one is one of the failed students with a higher activity, c1e8. On the other hand, ECODB detected three students as failed who were not detected by our method: c1ef12, c1ef13 and c1ef14. It must be pointed out that these three students performed a very low activity in the course. In particular, c1ef12 was not misclassified by any of the classifiers in the ensemble, and c1ef13 and c1ef14 were misclassified by only two of them. Modifying the "k" value barely changes the results. In the same way, increasing the number of outliers detected by ECODB only relies in more

Table 5. Students detected by LOF on d1a1

Id	Act	v-re	v-co	v-da	Grade
c1pe1	1.00	0.63	1.00	0.74	Pass
c1pe2	0.76	0.84	0.73	0.98	Pass
c1e6	0.91	1.00	0.51	0.72	Fail
c1pe3	0.74	0.49	0.88	0.59	Pass
c1pe4	0.72	0.65	0.61	0.59	Pass
c1fe10	0.00	0.00	0.00	0.00	Fail
c1fe11	0.00	0.00	0.00	0.00	Fail
c1fe12	0.00	0.00	0.00	0.00	Fail

Table 6. Students detected by ECODB on d1a1

Id	Act	v-re	v-co	v-da	Grade
c1e2	0.20	0.16	0.16	0.15	Fail
c1ef12	0.13	0.40	0.16	0.022	Fail
c1e6	0.91	1.00	0.51	0.72	Fail
c1e5	0.45	0.35	0.27	0.48	Fail
c1e9	0.37	0.30	0.44	0.50	Fail
c1pe16	0.23	0.35	0.26	0.04	Pass
c1ef13	0.12	0.23	0.20	0.13	Fail
c1ep15	0.40	0.19	0.30	0.39	Pass
c1ef14	0.12	0.19	0.09	0.17	Fail
c1e7	0.50	0.51	0.381	0.20	Fail
c1e4	0.31	0.16	0.331	0.24	Fail
c1e3	0.23	0.30	0.178	0.24	Fail

students with non-irregular behaviour being detected and, on the other hand, decreasing it results in a lower detection of students with irregular behaviour.

Similar results are achieved on the "d2a1" data set. Table 7 displays the results obtained by applying ECODB with k = 3 and number of outliers = 4. One of the failed students, c2e2, was also detected by our ensemble, and had a high activity. However, the other failed student detected, d2ef7, carried out a low activity in the course. Moreover, c2e1, who performed the highest activity among all students who failed, was detected by our proposal but not by ECODB.

Table 7. Students detected by ECODB on d2a1

Id	Act	v-re	v-co	a-di	v-fo	Grade
d2ep2	0.02	0.05	0.02	0.00	0.01	Pass
d2ep3	0.00	0.00	0.01	0.00	0.00	Pass
c2e2	0.51	0.44	0.48	0.00	0.28	Fail
d2ef7	0.02	0.03	0.02	0.00	0.02	Fail

In sum, we can state that the ensemble better detects the learners with an irregular behaviour. Furthermore, the technique only requires a parameter to be set, the number of classifiers that must misclassify an instance, which can be considered as a rough parameter. That means, all the instances detected correspond to irregular behaviours, and the higher we set up a value, the higher probability of being our students at risk. Thus, the setting of this parameter affect less to the accuracy of the results than the tuning of algorithm parameters. Outliers techniques, as shown, are less accurate than our proposal.

6 Conclusions and Future Work

Student dropout is a critical problem, particularly in distance education systems. Although there are many research works which address this issue, these have been generally focused on detecting learners whose activity decays along the course. But there are another important group of students who are at risk. They are those who work and participate actively in the course, but their performance is low in comparison with those peers who carry out a similar activity and pass.

This paper contributes with a method addressed to detect these learners at run-time of the course, using the activity performed in the learning platform during the period established for each task and the grade achieved. These students might obtain a feedback with an encouraging message that helps them to keep active, and some notes that guide them towards the main resources of the course. This technique has been compared with two techniques from the outlier detection domain, which shows our ensemble-based method to be more accurate.

There are some tasks and challenges to be dealt with in a near future, such as getting a large and diverse collection of virtual courses, or wrapping the proposed method so it can be installed in LMS platforms.

Acknowledgements. This work has been partially funded by the Spanish Government under grant TIN2014-56158-C4-2-P (M2C2).

References

1. Hara, N., Kling, R.: Student distress in web-based distance education. Educause Q. **24**(3), 68–69 (2001)
2. Giesbersa, B., et al.: Investigating the relations between motivation, tool use, participation, and performance in an e-learning course using web-videoconferencing. Comput. Hum. Behav. **29**(1), 285–292 (2013)
3. Romero, C., Ventura, S.: Data mining in education. Wiley Interdisc. Rew. Data Mining and Knowl. Disc. **3**, 12–27 (2013)
4. Wolff, A., Zdrahal, Z., et al. : Developing predictive models for early detection of at-risk students on distance learning modules. In: 4th International Conference on Learning Analytics and Knowledge, pp. 24–28. Indianapolis (2014)
5. Xing, W., Guo, R., et al.: Participation-based student final performance prediction model through interpretable genetic programming: integrating learning analytics, edm and theory. Comput. Hum. Behav. **47**, 168–181 (2015)
6. Koprinska, I., Stretton, J., Yacef, K.: Predicting student performance from multiple data sources. In: International Conference on Artificial Intelligence in Education, pp. 678–681 (2015)
7. Frenay, B., Verleysen, M.: Classification in the presence of label noise: a survey. IEEE Trans. Neural Netw. Learn. Syst. **25**(5), 845–869 (2014)
8. Rokach, L.: Ensemble-based classifiers. Artif. Intell. Rev. **33**, 1–39 (2010)
9. Smith, M., et al.: The robustness of majority voting compared to filtering misclassified instances in supervised classification. Artif. Intell. Rev. **49**, 105–130 (2017)
10. Breunig, M.M., Kriegel, H.P., et al.: LOF: identifying density-based local outliers. In: ACM International Conference on Management of Data, Dallas, pp. 93–104 (2000)
11. Saad, M.K., Hewahi, N.M.: A comparative study of outlier mining and class outlier mining. Comput. Sci. Lett. **1**(1) (2009)

The Use of Gamification as a Teaching Methodology in a MOOC About the Strategic Energy Reform in México

J. Mena[1(✉)], E. G. Rincón Flores[2], R. Ramírez-Velarde[2], and M. S. Ramírez-Montoya[2]

[1] University of Salamanca, Pº Canalejas, 169, Salamanca, Spain
juanjo_mena@usal.es
[2] Tecnológico de Monterrey, Avda. Av. Eugenio Garza Sada 2501 Sur, Tecnológico, 64849 Monterrey, NL, Mexico

Abstract. The arrival of online programs in education is pushing educators to promote new ways of teaching to engage students. Nonetheless, most higher education teachers are not trained in the practices of e-learning. In this paper, our purpose is to study whether the use of gamification better promotes learning in online courses. Over 6,000 participants enrolled in the MOOC "Conventional and green energy sources" as a part of the activities of the "Binational Laboratory on Smart Sustainable Energy Management and Technology Education" project. About 1,000 eventually completed it. Main results indicate that for all participants' profiles (i.e., gender, age, and educational level) the completion of a gamification challenge favored higher final test scores on the contents of the course. This lead us to think that gamification improves students' performance in online teaching. However, there are technical limitations associated to the courses platforms that need to be solved for teachers to be able to implement non-traditional learning approaches.

Keywords: Gamification · Teacher education · MOOCs

1 Introduction

The use of technical innovation and communication tools (ICT) in classroom has pushed schools and education systems to promote new ways of teaching based on multiple interactive strategies (e.g., synchronic and asynchronic) of instruction and assessment [1, 2].

Furthermore, incentives to move education and teacher education course contents to online formats (e-learning) are rapidly spreading worldwide. This global scenario has been referred to the "digital turn". The digital turn is defined as "an analytical strategy to discuss the digitalization process affecting society, and as a description of the digitalization process itself. This process leads from the 'book culture' of the so-called Gutenberg Galaxy to a digital age" [3].

T. Di Mascio et al. (Eds.): MIS4TEL 2018, AISC 804, pp. 29–36, 2019.
https://doi.org/10.1007/978-3-319-98872-6_4

Our main purpose is to understand to what extent the use of gamification techniques facilitates learning in a MOOC on energy sources.

The following pages are organized as follows: a brief theoretical background on MOOCs and gamification, the description of our study methodology and main results.

2 Theoretical Background

It has been long reported that most higher education teachers and teacher educators are not trained in the practices of e-learning. Thus, educators can be pressured to pursue online (or blended) course implementation, such as Massive Open Online Courses (MOOCs), without the necessary e-learning skills and tools [4].

2.1 The Use of MOOC in Higher Education Programs

Massive Open Online Courses (MOOCs) arises as a new model of online teaching, particularly demanded in higher education that push teachers in a position of being digitally competent [5]. They are typically divided into XMOOCs (based on traditional teaching) and cMOOCs (based on connectivist teaching [6].

Most courses to date are offered as xMOOCS because they allow unlimited numbers of participants at one time from different locations. Its uniqueness relies on the fact that they are "massive" -which have no precedent in distant education- but its spreading is basically driven by two factors: (1) rapid interchange of knowledge and (2) low or no cost tuition fees. However, there are some limitations: low completion rates that account of less than 10% [7], weaker intensity of interactions between teachers and students, lack of immediate feedback from instructors [8].

2.2 Gamification and MOOCs

Gamification is a relatively new methodology in MOOCs. It is an innovative educational strategy that borrows elements from games to be used in non-traditional game-based contexts. Its major purpose is to transpose students to virtual simulated scenarios that engage them in the school activities. Gamification is traditionally associated to elements such as badges, ranks and scores that serve to reward students' performance. Initially developed in the military sector and then expanded to other fields such as marketing, businesses, health-care, education.

According to [9]'s gamification model, there are three major components related to this teaching strategy: dynamics, mechanics and components. Dynamics are the contexts in which the gamification occurs, mechanics are the activities carried out within the dynamics and components are the objects used for the mechanics. In addition to that three dimensions should be added [10]: cognitive (i.e., problem solving, logical solutions, etc.), social (cultural skills, values, leadership, etc.) and affective (i.e., emotions, belief systems and attitudes). The components and dimensions combined result in an array of different uses of gamification that lead to different types of learning. For instance, an affective dynamic based on challenge mechanics and using appealing components such as badges, points or trophies foster students' learning by improving the engagement instructional episode within a class [11].

Although xMOOCs have expanded educational contents to wider populations, their behaviorist approach limits the inclusion of cognitive methodologies and teaching strategies that have been largely proven to have positive effects in students' learning. In this regard, in most MOOCs digital platforms it is technically challenging to use new methodologies such as gamification. This push to using traditional ways of teaching which, in turn, leads to boredom [12], the no differentiation of participants' profiles [13].

The main interest of this study is to analyze the effect of gamification-based teaching in the MOOC participants' performance depending on different participants' profiles. More specifically, our work aims at:

1. Determining the participants' profile in the MOOC.
2. Identify whether there are statistical differences between three major profilcs based on (1) age; (2) gender and (c) educational level and the completion of the gamification challenge in the MOOC.
3. Identify whether statistical differences might be found between three major profiles based on (1) age; (2) gender and (c) educational level and a performance test (final exam) about the contents of the MOOC.

3 Methodology

3.1 Course Design

The National Council for Science and Technology (CONACYT), the Mexican Secretary of Energy (SENER) and the Tecnológico de Monterrey (higher institution in Mexico) launched in 2015 a Strategic Energy Initiative Project to develop the Energy Reform in México which aims to impact academic, business and social communities in Mexico and Latin America by raising awareness through open innovation about the sustainable energy options available [11]. The project entitled "Binational Laboratory on Smart Sustainable Energy Management and Technology Education" (http://energialab.tec.mx/) between Mexico and the United States includes twelve subprojects. One of these subprojects aims at teaching energy sustainability through massive open online courses (MOOCs).

This research is based upon the results of one those MOOCs entitled "Conventional and green energy sources". The course included six themes: (1) Definition and types of energy, (2) Conventional energy sources, (3) Energy storage; (4) Eolic and hydric energy, (5) Solar and photovoltaic energy, (6) Biomass energy; six assignments and one final challenge where gamification was used. In the challenge the participants could win a prize in the form of a golden, silver and bronze cup depending on the correctness of their answer. The challenge was about an emergency in which the character takes shelter in a cave during a thunder storm. He should light-up the cave by making a lamp with several objects. The energy sources at his disposal are: ethanol, diesel or gas. Two criteria must be taken into the account: longer duration and less pollution.

3.2 Sample

A number of 6,022 participants enrolled in the MOOC entitled “Conventional and green energy sources”. It launched on late January 2017, and on September of the same year all the participants started it. Out of them 1,016 students eventually completed it.

3.3 Data Collection and Analysis

Descriptive data of participants’ profiles was obtained from a questionnaire that participants filled in during the first week of the course. For the second and third research questions (motivation and performance) we collected information provided by the analytics of the *MéxicoX* platform based on a validated Likert scale questionnaire (Cronbach’s Alpha of 0.8102) focus on participants’ perception on cognitive, social and emotional dimensions. The participants’ performance was measured by an individual test of twelve questions about the contents of the course. The questions were open ended and multiple choice.

The methodology applied had a quantitative approach [14]. Descriptive and inferential statistics were used to see different learning performances in the types of participants’ profile. The Minitab software and XLSTAT were used for factor analysis (Principal components analysis) logistic regression, best subset regression and ANOVA.

4 Results

4.1 Participants’ Profile

Out of 1,060 students, 1,019 passed the course (96%) and 41 failed (4%), whereas 1,025 students did the gamified challenge (97%) and 35 (3%) did not. Overall the completion rate was of 17% a percentage that was above the 4% to 10% average completion rate of this types of online courses [7]. Figure 1 shows the MOOC participant’s profile.

As Fig. 1 shows, four participants’ profiles were identified from the demographic analysis: (a) gender; (b) age; (c) educational level and (d) profession. Each of them made a difference in both completing the MOOC and in the overall learning of the contents about energy. Overall, 62% were men and 38% women. More than half of participants ranged from 21 to 30 years old (51%). The lowest frequency accounted for participants of 41 to 50 years old (16%). As for educational level, 37% of the students completed high school, 40% had Bachelor degrees and the rest (23%) upper degrees. The occupation of most of them was related to education (72%), 13% to managers and business (10% managers and 3% to companies) and 10% to administration (8% plus 2% of support activities).

4.2 Gamified Challenge Completion

The Principal Component Analysis (PCA1) were conducted for the variables Gender, Educational Level, Age and Gamified Challenge. The ANOVA analysis showed that

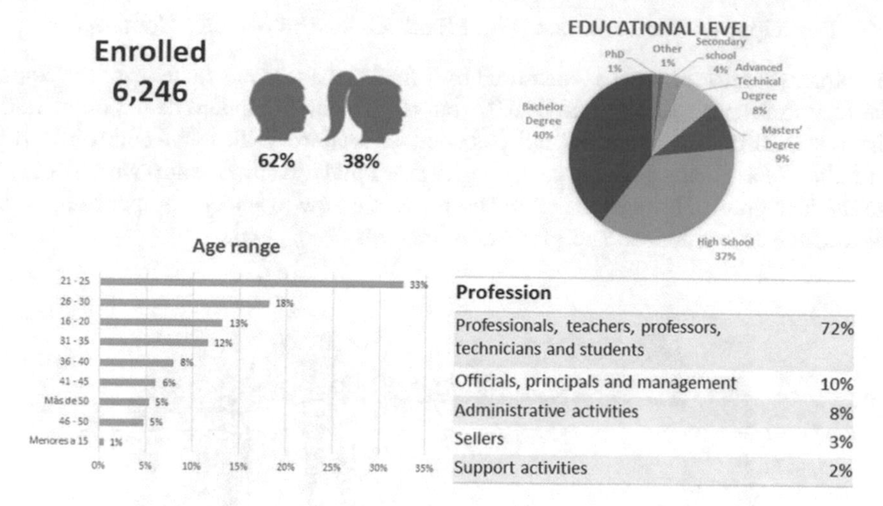

Fig. 1. Participants' profile in the "Conventional and green energies" MOOC [15].

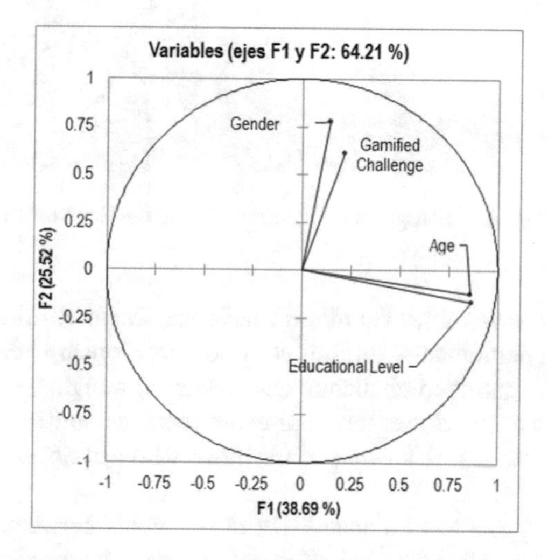

Fig. 2. PCA1. correlation biplot identifying causality relationships between variables.

there were no statistical differences between age or gender regarding the gamified challenge (p. = 0.274). Nevertheless, there were a slight tendency for males to be more likely to do the gamified challenge than females (see Fig. 2).

As for the educational level, we found that there were statistically significant differences (p = 0.024). The mean scores of the six types of educational levels are not statistically equal. The more advanced the educational level, the more likely it is the student to accomplish the gamified challenge.

4.3 Participants' Performance (Final Test About the MOOC Contents)

Participants' performance was measured by a final test about the contents of the course. The final exam has a mean score of 9.07 (out of 10) with a standard deviation of 1.304, with $p < 0.0001$. The exam did not pass neither Shapiro-Wilk nor Anderson-Darling normality tests. Figure 3 shows the histogram for Final test (final exam) with Normal fit and the histogram. This means that we do not know how to select a proper sample size nor to estimate confidence and prediction intervals.

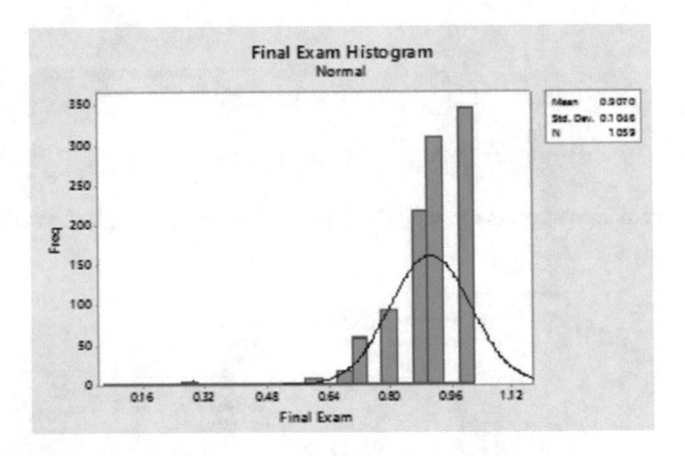

Fig. 3. Histogram and normal fit for the final exam

PCA2 included the variables Gamified Challenge, Gender, Educational Level, Age and Final Exam. All participants' profiles showed a relationship between succeeding in the completion of the gamified challenge and obtaining a higher score in the final test. With a $p < 0.0001$ we found that the final exam mean of students that completed the gamified challenge was $x = 9.1$ whereas for those who did not completed it correctly was $x = 7.8$.

The ANOVA analysis test indicated that mean scores between males and females on the final exam were slightly dissimilar but not statistically significant ($x = 9.1$ vs $x = 0.90$). However, there were significant differences between different ages ranks, specially between 21–25 and 41–50 ($p = 0.015$) in favor of the first group. As for Education Level, means scores were not the same ($p < 0.0001$). We found a noticeable trend in the dataset suggesting that final exam scores were higher as education level increases (see Fig. 4).

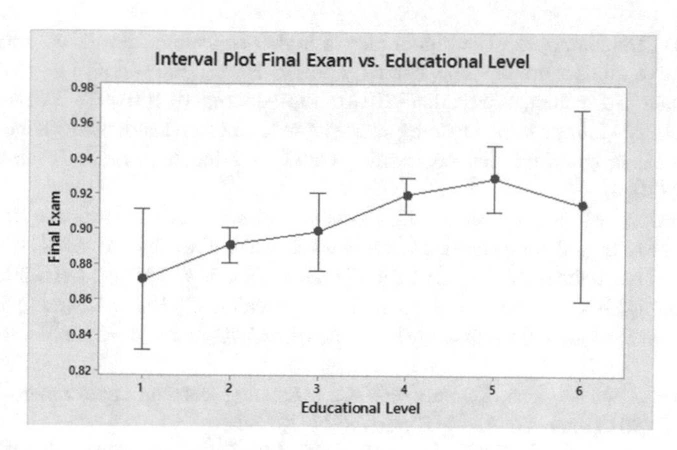

Fig. 4. Final exam interval plot for different education levels.

5 Discussion and Conclusion

The first result reveals that a relatively high percentage of participants (17%) finalized the MOOC course. According to [16] the average completion rate in Coursera is of 4% whereas [17] extend this rate around 10%. Our success may be due to the increasing interest on the energy sector among the Mexican population and the promotional work

Secondly, males and female participants showed the same interest in gamification. Their engagement to the gamified challenge is opposed to [18] who found that 99% of the male participants (against 94% females) were willing to play when accomplishing their class tasks.

Thirdly, our study shows that, in contrast to what it is widely believed, millennials (aged 26–30) are not more inclined to use gamification than older generations. This is not the case of the X generation (aged 21–25) who are keener to use this short of educational strategy.

The limitation of our study though was to technically adequate the gamified challenge to the *MexicoX* platform because the answer to it was based in a traditional answer type question (i.e., multiple choice). This leads us to think that MOOC interfaces nowadays greatly restrict the application of the gamification teaching methodology for online courses.

References

1. Rohatgi, A., Scherer, R., Hatlevik, O.: The role of ICT self-efficacy for students' ICT use and their achievement in a computer and information literacy test. Comput. Educ. **102**, 103–116 (2016)
2. Sánchez, A.B., Mena, J., He, G., Pinto, J.: Teacher development and ICT: the effectiveness of a training program for in-service school teachers. Procedia Soc. Behav. Sci. **92**, 529–534 (2013)

3. Kergel, D., Heidkamp, B.: The digital turn in higher education towards a remix culture and collaborative authorship. In: The Digital Turn in Higher Education, pp. 15–22. Springer Fachmedien Wiesbaden, Wiesbaden (2018). http://doi.org/10.1007/978-3-658-19925-8_2
4. Cutri, R.M., Whiting, E.F.: Opening spaces for teacher educator knowledge in a faculty development program on blended learning course development. Stud. Teach. Educ. **14**(2), 125–140 (2018)
5. Ramírez-Montoya, M.-S., Mena, J., Rodríguez-Arroyo, J.A.: In-service teachers' self-perceptions of digital competence and OER use as determined by a xMOOC training course. Comput. Hum. Behav. **77**, 356–364 (2017). https://doi.org/10.1016/J.CHB.2017.09.010
6. Oswal, S.: MMOC in the global context. In: Monske, E., Blair, K. (eds.) Handbook of Research on Writing and Composing in the Age of MOOCs, pp. 39–55. IGI Global, Hershey (2017)
7. Bartolomé, A.R., Steffens, K.: Are MOOCs promising learning environments? Comunicar **44**, 91–99 (2015). https://doi.org/10.3916/C44-2015-10
8. Xing, W., Chen, X., Stein, J., Marcinkowski, M.: Temporal predication of dropouts in MOOCs: reaching the low hanging fruit through stacking generalization. Comput. Hum. Behav. **58**, 119–129 (2016). https://doi.org/10.1016/j.chb.2015.12.007
9. Werbach, K., Hunter, D.: The Gamification Toolkit. Wharton Digital Press, Philadelphia (2015)
10. Lee, J., Hammer, J.: Gamification in education: what, how, why bother. Acad. Exch. Q. **15**(2), 146 (2011)
11. Rincón-Flores, E., Ramírez-Montoya, M.S., Mena, J.J.: Challenge-based gamification as a teaching' open educational innovation strategy in the energy sustainability area. In: Proceedings of the Fourth International Conference on Technological Ecosystems for Enhancing Multiculturality, TEEM 2016, Salamanca, Spain (2016a). http://hdl.handle.net/11285/620886
12. Hew, K.F.: Promoting engagement in online courses: what strategies can we learn from three highly rated MOOCS. Br. J. Edu. Technol. **47**(2), 320–341 (2016). https://doi.org/10.1111/bjet.12235
13. Kahan, T., Tal, S., Nachmias, R.: Types of participant's behaviors in a massive open online course. In: EDEN 2015 Annual Conference Book of Abstracts -Expanding Learning Scenarios, vol. 18, pp. 52–53, June 2015. https://doi.org/10.19173/irrodl.v18i6.3087
14. Kerlinger, F., Lee, H.: Investigación del Comportamiento. Métodos de investigación en Ciencias Sociales. Cuarta edición. MacGraw-Hill, México (2002)
15. Ramírez-Montoya, M.S., García, C., González, S., Aldape, P., Farías, S.: Estadísticas de Impartición de MOOC's en abril 2017 (Presentación del Proyecto 266632 Laboratorio Binacional para la Gestión Inteligente de la Sustentabilidad Energética y Formación Tecnológica). Tecnológico de Monterrey, Monterrey, México. Documento inédito. Disponible en (2017). http://hdl.handle.net/11285/622441
16. Armstrong, L.: 2013- The year of ups and downs for the MOOCs. Changing Higher Education (2014). http://goo.gl/SqwGWn
17. Liyanagunawardena, T., Williams, S., Adams, A.: The impact and reach of MOOCs: a developing countries' Perspective. eLearning Papers **1**(33), 1–8 (2013). http://elearning europa.info/sites/default/files/asset/In-depth_33_1.pdf
18. Lenhart, A., Kahne, J., Middaugh, E., Rankin Macgill, A., Evans, C., Vitak, J.: Teens, Video Games, and Civics: Teens' gaming experiences are diverse and include significant social interaction and civic engagement. Pew Internet & American Life Project, pp. 1–64 (2008). http://doi.org/10.1016/j.chembiol.2006.01.005

Real Marks Analysis for Predicting Students' Performance

Giulio Angiani, Alberto Ferrari, Paolo Fornacciari, Monica Mordonini, and Michele Tomaiuolo(✉)

Department of Engineering and Architecture, University di Parma, Parma, Italy
michele.tomaiuolo@unipr.it

Abstract. In the last few years the amount of electronic data in high schools has grown tremendously, also as a consequence of the introduction of electronic logbooks, where teachers store data about their students' activities: school attendance, marks obtained in individual test trials and the typology of these tests. However, all this data is often spread across multiple providers and it is not always easily available for research purposes. Our research project, named ELDM (Electronic Logbook Data Mining), focuses exactly on this information. In particular, we have developed a web-based system which is freely usable by school stakeholders; it allows them to (*i*) easily share school data with the ELDM project, and (*ii*) check the students' very different learning levels. On the basis of data collected from adhering schools, we have applied data mining techniques to analyze all the students' behaviours and results. Our findings show that: (*i*) it is possible to anticipate the outcome prediction in the first school months; and (*ii*) by focusing only on a small number of subjects, it is also feasible to detect serious didactic situations for students very early. This way, tutorship activities and other kinds of interventions can be programmed earlier and with greater effectiveness.

Keywords: Machine learning · Electronic logbooks
Student performance · Educational Data Mining

1 Introduction

Predicting students' performances is one of the most interesting challenges that an educational institution can face today. Being able to understand students' difficult situations very early paves the way for timely intervening with educational and didactical strategies, to prevent a negative outcome of the scholastic course. Data Mining (DM) techniques are a useful tool that can be also applied successfully to educational data. DM has been indeed used in several fields, as finance [10], healthcare [17], weather forecasting [16] and in social network analysis [1,8]. Application of Data Mining in the educational field is usually called Educational Data Mining (EDM) [7]: this is a still emerging interdisciplinary research area, that has received the attention of the scientific community in recent years.

T. Di Mascio et al. (Eds.): MIS4TEL 2018, AISC 804, pp. 37–44, 2019.
https://doi.org/10.1007/978-3-319-98872-6_5

In one of the first important works about EDM [4], Baker detects five fields of study in EDM: prediction, clustering, relationship mining, discovery within models, and extraction of data for human judgment. However, almost all EDM previous works are related to academic world or virtual learning [6] and only in a few cases to High School data [12]. In all cases, these works analyze data collected from surveys, or related to family status and economic position.

Our research instead focuses only on day-by-day students' scholastic life information, using data stored in electronic logbooks of ten Italian high schools. In these logbooks, indeed, teachers store data about their students' activities: school attendance, marks obtained in individual test trials and the typology of these tests. However, these logbooks are often spread across multiple different providers and are not readily available for analysis. In particular, we have developed a web-based system, named ELDM (Electronic Logbook Data Mining). The system allows school managers and authorised users to: (*i*) easily share school data with the ELDM project, exporting them from their own service provider, with just few clicks; and (*ii*) obtain some detailed results of data analysis for their own school.

It's important to highlight some previous related works, which describe useful results and experiences, for our research. Kapur *et al.* [11] performed a comparison of different classification algorithms, applied to educational data. They worked with 480 entries related to students enrolled Kalboard 360 e-learning platform. For each entry, they collected 16 features; part of these are strictly related to family factors. The research explains as J48 Decision Tree and Random Forest are the most effective algorithms for this purpose. In Veracano *et al.* [12], several experiments for predicting students' dropout are performed using real data from 419 students of one Mexican High School. This work shows different algorithms that use data related also to social conditions and to the marks of the previous school grade. Due to the small number of analyzed cases, the dataset is highly unbalanced. In [3], Asif *et al.* analyze university studies. Their work indicates that by focusing on a small number of courses, that are indicators of particularly good or poor performance, it is possible to provide prompt warning and support to low achieving students. In [15], Saarela *et al.* present a research allowing to learn the difficulty level of different math questions and to predict weather or not a student with a particular background profile will be successful in answering correctly. Daud *et al.* [5] show with experimental results that some outcomes can be predicted using information related to family expenditures and students' personal features. In [18], Xu *et al.* develop a novel algorithm that enables the progressive prediction of students' performance by adapting ensemble learning techniques and using education-specific domain knowledge. They prove the prediction performance of their algorithm and show its improvement against benchmark algorithms on a real-world student dataset from UCLA. Asif *et al.* [2] use data mining techniques for predicting the students' graduation performance in the final year of university using only pre-university marks and examination marks of early years at university. They identify two main groups of students using clustering and using only a reduced set of indicator subjects.

Pereira *et al.* [13] use decision trees in a comparative analysis for predicting students' performances. CHAID decision tree proves the best algorithm for this issue. Finally, Prasada Rao *et al.* [14] compare the performance of J48, Naïve Bayes and Random forest algorithms. Their dataset consists of 200 entries of undergraduate computer science and engineering students. Their result shows that Random Forest provides the best performances, for their dataset.

The rest of the manuscript is arranged in the following way: Sect. 2 describes the data collection and methodology used for this study; results and discussions are presented in Sect. 3; finally, Sect. 4 provides some concluding remarks.

2 Data and Metodology

Data. Data used for our research have been extracted from the electronic logbooks of 10 high schools, located in different parts of Italy. All data have been anonymized in accordance with current Italian privacy laws [9]. All information are related to marks obtained by students in their school tests and to their class attendance. Also end period marks (italian school year is usually divided into two periods) and end of year outcomes have been extracted, for each subject. The informations about marks and attendance have been used for training some classifiers to predict the final outcome.

Data Preparations. Since Italian high schools do not have standardized evaluation tests, marks are assigned by teachers in a 1–10 scale. Each teacher, for each test, can choose his/her way of evaluation. However, the assigned marks must belong to the [1–10] interval and are float values. Values outside this interval have not been used for our research. All the subjects have been clustered in six groups:

1. Italian Language (ita)
2. Mathematics (mat)
3. English Language (eng)
4. History (his)
5. Subjects strictly related to the student course (cou)
6. Other subjects (oth)

The clustering of subjects has been necessary for analyzing students belonging to different courses. The first four groups are common to all Italian high school courses, but each course has also its own subjects: these ones have been assembled in the fifth group (*cou*). The sixth group (*oth*) contains all the remaining subjects.[1]

Feature Selection and Transformation. Starting from daily raw-data, we have built the students' features, grouping data for each month with the following method: for each group, we calculated the average mark, collected in the

[1] E.g.: "Informatics" belongs to the 5th group for students of a course on ICT, but to 6th group for students of a course on foreign languages.

period from Sep, 15th (start of school) till the end of each month, from October to May. The name assigned to these features has the following format, for values related to a single student: *<subject>_<month>*; and the following format, for the average obtained using marks of all his/her classmates for the same period and for the same subject: *<subject>_<month>_grp*. The second group of features contains information about a student's school attendance and about the average attendance of his/her classmates. Each feature shows a student's school attendance in a certain period, as a number of days. Like in the previous case, these features have the following format: *abs_<month>*; and *abs_avg_<month>_grp*.

The third set of features is related to a student's trend in a certain period. For each student and for each subjects group, we have selected all marks related to that group, and we have calculated the linear regression line for these marks. The *trend* value is pointed out by the tuple $(m,\ c,\ dev)$[2] which contains the values used to populate this third set.

Also in this case, the computed value has been compared with the corresponding average value of all the classmates. The same features have been calculated also for each student's group.

The last set of features contains only data about the school, the year, the course-year, the study course and about some data of end-of-school subjects marks. The final feature set F contains 410 elements.

Final Dataset. The whole dataset contains 13151 different instances, related to 10342 different students attending 10 Italian high schools. There are more instances than students, because for some schools we have collected data of several years. Each instance includes 410 float value at most, one for each feature of the F set. Each instance has also the end-of-year outcome feature, which is mandatory in classification experiments. The outcome feature can assume one of the following values: "POSITIVE","NEGATIVE" or "SUSPENDED"[3]. In Table 1, the distribution of students according to the final results is shown. With such a number of instances, it has been possible to build an absolutely balanced dataset, differently from almost all the works presented in Sect. 1. First, we have trashed out all the instances with more than 50 not valid values (for example an average value calculated by only a mark) and then, for each of three classes, we have used 1000 correct instances which have been divided in equal parts between training and test set.

Therefore both the definitive training set and test set are composed with 1500 instances (500 instances for each of the three classes).

3 Results and Discussion

In our work, we performed several experiments. In the first one we focused in forecasting the final outcome.

[2] m is the slope and c is the y-intercept of the linear regression line, while *dev* is the standard deviation for the selected marks set.

[3] The "SUSPENDED" value indicates that the student must pass another exam at the end of August, for accessing the next class.

Table 1. Distribution of students' final results.

Final result	Number of students	Percentage
POSITIVE	10609	80.7%
NEGATIVE	1100	8.6%
SUSPENDED	1442	11.0%

Comparison Among Different Classification Methods. For this purpose we used three standard classification techniques. We compared classification results using the first features, containing data about the first three months of school lessons (i.e. October, November and December), with results obtained from the next three months (January, February and March). In such a case, each instance consists of 148 features. Table 2 shows the classification result calculated, for the first and the second period, with three different algorithms: J48, SVM and Random Forest. It is not surprising to find a great increase of accuracy, using data of the second period: March marks are absolutely valuable for understanding the real situation of students. We notice that, in both cases, the Random Forest algorithm obtains better results in terms of accuracy for classification. However, it's very valuable to predict the final year outcome with an accuracy of 79%, using only the marks of the first three months. It is very interesting also that almost all the classification errors are between the "SUSPENDED" class and the others, as shown in Table 3. Only for 45 instances, the POSITIVE and NEGATIVE outcomes have been confused by the classifier (the error is slightly greater than 3%).

Table 2. Classification results with different algorithms with balanced data for the first and second period (three months), calculated on the test set.

Algorithm	First period		Second period	
	Accuracy	F-Measure	Accuracy	F-Measure
J48 Tree	71.00%	0.710	80.00%	0.800
SVM	73.13%	0.732	76.33%	0.765
Random Forest	**79.46%**	0.795	**82.60%**	0.827

Table 3. Confusion matrix, calculated on the test set with the Random Forest algorithm.

	Positive	Negative	Suspended
Positive	391	20	89
Negative	25	422	53
Suspended	37	84	379

Prediction of Last Year Students' Outcome. The second experiment is focused only on students attending their last course year (Italian high schools study courses usually are 5 years long). At the end of the last year, only two outcomes are possible: POSITIVE, if a student can access the final high school examination, and NEGATIVE, otherwise. Using only instances related to students in their last year we reduced our dataset to 3707 instances, with only 100 instances belonging to the NEGATIVE class. Hence, for obtaining a balanced train and test sets, we used only 200 instances equally divided between the training set and the test set. The classification results about this reduced dataset have an accuracy level of **94%**. Also in this experiment, the Random Forest algorithm has obtained the best results, in accordance with other cited works [11,14].

Relation Between Subjects and Accuracy. In the last experiment we have investigated how the accuracy could depend on subjects or periods. For this issue we have calculated the classification accuracy using a different number of relevant features. To determine which are the most important features for classification, we have applied the Information Gain algorithm to the training set. Table 4 shows the ranking of the 30 most important features. According to these results, we have calculated the accuracy using 5, 10, 20, 30, 50, 75, and 100 features. A comparison between the results obtained in this experiment is shown in Fig. 1. Also in this case, all classifications are performed with the Random Forest algorithm using all the attributes in the random selection. The results show that the highest value of accuracy is obtained with only 50 features (of 132 available using data of the first 3 months). We observe that a good level of accuracy (more than 75%) can be reached just with the first 20 attributes.

Table 4. Features relevance ranking calculated by Information Gain Ranking algorithm.

Rank	Feature	Rank	Feature	Rank	Feature
1	cou_dec	11	ita_dec	21	mat_oct
2	cou_nov	12	trend_c_cou_oct	22	Gft.trend_c_oth_dec
3	cou_oct	13	trend_c_his_nov	23	gft.trend_c_oth_nov
4	trend_c_cou_dec	14	trend_c_mat_dec	24	gft.trend_dev_mat_dec
5	his_dec	15	ita_nov	25	trend_c_eng_nov
6	mat_dec	16	gft.trend_dev_oth_nov	26	trend_c_mat_oct
7	trend_c_cou_nov	17	eng_dec	27	trend_c_eng_dec
8	mat_nov	18	gft.trend_dev_cou_dec	28	gft.trend_m_ita_dec
9	his_nov	19	trend_c_mat_nov	29	trend_dev_cou_dec
10	trend_c_his_dec	20	eng_nov	30	gft.trend_m_oth_dec

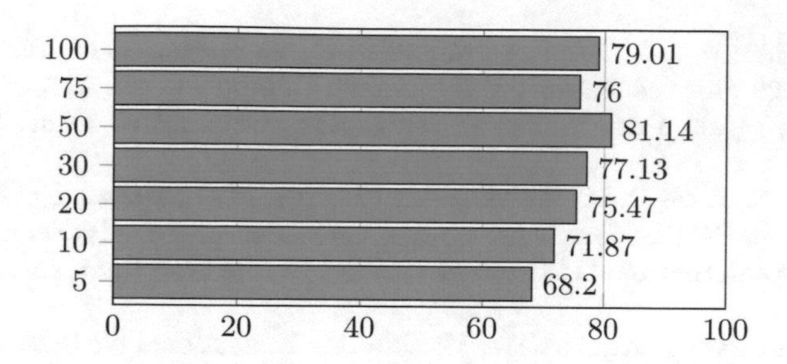

Fig. 1. Accuracy in prediction using different number of attributes.

4 Conclusions

On the basis of their own empirical experience, teachers and school managers tend to form an idea of the probability of a student's success or failure, already after a few months from the beginning of the lessons and after the initial evaluation tests. This evaluation is based on the results of the tests they have given but above all on observing the behavior and attitudes of students measured according to their years-long teaching experience. Our aim is to provide additional help by trying to extract some knowledge which may be hidden in the large volume of data, without using data related to social or familiar condition, neither to previous school study grades, and focusing only on students real marks and school attendance. The goal is to predict difficulties already in the first part of the school year. This would allow schools to intervene with recovery tools far ahead of the usual practices, that are typically driven by the results, after finding an obvious evidence of an existing problematic situation.

Although the results are very encouraging, our research can not indicate the teaching techniques to help students in difficulty, but can only highlight their situations. A possible future direction of investigation may regard the interventions applied by teachers for students with similar situations. The analysis may focus on the progress of students during the entire school year, to identify similar behaviors – for example, a constantly positive, discontinuous, or constantly negative trend – and study their correlation with overall performances.

Acknowledgements. This research is supported by Emilia-Romagna Assemblea Legislativa. The authors, therefore, acknowledge technical support of the Spaggiari Group and to ARGO which have developed software procedure for extracting data from school databases. Special thanks to all the schools which have joined our research project.

References

1. Angiani, G., Fornacciari, P., Iotti, E., Mordonini, M., Tomaiuolo, M.: Models of participation in social networks, pp. 196–224. IGI Global (2017)
2. Asif, R., Hina, S., Haque, S.I.: Predicting student academic performance using data mining methods. Int. J. Comput. Sci. Netw. Secur. **17**(5), 187–191 (2017)

3. Asif, R., Merceron, A., Ali, S.A., Haider, N.G.: Analyzing undergraduate students' performance using educational data mining. Comput. Educ. **113**, 177–194 (2017)
4. Baker, R., et al.: Data mining for education. Int. Encycl. Educ. **7**(3), 112–118 (2010)
5. Daud, A., Aljohani, N.R., Abbasi, R.A., Lytras, M.D., Abbas, F., Alowibdi, J.S.: Predicting student performance using advanced learning analytics. In: Proceedings of the 26th International Conference on World Wide Web Companion, pp. 415–421 (2017)
6. Ducange, P., Pecori, R., Sarti, L., Vecchio, M.: Educational big data mining: how to enhance virtual learning environments. In: International Joint Conference SOCO 2016-CISIS 2016-ICEUTE 2016, pp. 681–690. Springer, Cham (2016)
7. Ferguson, R.: Learning analytics: drivers, developments and challenges. Int. J. Technol. Enhanc. Learn. **4**(5–6), 304–317 (2012)
8. Fornacciari, P., Mordonini, M., Tomaiuolo, M.: Social network and sentiment analysis on twitter: towards a combined approach. In: KDWeb 2015. CEUR Workshop Proceedings, vol. 1489. CEUR-WS.org (2015)
9. Gazzetta: Codice di deontologia e di buona condotta per i trattamenti di dati personali per scopi statistici e scientifici. Gazzetta Ufficiale della Repubblica Italiana 2004(190) (2004)
10. Jadhav, S., He, H., Jenkins, K.W.: An academic review: applications of data mining techniques in finance industry. Int. J. Soft Comput. Artif. Intell. **4**(1), 79–95 (2017)
11. Kapur, B., Ahluwalia, N., Sathyaraj, R.: Comparative study on marks prediction using data mining and classification algorithms. Int. J. Adv. Res. Comput. Sci. **8**(3) (2017)
12. Márquez-Vera, C., Cano, A., Romero, C., Noaman, A.Y.M., Mousa Fardoun, H., Ventura, S.: Early dropout prediction using data mining: a case study with high school students. Expert Syst. **33**(1), 107–124 (2016)
13. Pereira, B.A., Pai, A., Fernandes, C.: A comparative analysis of decision tree algorithms for predicting student's performance. Int. J. Eng. Sci. **7**(4), 10489–10492 (2017)
14. Rao, K.P., Rao, M.C., Ramesh, B.: Predicting learning behavior of students using classification techniques. Int. J. Comput. Appl. **139**(7) (2016)
15. Saarela, M., Yener, B., Zaki, M.J., Kärkkäinen, T.: Predicting math performance from raw large-scale educational assessments data: a machine learning approach. In: JMLR Workshop and Conference Proceedings, vol. 48. JMLR (2016)
16. Venkatesh, S., Chandrakala, D.: A survey on predictive analysis of weather forecast. Weather (2017)
17. Vidhu, R., Kiruthika, S.: A survey on data mining techniques and their comparison approaches for healthcare. Data Min. Knowl. Eng. **9**(1), 14–19 (2017)
18. Xu, J., Han, Y., Marcu, D., Van Der Schaar, M.: Progressive prediction of student performance in college programs. In: Thirty-First AAAI Conference on Artificial Intelligence, pp. 1604–1610 (2017)

Multisensory Educational Materials: Five Senses to Learn

Michela Ponticorvo[1(✉)], Raffale Di Fuccio[1], Fabrizio Ferrara[1], Angelo Rega[2], and Orazio Miglino[1]

[1] Department of Humanistic Studies, University of Naples "Federico II", Naples, Italy
michela.ponticorvo@unina.it
[2] IRFID, Institute for Research, Formation and Innovation on Disabilities, Ottaviano, Italy

Abstract. The digital revolution has deeply transformed the educational materials, especially for children, in a pathway going from physical objects to digital ones. In this paper, after delineating some relevant milestones in this route, one further step is delineated which proposes multisensory materials for learning. These kind of materials keep the main advantages of physical educational materials and marry them with the ones derived from the digital world. Then, an example of multisensory materials for learning is introduced: STTory, which is dedicated to multisensory storytelling, together with results about one study on this material which indicate that introducing multisensory elements in digital materials has positive effects on learning in children. Multisensory educational materials provide multimodal input for learning individual, enriching learning environments and educational scenarios.

Keywords: Multisensory educational materials · Training tools
Intelligent learning environment

1 Introduction

The five senses are the channels through which humans can know the world around them to get information about and to better adapt. Since birth, and more and more as development goes by, children refine and exploit their senses harmonically with action which allows to select relevant information from the environment. Let us imagine a eight-months years old boy, sitting on a carpet with many toys. With sight he will appreciate colours and will grasp the rattle he prefers, with hearing he will listen to the sounds it will produce, shaking and beating it against the floor, with touch he will experience if it is soft or hard. He will probably also taste and smell it, exploring features related to these senses. Hands play a crucial world, in this exploration process, along the whole human life-cycle. This explains, on the technological hand, why it is so easy to learn to use devices which exploits pointing and touching such as the mouse and the tablet and, on the educational hand, why so many educational materials that can

T. Di Mascio et al. (Eds.): MIS4TEL 2018, AISC 804, pp. 45–52, 2019.
https://doi.org/10.1007/978-3-319-98872-6_6

be manipulated are used at school. These physical objects, that are specifically designed to foster learning are called manipulatives, are mainly used to learn and teach mathematics [2,14] and are fundamental in influential pedagogical approaches.

The first manipulatives specifically conceived for education were introduced by Froebel [7], in the 19th century who used them in his kindergarten. He developed different types of objects to favour pattern recognition and geometric forms identification in nature. At the beginning of 1900, Maria Montessori [11] made a great use of manipulatives, further advancing their importance in education. She indeed designed many materials to help children to actively discover and learn basic ideas, not only in math, but also in other subjects. Froebel and Montessori proposed a wide variety of materials that favour active exploration and try to stimulate all children senses. Zuckerman and colleagues [16], argue that Froebel-inspired manipulatives foster modelling of real-world structures, whereas Montessori-inspired manipulatives foster modelling of more abstract structures; nonetheless they share the relevant feature to appeal to all the different senses. For example, Froebel gifts are play materials for young children designed for the original Kindergarten that indeed allowed children to learn using their senses, and Montessori used specific materials, named sensorial materials and widely used in the Montessori classroom to help children to develop and refine their five senses. They offer the chance to engage the learning child sight, touch, sense of smell, taste and hearing, also promoting action which is especially important in the first phases of human development [9]. The digital revolution has deeply transformed the educational materials, especially for children, and the physical objects, the manipulatives have undergone a deep transformation, which will be outlined in the following section.

2 From Manipulatives to Digital and Virtual Manipulatives

The traditional manipulatives, introduced in the previous section, have experienced a deep transformation due to the coming of ICT. Manipulatives have become a tangible interface for education, in informatics language a Tangible User Interface TUI [1,8]. Digital, augmented, virtual version of manipulatives have been proposed: Resnick and colleagues [15] developed a new generation of digital manipulatives, computationally-enhanced versions of traditional children's toys which enable children to explore a new set of concepts, namely blocks, beads, balls, and badges. Zuckerman and colleagues proposed a computationally enhanced versions of manipulatives, derived from Montessori materials, in the form of enhanced building blocks: physical, modular interactive systems that serve as general-purpose modelling and simulation tools for dynamic behaviour. In a similar vein, virtual manipulatives have been proposed [12]. They are a new class of manipulatives for computer programs that use visual representations and include static and dynamic visual representations of concrete manipulatives. This kind of materials keeps some features of traditional manipulatives, such as the chance to attract children and favour learning, for example, in mathematics and have specific advantages as the flexibility or the opportunity to record data. But they lose an important feature underlined above: they mainly rely on sight, sometimes on hearing and touch, whereas smell and taste are completely lost. This constitutes, in our opinion, a drawback of digital materials, because, smell and taste, the so-called chemical senses, are indeed important in everyday life, (and have been important in human phylogenetic story), have specific neuro-cognitive features and can help the learning process. Olfaction, for example, has strong links with emotions and can therefore affect behaviour, thanks to emotional associative learning to odours. Moreover, the olfactory network has the uniqueness to do not pass through the thalamus and go directly to the cortex, thus providing the neural basis for the strong connection between olfactory stimuli and emotional memory. For this reason, in the following sections, a platform to design and implement multisensory learning materials is introduced, an example of learning system is provided, together with data, to show the effectiveness of this approach on cognitive functions as learning and memory.

3 STTory: A Tool for Multisensory Storytelling

In this section, STTory [3,4,6,13] is introduced. It is a tool for digital and multisensory storytelling that allows a narration with smell, taste and touch and it allows the learner to jump inside a story through all the senses. The learner is not a passive listener, but has the chance to intervene directly in the multisensory narration, affecting some story elements and characters' behaviours. The learner interacts with a manipulative, TUI, as described above, made of a series of physical learning materials equipped by a passive RFID, which is pasted in a hidden place, namely:

- 3 dolls representing a mouse, a princess and a duck;
- 6 smelling jars (the smells are apple, soap, rose, sea, burnt, mint);
- 6 pieces of fabric representing 3 shoes in three different colours (red, blue and yellow) and 3 skirts in three different colours (red, blue and yellow);
- 2 flavours contained in two different little red jars (strawberry and blackberry candies).

The materials are freely at disposal on a table, and the learners are able to manipulate and interact with them. STTory proposes three different stories: the Mouse Story, the Princess Story and the Duck Story. The stories have a parallel structure and the required tasks are in common (six steps in total, one doll manipulation, one doll with dress, three interaction with odours, one interaction with flavours). At every step of the story, STTory asks for one or more materials and the learner replies by placing it/them on the active table. The request is made by a visual and an aural message.

Primarily, STTory asks to find the doll showed on the screen by the system, selecting the story assigned. After the doll selection, the story requires to choose a dress for the related doll (the shoes for the stories of Mouse and Duck, the skirt for Princess story). The requested task is to select the dress with the favorite colour and to put it directly on the doll, dressing it (Fig. 1). The active table is able to recognize the couple formed by the doll and the selected dress, so the character will wear the selected dress during the next steps, as a result of the choice.

Fig. 1. A screenshot from the digital side of STTory and the doll dressing on the physical side

As the storytelling goes on, the learner is involved in the story by sensing the stimulation related to a particular event in the story. For instance, in the case of the princess walking in a garden, when she feels a nice smell of grass, the learner has to find the grass odour in order to go on in the story. After the learners finds the right smell, the story can continue with a new event, again involving smell.

The last step is similar to smelling steps, but it involves taste. If the character buys a strawberry ice-cream to his girlfriend, the learners has to find the strawberry flavour between the proposed candies.

Every time the learners fails the task, selecting the wrong smell/taste (in the case of dresses all solutions are correct), STTory asks to retry with another jar; when it is correct, the learners gets a congratulation feedback and the story proceeds to the next frame. A study was run to test STTory effectiveness in learning.

4 Materials and Method

The experimentation aims at evaluating the acceptability and thememorization of story with the STTory system and the multisensory manipulative materials, as the hyphotesis is that using multisensory manipulatives can improve memorization. For this reason, the experimental design include 3 conditions with symmetric stories (Fig. 2): multisensory materials; only digital with touch-screen technology, a traditional book (a game-book). Therefore each story has three variants:

a. **Multisensory**: with the STTory system, where the user has to interact with the physical and digital material by the Active Board (as described before);
b. **Touch-screen**: with the touch-screen technology (tablet), where the user interacts by clicking with the fingers on a screen;
c. **Book**: with the book, where the user interacts with pages, by browsing in them.

Fig. 2. The materials for the 3 versions of the story: Multisensory, Touch-screen and Book

The Multisensory version is implemented using a tool STELT, developed by authors, which connect physical and digital materials that allow to write a wider variety of scenarios. For the Multisensory version, the learners interact with the TUIs and with Activity Board 1.0 [5] with a Mediacom WinPad W801 Tablet. The connection between the tablet and the Activity Board 1.0 is by USB cable. For the Touch-screen version, the learners play with a Mediacom WinPad W801 Tablet. The software applications for the Multisensory and Touch-screen versions, were developed by using the software STELT [10]. For the game-book version, the user browses some printed sheets (A4) in the form of a book.

The participants are 24 students, 14 males and 10 females, of the primary school (second and third class: 7–9 years old), in particular from the Istituto Comprensivo of Viggianello (Viggianello, Basilicata, Italy) and the Piaget- Majorana Institute of Rome, Italy. The participants were divided in little groups of three users and each group interacted with three conditions (Multisensory, Touch-screen and Book). The matching between the stories (Mouse, Princess and Duck) and the technologies were randomized as well the execution order.

The setting was a classroom where there were only the group of three learners and two researchers, one with the role of assistance (when requested) and another as qualitative observer of the intra-group interactions and user- technologies interactions. At the end of the session, after an elapsed time of about two hours, the researchers asked to the learners, one by one:

(1) which type of technological support they prefer. They could see the three technological tools to avoid confusion. In particular the question was: Which tool do you prefer?
(2) some questions about the story; the questions for the three stories are symmetric with the same structure.

The crosses between stories and technologies where equally covered by 24 participants. The first questions aims to explore the acceptability of the technologies, whereas the other questions focuses on learning and memorization about stories with different approaches. We expect that the system are highly accepted and that manipulatives help improving memorization.

After playing with one of the three books, depending on experimental conditions, the children were asked questions, with a structured tool, to assess acceptability and memorization of story contents.

4.1 Acceptability Study

The acceptability study embraces the entire group of participants in this track of the trial. After the interaction with the three story with different and randomized technologies (Multisensory, Touch-screen and Book), the learner answered individually to the question: Which tool do you prefer?.

The majority of the students prefers the tangible materials; in particular 66% of the participants preferred the tangible materials (16 participants). The 25% (6 participants) preferred the touch-screen technology on the tablet, and only 2 children (8%) chose the book mode. The gender does not affect the preference.

4.2 Memory Study

At the end of the trials with the three conditions and related stories, the children answer individually some questions about the story content. Each child freely replies to the questions, there are no pre-defined choices. The answer is considered as incorrect in the case it does not adhere to the story, or the learner does not remember the reply.

5 Results

Children remember better the stories when they use the Multisensory materials with a 60,4% of correct replies. In the case of Touch-screen the correct replies were (32,3%). With the Book interaction the situation is intermediate.

This distribution is significantly different from chance ($chi^2 = 10.03$; prob. below 0.05). The average of correct replies is higher in the Multisensory condition (av.= 2.42, st.dev. = 1.28), followed by the Book (av.= 1.87, st.dev. = 1.11) and by the Tablet (av.= 1.29, st.dev. = 1.33).

6 Qualitative Observations

The data and the observations of the researchers during the experimentation converge. The answers in the case of the Multisensory story are very quick and without doubt, in the case of touch-screen the replies are uncertain. However it is very common that children try to reply to the question even if they do not remember, by using their common sense.

These data indicate that using STTory multisensory materials is effective for learning. It seems that the tablet offers an interaction which is too fast, thus preventing memorization. In fact children tend to be attracted by the interaction in itself, without focusing on the story. On the contrary, experience with hands and all senses, activates memorization, keeping the mnestic trace. The traditional book is the best-known materials and reading allows to focus on the story better than the tablet by itself. The STTory system that puts together physical and digital attract children and are strongly appreciated by them, favouring learning and memorization.

7 Conclusions

Learning takes place in scenarios that change continuously according to technology development, which has been dramatic in the last years. The digital side has become very relevant in people life and this issue stimulates vibrant debates. For learning it has led to lose something relevant, namely the interaction with the physical world. But technology, thanks to augmented reality and Internet of Things approaches, offers also the chance to recover this aspect, with tangible interfaces.

Children and adults, this way, can lift their eyes from the screen and use their hands, that are, as Montessori asserted, instruments of intelligence, an extension of thought, just like their bodies that allow to grab, write, shake, smell and taste.

Moreover, the technological gap which children experience between school and home context, represents a rift between educational contexts that, on the contrary, can work together to educate children.

The data reported above, indicate that touch-screen technology favours fast and immediate interaction between children and story, but this can be counterproductive if the learning task requires reflection and memorization, which is favoured by an interaction mediated by physical tools. This argument would be in favour of anti-technological coalition in the cited debate.

But if we consider the Multisensory system, the performance is better than the traditional books.

Putting physical and digital together, keeps the advantages of both: flexibility, manageability, data recording, immediate and personalized feedback for the digital; manipulation, active exploration, stimulation of all senses for the physical.

Designing and implementing multi-sensory materials opens the challenge to re-conceive learning environments and educational scenarios, to re-thinking the multimodal input for learners, to re-invent traditional educational approaches with multimodal materials, in one word, to re-shape educational approaches.

References

1. Billinghurst, M., Kato, H., Poupyrev, I.: Tangible augmented reality, p. 7. ACM SIGGRAPH, ASIA (2008)
2. Boggan, M., Harper, S., Whitmire, A.: Using manipulatives to teach elementary mathematics. J. Instr. Pedagogies **3**, 1 (2010)
3. di Ferdinando, A., di Fuccio, R., Ponticorvo, M., Miglino, O.: Block magic: a prototype bridging digital and physical educational materials to support children learning processes. In: Smart Education and Smart e-Learning, pp. 171–180. Springer, Cham (2015)
4. Di Fuccio, R., Ponticorvo, M., Di Ferdinando, A., Miglino, O.: Towards hyper activity books for children. connecting activity books and montessori-like educational materials. In: Design for Teaching and Learning in a Networked World, pp. 401–406. Springer, Cham (2015)
5. Di Fuccio, R., Siano, G., De Marco, A.: The activity board 1.0: RFID-NFC WI-FI multitags desktop reader for education and rehabilitation applications. In: World Conference on Information Systems and Technologies, pp. 677–689. Springer, Cham (2017)
6. Di Fuccio, R., Ponticorvo, M., Ferrara, F., Miglino, O.: Digital and multisensory storytelling: narration with smell, taste and touch. In: European Conference on Technology Enhanced Learning, pp. 509–512. Springer, Cham (2016)
7. Froebel, F.: The education of man. A. Lovell and Company, New Orleans (1885)
8. Ishii, H.: The tangible user interface and its evolution. Commun. ACM **51**(6), 32–36 (2008)
9. Kontra, C., Goldin-Meadow, S., Beilock, S.L.: Embodied learning across the life span. Top. Cogn. Sci. **4**(4), 731–739 (2012)
10. Miglino, O., Di Ferdinando, A., Di Fuccio, R., Rega, A., Ricci, C.: Bridging digital and physical educational games using RFID/NFC technologies. J. e-Learning Knowl. Soc.**10**(3), (2014)
11. Montessori, M.: The montessori method. Transaction publishers, New Jersey (2013)
12. Moyer, P.S., Bolyard, J.J., Spikell, M.A.: What are virtual manipulatives? Teach. Child. Math. **8**(6), 372 (2002)
13. Ponticorvo, M., Di Fuccio, R., Di Ferdinando, A., Miglino, O.: An agent-based modelling approach to build up educational digital games for kindergarten and primary schools. Expert Systems (2017)
14. Puchner, L., Taylor, A., O'Donnell, B., Fick, K.: Teacher learning and mathematics manipulatives: a collective case study about teacher use of manipulatives in elementary and middle school mathematics lessons. Sch. Sci. Math. **108**(7), 313–325 (2008)
15. Resnick, M., Martin, F., Berg, R., Borovoy, R., Colella, V., Kramer, K., Silverman, B.: Digital manipulatives: new toys to think with. In: Proceedings of the SIGCHI Conference on Human Factors in Computing Systems, pp. 281–287. ACM Press/Addison-Wesley Publishing Co., New York, January 1998
16. Zuckerman, O., Arida, S., Resnick, M.: Extending tangible interfaces for education: digital montessori-inspired manipulatives. In: Proceedings of the SIGCHI Conference on Human Factors in Computing Systems, pp. 859–868. ACM (2005)

Human Expert Labeling Process: Valence-Arousal Labeling for Students' Affective States

Sinem Aslan[1], Eda Okur[1], Nese Alyuz[1](✉), Asli Arslan Esme[1], and Ryan S. Baker[2]

[1] Intel Corporation, Hillsboro, OR 97124, USA
{sinem.aslan, eda.okur, nese.alyuz.civitci, asli.arslan.esme}@intel.com
[2] University of Pennsylvania, Philadelphia, PA 19104, USA
rybaker@upenn.edu

Abstract. Affect has emerged as an important part of the interaction between learners and computers, with important implications for learning outcomes. As a result, it has emerged as an important area of research within learning analytics. Reliable and valid data labeling is a key tenet for training machine learning models providing such analytics. In this study, using Human Expert Labeling Process (HELP) as a baseline labeling protocol, we investigated an optimized method through several experiments for labeling student affect based on Circumplex Model of Emotion (Valence-Arousal). Using the optimized method, we then had the human experts label a larger quantity of student data so that we could test and validate this method on a relatively larger and different dataset. The results showed that using the optimized method, the experts were able to achieve an acceptable consensus in labeling outcomes as aligned with affect labeling literature.

Keywords: Affective state labeling · Circumplex Model of Emotion Inter-rater agreement · Intelligent tutoring systems · Affective computing

1 Introduction

Affect has emerged as an important part of the interaction between learners and computers, with important implications for learning and learner outcomes. As a result, it has emerged as an important area of research within learning analytics [1–3]. Reliable and valid data labeling is a key tenet for training machine learning models providing such analytics.

However, there is still considerable disagreement on key aspects of the study of affect - including even how affect itself is conceptualized. Two key paradigms have emerged for how affect is represented by researchers: (1) Affect as a set of discrete states [4–9] and (2) affect as a combination of a two-dimensional space of attributes. Several models exist that represent affect and emotion as a set of discrete states; perhaps the most widely-known such model is Ekman's set of six basic emotions [10], but other key models include the OCC model of the cognitive structure of emotions [11], and the set of

T. Di Mascio et al. (Eds.): MIS4TEL 2018, AISC 804, pp. 53–61, 2019.
https://doi.org/10.1007/978-3-319-98872-6_7

affective states studied by D'Mello and Graesser and their colleagues [12]. On the other hand, various models exist that represent affect as a two-dimensional structure [13]. The most widely-used model, however, both in education and other domains, is Russell's Circumplex Model of Emotion [14], which represents affect as a 2×2 combination of Valence (Negative to Positive) and Arousal (Calm to Excited).

Human Expert Labeling Process (HELP) is a labeling protocol [15], which was originally developed to enable affect labelers (i.e., human experts) with backgrounds in Psychology or Educational Psychology label students' discrete affective states (i.e., Satisfied, Confused, and Bored) occurring in a 1:1 digital learning scenario. In this study, using HELP as a baseline labeling protocol, we investigated an optimized method for labeling student affect based on Circumplex Model of Emotion (Valence-Arousal). Therefore, there is one major research question that this study aims to address: What is an optimized method for labeling students' affect in terms of valence and arousal? Identification of such a method will be critical for obtaining ground truths necessary for generation of analytics based on machine learning techniques.

2 Data Collection

The student data used in this study is a part of a larger dataset previously collected through authentic classroom pilots of an afterschool Math course in an urban high school in Turkey [16]. During the pilots, the students used an online learning platform for watching instructional videos and solving assessment questions. Our data collection application running in the background collected two streams of videos from the students: (1) Student appearance videos from the camera, to enable monitoring of observable cues available in the individual's face or upper body; and (2) student desktop videos, to enable observation of contextual information.

3 Labeling Tool, Labelers, and Training

We developed a labeling tool customized to various labeling experiments (see Fig. 1).

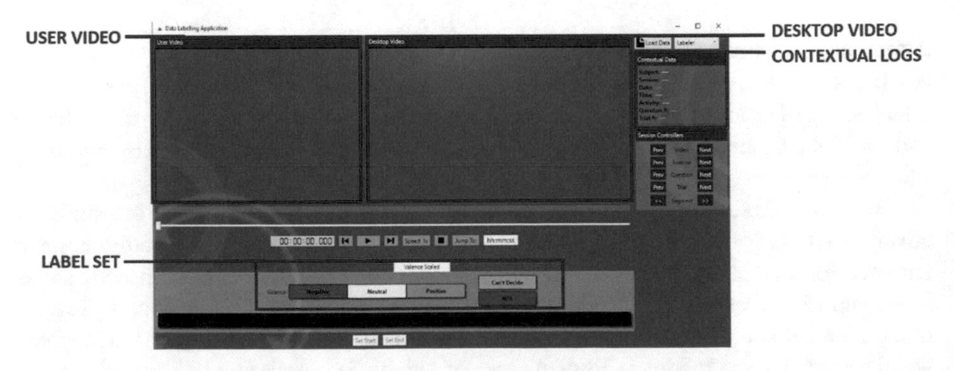

Fig. 1. Customized labeling tool (sample view), for labeling valence.

We recruited and trained six human experts with backgrounds in Psychology/ Educational Psychology. The training process took around eight hours which included instruction and demonstration, practice with feedback, as well as reviewing each other's labels and discussing differences in labeling outcomes. For all labeling tasks, based on observed state changes, the experts provided their Valence-Arousal labels using all available cues (e.g., student video/audio, desktop recording with mouse cursor locations, as well as any relevant contextual information from the device and content platform).

4 Experimental Conditions

To find an optimized method for Valence-Arousal labeling on the student data, we experimented with two variables: (1) Selection of Labels - which labels to use; and (2) Labeling Method - how to label.

For Selection of Labels, we had two conditions: Binary Labeling vs. Scaled Labeling. Binary Labeling had two levels of states: Positive vs. Negative for Valence, and Low vs. High for Arousal. Scaled Labeling had a scale of three levels: Negative, Neutral, and Positive for Valence; and Low, Medium, and High for Arousal.

For Labeling Method, we had three conditions: (1) Separate Labeling, (2) Combined Labeling, and (3) Separate Labeling with Displayed Labels. In the Separate Labeling condition, the human experts were asked to label either Valence or Arousal - one at a time. In Combined Labeling condition, they were asked to label Valence and Arousal simultaneously. In Separate Labeling with Displayed Labels condition, the experts labeled one construct first, and then labeled the other construct with the first construct's labels displayed - i.e., the experts were asked to label either Valence with their previous label of Arousal for the same data displayed or label Arousal with their previous labels of Valence for the same data displayed.

For Selection of Labels, we assigned the human experts to the conditions (Binary vs. Scaled) at the beginning of the study so that we could train them based on the specific labels for their assigned condition: We randomly assigned three experts to the Binary Labeling condition, and the other three to the Scaled Labeling condition.

For all three Labeling Method conditions (i.e., Separate Labeling, Combined Labeling, and Separate Labeling with Displayed Labels), the human experts in both Binary and Scaled Labeling individually labeled the same student data (around seven hours collected from five students in two sessions - each session was 40 min). In summary, all six experts followed the procedures outlined below (each expert labeled the same student data five times so that we could have comparative results):

1. Valence labeling only (Separate Labeling).
2. Arousal labeling only (Separate Labeling).
3. Valence and Arousal labeling together (Combined Labeling).
4. Arousal labeling with Valence labels displayed (Separate Labeling with Displayed Labels).
5. Valence labeling with Arousal labels displayed (Separate Labeling with Displayed Labels).

Note that we randomized the order of the first three procedures to minimize the effect of time and familiarity of the student data being labeled. After the experts completed (1–3), they conducted (4) first, and then (5). Note that (4) and (5) were conducted after (1–3), since we needed either Valence or Arousal labels gathered from the individual experts so that we could display them during labeling.

5 Valence-Arousal Labels

In this research, Valence is defined as the direction of a student's affect and Arousal as the level of activation in physical response of the student during the learning process. For Valence, we had three possible labels:

1. Negative: The student seems to experience negative affect (e.g., getting frustrated, stressed, agitated, bored, etc.). Any negative affect is placed within this category.
2. Neutral: The student's affect seems to be neutral. One cannot observe any clear direction towards negative or positive affect (e.g., calm).
3. Positive: The student seems to experience positive affect (e.g., feeling satisfied, excited, etc.). Any positive affect is placed within this category.

For Arousal, we had three possible labels:

1. Low: The student does not seem to be emotionally activated, dynamic, reactive, or expressive of his/her affect.
2. Medium: The student seems to be emotionally somewhat dynamic, reactive, and expressive of his/her affect.
3. High: The student seems to be emotionally very dynamic, reactive, and expressive of his/her affect.

In Table 1, we summarized the final list of Valence-Arousal labels as customized for the Binary and Scaled Labeling conditions. In addition to Valence and Arousal labels, we also had control labels that apply to both of these conditions: Can't Decide (if the human expert cannot decide on a final label) and N/A (if data cannot be labeled - e.g., there is no one in front of the camera).

Table 1. Binary and scaled Valence-Arousal labels

	Binary labels	Scaled labels
Valence	Negative vs. Positive	Negative–Neutral–Positive
Arousal	Low vs. High	Low–Medium–High

6 Analysis of the Labeled Data

Upon completion of labeling, we preprocessed the labeled data prior to analysis: We first aligned label-sets of all experts to each other. Then, we applied windowing over each expert's labeling outputs to obtain the corresponding instance-wise labels. For this, we utilized a sliding window of 8 s with an overlap of 4 s. Hence, after preprocessing, we obtained instance-wise label sets that were timely synchronized with each other.

To compare labeling results for different experimental conditions, we calculated inter-rater agreement among multiple human experts. For inter-rater agreement, we used consensus measures which are designed to estimate the degree of agreement among multiple experts [18]. In this study, we used Krippendorff's alpha [19], as it is robust against incomplete data and is suitable for multiple raters. Despite of disagreements for acceptable value in the related literature, a value above 0.4 is often considered moderate agreement for affect labeling [20].

To investigate the differences among different experimental conditions (i.e., Set of Labels and Labeling Method), inter-rater agreement measures were calculated for the given Valence-Arousal labels using the indicated labeling method:

- Valence Labels with Separate Labeling
- Arousal Labels with Separate Labeling
- Valence Labels with Combined Labeling
- Arousal Labels with Combined Labeling
- Arousal Labels with Arousal Labeling with Valence Labels Displayed
- Valence Labels with Valence Labeling with Arousal Labels Displayed

All these analyses were conducted for the Binary and Scaled label sets separately. Furthermore, we conducted additional analysis, to provide a comparison between Binary and Scaled results: We post-processed Scaled label sets, converting Neutral/Medium labels to either extreme (checking both possibilities), to obtain pseudo-Binary labels. See below for how we converted these labels and their acronyms as used in the Results section:

- For Valence: **NN**: Negative and Neutral merged. | **NP**: Neutral and Positive merged.
- For Arousal: **LM**: Low and Medium merged. | **MH**: Medium and High merged.

7 Results

The inter-rater agreement results for each experimental conditions are summarized in Table 2. These results show that for Valence, the highest consensus among the human experts was achieved in the Separate Labeling with Binary Labels condition (0.495). However for Arousal, the best consensus was obtained when the experts used Scaled Labels (in the Arousal Labeling with Valence Displayed condition), which was followed by converting those Scaled Labels into LM Binary Labels (Low and Medium merged), obtaining an alpha of 0.602.

The findings in Table 2 also show that Valence labeling resulted in higher consensus among the experts than Arousal labeling (before any conversions into pseudo-Binary labels), regardless of whether the Binary or Scaled label set was used, for both Separate and Combined labeling. However, an exception to this finding was labeling Arousal after having previously labeled Valence, with the Valence labels displayed.

Table 2. Consensus (Krippendorff's alpha) among the human experts

		Binary labels	Scaled labels	LM/NN*	MH/NP*
Separate Labeling	Arousal	0.382	0.189	0.405	0.251
	Valence	0.495	0.225	0.266	0.393
Combined Labeling	Arousal	0.235	0.220	0.558	0.221
	Valence	0.355	0.237	0.333	0.388
Separate Labeling with Displayed Labels	Arousal with valence displayed	0.495	0.378	0.602	0.407
	Valence with arousal displayed	0.467	0.200	0.333	0.355

* Merging rules: LM: Low-Medium, NN: Negative-Neutral, MH: Medium-High, NP: Neutral-Positive

When comparing consensus among the human experts, we also found that consensus was always higher for the Binary Labeling conditions than for the Scaled conditions. This suggests that Binary Labeling was easier for the human experts. Additionally, when Scaled label sets are converted to pseudo-Binary labels, LM is always better than MH for Arousal agreements in all cases. This suggests that the experts found it more difficult to distinguish Low vs. Medium Arousal, than Medium vs. High Arousal. Similarly, when Scaled label sets are converted to pseudo-Binary labels, NP is always better than NN for Valence agreements in all cases. This implies that the experts found it more difficult to distinguish Neutral vs. Positive Valence, than Negative vs. Neutral Valence.

In addition to these quantitative results, we asked the human experts about their preferences for how to label, and which methods were easier to use, at the end of the experiments. The feedback we got from the six experts can be summarized as follows:

- The majority of the experts in the Binary labeling condition found Valence easier to label than Arousal, matching our quantitative findings.
- All experts in the Scaled labeling condition found Arousal easier than Valence to label. (Note, however, that inter-rater agreement was actually lower for Arousal in several cases, within this condition).
- 5 of the six experts indicated that they preferred to first label Valence, and then Arousal, in line with the quantitative findings.

Leveraging the quantitative results and considering the feedback from the human experts, it appears that the most optimized method for Valence-Arousal labeling on the student data, at least as far as our study is concerned, would be as follows:

1. Obtain binary Valence labels (Positive vs. Negative);
2. Displaying the binary Valence labels, obtain the scaled Arousal labels (Low-Medium-High);
3. Merge the Low and Medium Arousal scales to obtain final binary Arousal labels (Low vs. High).

Once this optimized method was identified, the next step was to test and validate this method using a relatively larger dataset from more students. Towards this end, we had the experts label around 104 h of student data in total (from 17 students in 13 sessions) using the optimized method. The results obtained with the optimized labeling method are summarized in Table 3, both for the complete dataset and the subset of data utilized in the previous experiments (i.e., experimental data). As the results in Table 3 indicate, consensus among the experts is even higher for the complete dataset (Valence: 0.549; Arousal: 0.610) than for the subset previously studied.

Table 3. Consensus measures with the optimized method (Experimental vs. Complete data)

Name	Dataset details		Consensus measures	
	Student count	Total number of hours	Valence	Arousal
Experimental	5	7	0.495	0.602
Complete	17	104	0.549	0.610

8 Conclusions

To enable the human experts conduct Valence-Arousal labeling on the student data, we used HELP as a baseline labeling protocol. However, as the protocol was originally developed for labeling discrete affective states only, we needed to identify an optimized method for Valence-Arousal labeling on the student data through several experiments. Empirically, the optimized labeling methodology was found to be consisting of the following steps: (1) Obtaining binary Valence labels, (2) obtaining scaled Arousal labels when binary Valence labels are displayed, and (3) merging Low and Medium Arousal scales to obtain final binary Arousal labels. Additionally, the results of our experiments suggest that, before pseudo-Binary conversions, Valence labeling was easier than Arousal labeling for the human. The only exception to this finding was labeling Arousal with the Valence labels displayed, and this may be because labeling Arousal was generally more difficult, and labeling Valence first could have helped the experts isolate their thinking about Arousal, not taking Valence into account.

Using the optimized method, we then had the human experts label a larger quantity of student data so that we could test and validate this method on a relatively larger and different student dataset. The results showed that using the optimized method, the experts were able to achieve an acceptable consensus in labeling outcomes as aligned with relevant affect labeling literature [15, 17]. The researchers and practitioners can leverage the results of this study to design and implement similar data labeling tasks with a consideration of some limitations of the study (e.g., limited number of students, education-context dependency).

References

1. Sabourin, J., Mott, B., Lester, J.C.: Modeling learner affect with theoretically grounded dynamic Bayesian networks. In: Proceedings of International Conference on Affective Computing and Intelligent Interaction, pp. 286–295. Springer, Heidelberg (2011)
2. Jaques, N., Conati, C., Harley, J.M., Azevedo, R.: Predicting affect from gaze data during interaction with an intelligent tutoring system. In: Proceedings of the International Conference on Intelligent Tutoring Systems, pp. 29–38. Springer, Cham (2014)
3. Pardos, Z.A., Baker, R.S., San Pedro, M.O.C.Z., Gowda, S.M., Gowda, S.M.: Affective states and state tests: investigating how affect and engagement during the school year predict end of year learning outcomes. J. Learn. Anal. **1**(1), 107–128 (2014)
4. Kapoor, A., Picard, R.W.: Multimodal affect recognition in learning environments. In: Proceedings of the International Conference on Multimedia, pp. 677–682. ACM (2005)
5. Kapoor, A., Burleson, W., Picard, R.W.: Automatic prediction of frustration. Int. J. Hum.-Comput. Stud. **65**(8), 724–736 (2007)
6. Hoque, M.E., McDuff, D.J., Picard, R.W.: Exploring temporal patterns in classifying frustrated and delighted smiles. Trans. Affect. Comput. **65**(8), 323–334 (2012)
7. Grafsgaard, J.F., Wiggins, J.B., Boyer, K.E., Wiebe, E.N., Lester, J.C.: Automatically recognizing facial indicators of frustration: a learning-centric analysis. In: Proceedings of the International Conference on Affective Computing and Intelligent Interaction, pp. 159–165. IEEE (2013)
8. Bosch, N., D'Mello, S., Baker, R., Ocumpaugh, J., Shute, V., Ventura, M., Zhao, W.: Automatic detection of learning centered affective states in the wild. In: Proceedings of the International Conference on Intelligent User Interfaces, pp. 379–388. ACM (2015)
9. Arroyo, I., Cooper, D.G., Burleson, W., Woolf, B.P., Muldner, K., Christopherson, R.: Emotion sensors go to school. In: Proceedings of the International Conference on Artificial Intelligence in Education, vol. 200, pp. 17–24 (2009)
10. Ekman, P.: An argument for basic emotions. Cogn. Emot. **6**(3–4), 169–200 (1992)
11. Ortony, A., Clore, G.L., Collins, A.: The Cognitive Structure of Emotions. Cambridge University Press, Cambridge (1988)
12. D'Mello, S., Picard, R.W., Graesser, A.: Toward an affect-sensitive AutoTutor. Intell. Syst. **22**(4), 53–61 (2007)
13. Barrett, L.F., Russell, J.A.: The structure of current affect: controversies and emerging consensus. Curr. Dir. Psychol. Sci. **8**(1), 10–14 (1999)
14. Russell, J.A.: A circumplex model of affect. J. Pers. Soc. Psychol. **39**(6), 1161 (1980)
15. Aslan, S., Mete, S.E., Okur, E., Oktay, E., Alyuz, N., Genc, U., Stanhill, D., Arslan Esme, A.: Human expert labeling process (HELP): towards a reliable higher-order user state labeling process and tool to assess student engagement. Educ. Technol. **57**(1), 53–59 (2017)
16. Okur, E., Alyuz, N., Aslan, S., Genc, U., Tanriover, C., Arslan Esme, A.: Behavioral engagement detection of students in the wild. In: Proceedings of the International Conference on Artificial Intelligence in Education, pp. 250–261. Springer, Cham (2017)
17. Ocumpaugh, J., Baker, R., Rodrigo, M.M.T.: Baker Rodrigo Ocumpaugh monitoring protocol (BROMP) 2.0 technical and training manual. New York, NY and Manila, Philippines: Teachers College, Columbia University and Ateneo Laboratory for the Learning Sciences (2015)

18. Stemler, S.E.: A comparison of consensus, consistency, and measurement approaches to estimating interrater reliability. Pract. Assess. Res. Eval. **9**(4), 1–19 (2004)
19. Krippendorff, K.: Computing Krippendorff's alpha-reliability. Departmental Papers (ASC), 43. Retrieved from http://repository.upenn.edu/asc_papers/43 (2011)
20. Siegert, L., Böck, R., Wendemuth, A.: Inter-rater reliability for emotion annotation in human–computer interaction: comparison and methodological improvements. J. Multimodal User Interfaces **8**(1), 17–28 (2014)

A Week of Playing with Code, the Object-Oriented Way

Michele Tomaiuolo(✉), Giulio Angiani, Alberto Ferrari, Monica Mordonini, and Agostino Poggi

Department of Engineering and Architecture, University of Parma, Parma, Italy
michele.tomaiuolo@unipr.it

Abstract. This work presents the results of some summer stages for high school students, for the first introduction to computational thinking and programming. Differently from other approaches, our experiences are characterized by: (*i*) a full week of lessons and exercises, for gradually developing a small but complete and original project, (*ii*) an objects-early methodology, (*iii*) the choice of a dynamic videogame as the main programming project, and (*iv*) the availability of few small and focused examples, as source files, for students to start building their own application. At the end, the students' anonymous opinions about the stage, and their own projects, were collected and analyzed, for improving future similar activities.

Keywords: Computer games · Computational thinking
Computer science education · Computer programming · Python

1 Introduction

The theme of coding is generating much interest. In particular, many projects aim to facilitate the introduction of young students and other novices to computer programming; moreover, there is a long trail of debates about the best language for learning to code. Also about the choice of the application context, there are diverging opinions [8,10], though computer games have been proposed by multiple authors [4,20] and teachers [2,12,14,19,21].

This article discusses some experiences we gained in our university, in particular about a series of summer stages for high school students. These stages follow the changes which we have introduced in our first university course, where teaching has been focused more and more on problem solving skills and computational thinking [22]. The aim is to introduce the basic concepts of object-oriented programming, with an objects-early approach. Both in the summer stage and in the university course, the learning process has been modeled as a sequence of levels of growing difficulty. Progress is obtained through the development of various parts of a project, which at first is a dynamic videogame, with various interacting characters; i.e., progress corresponds to the basic acquisition of new concepts and skills in computer programming.

T. Di Mascio et al. (Eds.): MIS4TEL 2018, AISC 804, pp. 62–69, 2019.
https://doi.org/10.1007/978-3-319-98872-6_8

The rest of the paper is organized in the following way. Section 2 presents some background material, including debates about using computer games in education. Section 3 discusses our educational framework for the introduction of high school students to Computational Thinking and programming. Section 4 describes the results of a stage organized in summer 2017. Finally, Sect. 5 provides some concluding remarks.

2 Background

In the scientific literature, there are many studies concerning the use of gaming in the didactic context [15], also for courses oriented at computational concepts and coding [11]. While traditional schooling is often perceived as ineffective and boring by many students, these educational experiences show that the learning process can be improved if adequate facilities and methodologies are adopted. Conceptually, game development and gamification [1,5,9] are different and orthogonal aspects, but in practice they are often mixed. In this work, we discuss how the concepts of basic coding and object-oriented programming can be introduced and organized like an advancement game, with levels of growing difficulty; each level has to be completed by the invention of a personal solution to the task at hand, toward the development of a simple but functioning computer game [3,13,17].

The creation of computer games is strictly connected with objects and events: a game can be easily seen as composed by objects with different properties and behaviors that deal with events; understanding how inheritance works comes naturally with the need to define specialized characters from basic ones. This naturally guides students towards the use of the objected-oriented paradigm. The debate about objects-early or objects-late approach in teaching coding is presented in [11]. After discussing the problem of paradigm shift, the conclusion is that learning to program in an object-oriented style seems to be very difficult after being used to a procedural style. A radical approach states that a successful methodology to teaching introductory programming has not to be based on the objects-early or late approaches, on the procedural or object-oriented paradigms, but on the "game first" approach [12]. A line of research focuses on easy to use environments [6,7,11,16,21] and is interested in the development and learning process, more than the products. In this sense, game creation is not the purpose, but the instrument for driving users to learn the key concepts of programming.

Strategies concerning the teaching of code are set in a range that have as extremes: intensive learning, that is didactic and concentrated, and extensive learning, a sort of process that occurs without very specific intent. Our idea is to find the middle path, adopting some principles of extensive learning projects, e.g., "code to stay alive", borrowed from "fight to be alive", which is the main concept of many game. We propose more practical activities, "playing" with the code, but without losing sight of fundamental theoretical aspects. In our opinion, it is necessary to introduce theoretical concepts, while at the same time practicing with them, through hands-on exercises.

3 Teaching Material and Methodology

3.1 Proposed Projects and Frameworks

The proposed project for our summer stages is about the creation of a dynamic game (e.g., Frogger, Space Invaders, Snake, Bubble Bobble, etc.). In fact, such projects are proposed in both the summer stage and our first university course.

The nature of these dynamic games is based on various types of *actors* (or characters) moving in the gaming *arena*; it allows to test students in the design and construction of a hierarchy of classes from which to instantiate the various actors. Encapsulation, inheritance and polymorphism are intrinsic in these types of game, that also are particularly appropriate for the development of advanced projects for the best students (Fig. 1).

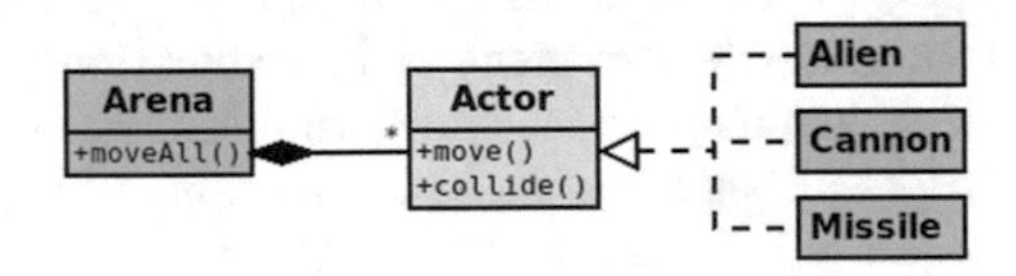

Fig. 1. The proposed framework.

A very simple framework has been used. With respect to more complete and complex frameworks for game development, it is just made of few dozens of lines. This way, it can be studied and grasped confidently, without much effort, and it brings the attention to the principles of abstraction and polymorphism, instead of game details[1]. It is built around two main abstractions: `Arena` and `Actor`.

- **`Actor`** is essentially an interface, which has to be implemented by all objects participating in the game. Each actor has to provide a `move` method, for acting at its own turn, and a `collide` method, for handling collisions with other actors.
- **`Arena`** is essentially a container for actors. It manages all actors according to a turn-based policy and it is completely agnostic about the types of actors that are introduced. When the main application calls the `move_all` method of its arena, the `move` method of each contained actor is called, sequentially. After each single turn, collisions are detected, and the `collide` method of both colliding actors is called.

3.2 Objects-Early Approach

The advantage of the proposed framework is that students are guided to use relatively advanced programming techniques (as polymorphism is often regarded),

[1] The framework and all teaching material is available at http://www.ce.unipr.it/stage/.

in a quite intuitive way. In fact, it is easy to acknowledge that each actor has to act in the arena, though each type of actor performs its actions according to a different logic.

The whole learning process is proposed as a sequence of levels, each one to be completed before advancing to the next one; i.e., progress corresponds to the basic acquisition of new concepts and skills in computer programming.

1. **Level 1 - Cycles and conditions**, to draw multiple shapes.
2. **Level 2 - Lists**, to store input data and for example draw an histogram in the available screen space.
3. **Level 3 - Functions**, to handle the periodical update of a simple scene (e.g. a single bouncing ball).
4. **Level 4 - Objects**, to encapsulate data and manage multiple moving shapes, together with the logic of their own movements (advancing and bouncing, for example).
5. **Level 5 - Polymorphism**, to handle the movement of a list of different actors homogeneously (introducing an intuitive notion of Liskov's substitution principle).
6. **Level 6 - Composition**, to begin building up a game, with the presence and interaction of multiple actors.

That is, object oriented concepts can be introduced gradually, as advancements in a gamified learning process, that is actually also about game development. This way, the utility of these various development concepts becomes apparent, while the complexity of the application grows. In our experience, all this development can be made also in a short time, together with students (i.e., providing suggestions and guidance, but preserving and encouraging their autonomy in finding original solutions), and starting with an existing framework and useful examples. In the summer stages, this development and learning process is encompassed in five days (40 h in total, in lab, using the option to stay in the afternoons, with the assistance of a tutor).

4 The Space Invaders, in a Week of Code

4.1 Students

In recent years, our department has organized each year a summer internship for a hundred high school students. Though the teaching methodology has been similar in all these stages, for clarity we present here the results of the last stage, only. In summer 2017, we hosted 115 students, from many differently specialized high schools: around 15 of them specializing in Humanities, Arts, or Languages; around 50 specializing in Sciences (which unfortunately often do not include specific courses about Computer Science); around 50 from Technical high schools. In fact, Fig. 2 show a quite varied situation, with different levels of initial skills.

The stage lasts 40 h spanning 5 days, at maximum, which include some optional hours in the afternoon, with tutors available for helping with exercises.

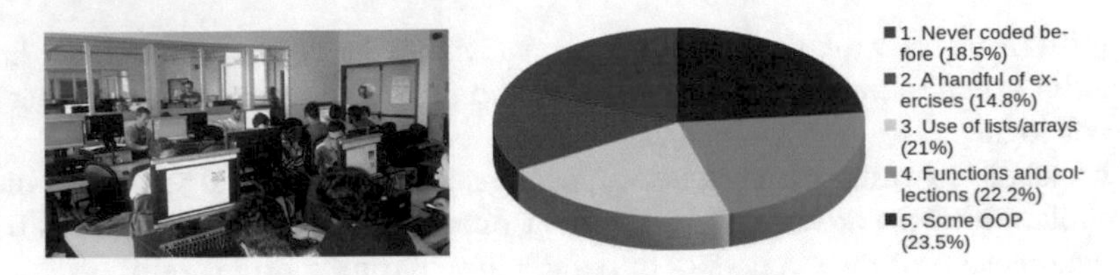

Fig. 2. A photo of students participating in stage activities (left); their previous programming skills (right).

4.2 Opinions

A survey is proposed to students in the last day and a single answer is requested for each question. Answering to the survey is not mandatory. In total, 81 students accepted the survey, anonymously (through Google Forms). Here, we'll report the most significant values.

Figure 3 highlights that OOP is the most difficult topic of the stage. For many students, a source of difficulty, which deserves attention, is the confusion among object fields, method parameters, and local variables.

As Fig. 4 shows, students have provided very positive feedback about developing a longer project, in contrast with solving smaller problems, and about computer games in particular. No one dislikes the specific application domain of computer games, and 91.5% of students have said to be motivated or strongly motivated in developing computer games. In general, the development of a longer project has been found very gratifying, or at least interesting, by 86.5% of students.

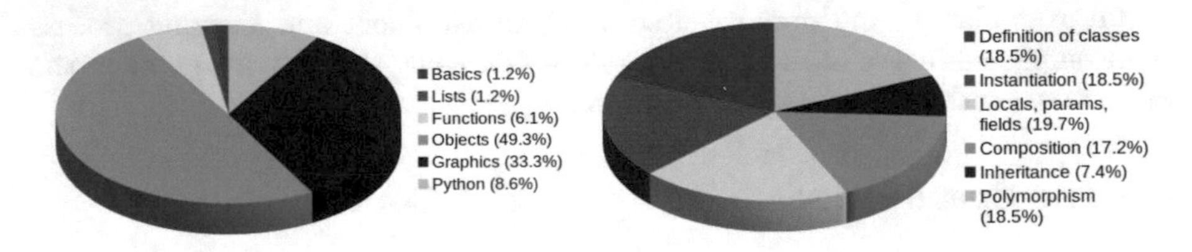

Fig. 3. Main difficulties reported, in general (left), and about OOP in particular (right).

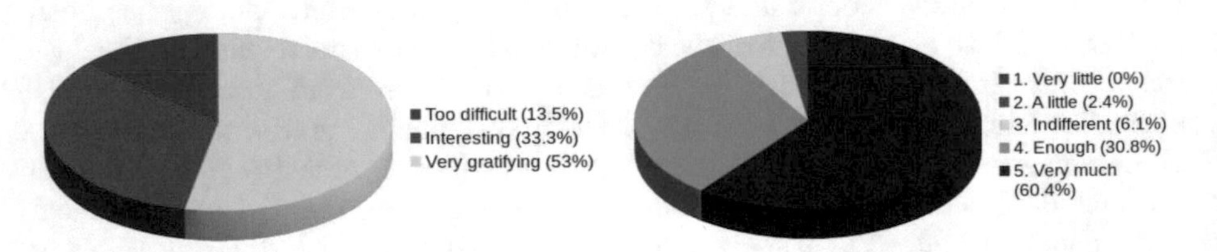

Fig. 4. Motivation for working on a longer project in general (left), and on a videogame in particular (right).

The programming language used to introduce programming, and object-oriented programming in particular, is important [18]. In general, students appreciate Python: only 16% would prefer a different language, for the stage. The most useful aspect of Python is its readability/terseness (according to 40.7% of students). Others (19.8%) instead appreciate above all its interactive shell. The access to fields and the use of `self` is considered the major annoyance of Python by 37% of students.

Finally, we have asked students about their self-perceived level of autonomy in problem solving (Fig. 5), at the end of the stage, and their own satisfaction about this experience (Fig. 6). Answers are shown in their aggregate form (at the left, in both figures) and then in association with the initial coding skills of students (at the right, in both figures).

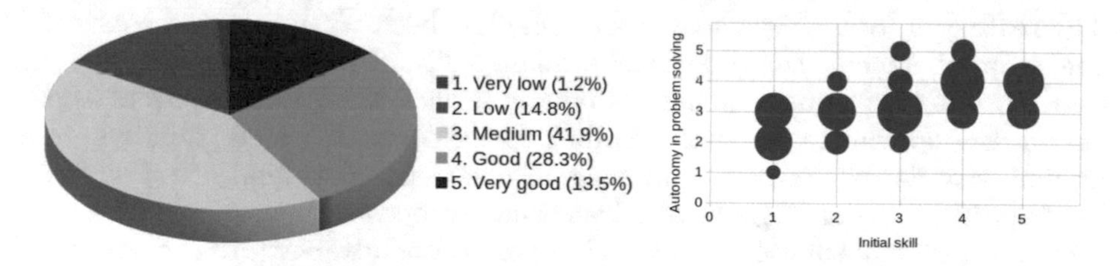

Fig. 5. Self-perceived autonomy in problem solving, at the end of the stage.

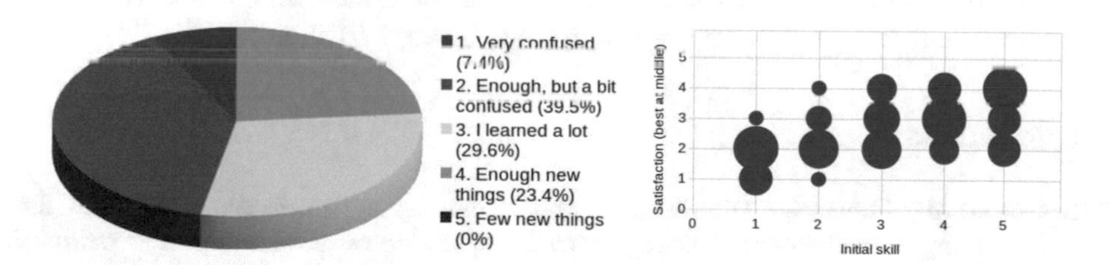

Fig. 6. Satisfaction about the stage. Middle values for learning enough, or much (at 3). Low values for confusion. High values for boredom.

4.3 Delivered Works

The proposed project allowed all students, also those without any previous programming experience, to achieve a functioning project. However, realized programs differ from the point of view of the actors' characteristics, the way they operate in the arena and the multimedia aspects of the game itself[2]. During the stage, some students programmed additional game logics and actors, like motherships, etc. During the university course, projects of this kind allow some of the best students to deepen the motion study and also to implement more interesting strategies (e.g., ghosts' own behaviors in the Pacman game, etc.) (Fig. 7).

[2] The delivered projects, in their original form, are also available at http://www.ce.unipr.it/stage.

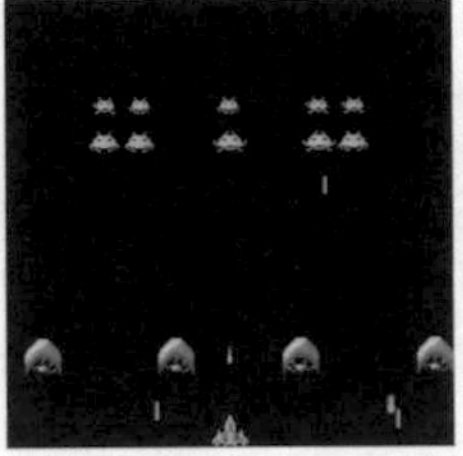

Fig. 7. Screenschots of some games developed by students during the stage.

5 Conclusions

The realization of a simple computer game has been often reported as a valid choice for introducing computational thinking and coding. In our experience, the development of a game stimulates students to face the typical issues of application development in general, according to the object-oriented paradigm, in a gradual way and with their continuous involvement. This kind of activity can be mapped to a sort of gamified advancement process, where students have to provide a personal solution to a certain task, before advancing to the next level of development. The analysis of students' answers to a poll proposed in summer 2017 has mostly confirmed the applicability of this methodology for a full-week introductory course. Results also highlight specific issues which deserve particular attention and possibly more examples dedicated to them.

References

1. Caton, H., Greenhill, D.: Rewards and penalties: a gamification approach for increasing attendance and engagement in an undergraduate computing module. Int. J. Game-Based Learn. **4**(3), 1–12 (2014)
2. Ceder, V., Yergler, N.: Teaching programming with python and pygame. Apresentado na PyCon (2003)
3. Cooper, S., Dann, W., Pausch, R.: Alice: a 3-D tool for introductory programming concepts. J. Comput. Sci. Coll. **15**(5), 107–116 (2000)
4. Dawson, M.: Python Programming for the Absolute Beginner. Cengage Learning, Boston (2010)
5. Deterding, S., Dixon, D., Khaled, R., Nacke, L.: From game design elements to gamefulness: defining gamification. In: Proceedings of the 15th International Academic MindTrek Conference: Envisioning Future Media Environments, pp. 9–15. ACM (2011)
6. Ferrari, A., Lombardo, G., Mordonini, M., Poggi, A., Tomaiuolo, M.: OOPP: tame the design of simple object-oriented applications with graphical blocks. In: Lecture Notes of the Institute for Computer Sciences, Social-Informatics and Telecommunications Engineering, LNICST, vol. 233, pp. 279–288 (2018)
7. Ferrari, A., Poggi, A., Tomaiuolo, M.: Object oriented puzzle programming. Mondo Digitale **15**(64) (2016)

8. Greenberg, I., Kumar, D., Xu, D.: Creative coding and visual portfolios for CS1. In: Proceedings of the 43rd ACM Technical Symposium on Computer Science Education, pp. 247–252. ACM (2012)
9. Howland, K., Good, J.: Learning to communicate computationally with flip: a bi-modal programming language for game creation. Comput. Educ. **80**, 224–240 (2015)
10. Kelleher, C., Pausch, R.: Lowering the barriers to programming: a taxonomy of programming environments and languages for novice programmers. ACM Comput. Surv. (CSUR) **37**(2), 83–137 (2005)
11. Kölling, M.: The problem of teaching object-oriented programming, part 1: languages. J. Object-Oriented Program. **11**(8), 8–15 (1999)
12. Leutenegger, S., Edgington, J.: A games first approach to teaching introductory programming. ACM SIGCSE Bull. **39**(1), 115–118 (2007)
13. Overmars, M.: Teaching computer science through game design. Computer **37**(4), 81–83 (2004)
14. Perkel, J.M.: Pick up python. Nature **518**(7537), 125 (2015)
15. Prayaga, L., Coffey, J.W., Rasmussen, K.: Strategies to teach game development across age groups. In: Design, Utilization, and Analysis of Simulations and Game-Based Educational Worlds, pp. 95–110. IGI Global (2013)
16. Price, T.W., Barnes, T.: Comparing textual and block interfaces in a novice programming environment. In: Proceedings of the Eleventh Annual International Conference on International Computing Education Research, pp. 91–99. ACM (2015)
17. Repenning, A.: Agentsheets®: An interactive simulation environment with end-user programmable agents. Interaction (2000)
18. Shein, E.: Python for beginners. Commun. ACM **58**(3), 19–21 (2015)
19. Silveira, I.F., Mustaro, P.N., Silva, L.: Using computer games to teach design patterns and computer graphics in CS and IT undergraduate courses: some case studies. In: Brazilian Symposium on Computers in Education, vol. 1, pp. 490–499 (2007)
20. Sweigart, A.: Making Games with Python and Pygame. CreateSpace, North Charleston (2012)
21. Vassilev, T.I., Mutev, B.I.: An approach to teaching introductory programming using games. In: Proceedings of the International Conference on e-Learning, vol. 14, p. 246 (2014)
22. Wing, J.M.: Computational thinking. Commun. ACM **49**(3), 33–35 (2006)

Emojis' Psychophysics: Measuring Emotions in Technology Enhanced Learning Contexts

Roberto Burro, Margherita Pasini, and Daniela Raccanello(✉)

Department of Human Sciences, University of Verona, Verona, Italy
{roberto.burro,margherita.pasini,
daniela.raccanello}@univr.it

Abstract. Emojis are pictorial representations of human facial expressions which are becoming particularly popular in computer-mediated communication. Within text-based messages, they function as surrogates of real nonverbal elements to convey emotional meanings, but in absence of a verbal label they can result ambiguous. In addition, a quantification of how much positive and/or negative valence is expressed by different emojis is still missing. We asked to 110 adults to evaluate 81 emojis on two scales, relating to positive and negative valence. Through Rasch models, we quantified the amount of positive and negative valence expressed by each emoji, deleting those emojis that were not scalable on the considered dimensions. This study is a preliminary step for the development of a set of scales formed by emojis representing discrete emotions, to be used in a variety of ways in psychological research as well as in technological learning contexts or for product evaluation purposes.

Keywords: Emotions · Emojis · Learning · Fundamental measurement
Rasch models · Psychophysical scaling

1 Introduction

Nowadays, the use of technologies for assessing how individuals perceive, elaborate, and react in everyday life, also in relation to learning, is increasing [34]. Within information and communication technology (ICT) contexts, a key role for measurement issues could be played by emojis. In this paper, we present the state-of-the-art on emojis and their measurement according to a psychophysical perspective, and we describe a preliminary study aiming at quantifying the valence of emojis of common use. This research is a preliminary step within a larger project which has the broader aim to develop an instrument including emojis to be utilized within learning technological environments to assess learning-related emotions. Currently the role of emotions within such environments is gaining particular attention [19, 27], even if the instruments deputed to assess them still need to be refined. Emojis are stimuli that are particularly appropriate for technological environments: They can for example be quickly used for assessing emotions also with methods such as the Experience Sampling [23], in which people are interrupted during a task to respond about how they are feeling. This enables to diminish memory distortions typical of self-report instruments presented at the end of a task. Nevertheless, in such cases it is essential that the

T. Di Mascio et al. (Eds.): MIS4TEL 2018, AISC 804, pp. 70–78, 2019.
https://doi.org/10.1007/978-3-319-98872-6_9

meaning of the proposed graphical stimuli is not ambiguous. In our specific case, quantifying the valence of emojis is preliminary to further develop a scale including emojis pertaining to different valences, emotions, and intensity, characterized by good psychometric properties.

1.1 Definition and Functions of Emojis

1.1.1 What Are Emojis and Who Uses Them?

Emoticons are pictorial representations of human facial expressions which are becoming particularly popular in computer-mediated communication (CMC) [1]. They were introduced during the eighties as formed uniquely by standard keyboard typographical characters, but more recently they have evolved into pictorial symbols with a higher degree of humanization, as it is the case for emojis [1]. The use of emojis in accompanying or substituting text in CMC is currently increasing [16], together with the rates of users of social media and text message platforms–e.g., for WhatsApp, one of the most popular mobile applications, the number of monthly active users worldwide has changed from 200 to 1300 million from April 2013 to July 2017 [33]. Data on mostly used emojis have also been recently published, for example identifying smileys and emojis with hearts or tears among the most popular 2016 emojis [14]. In addition, we know that emojis are utilized more frequently in conversations with friends versus strangers, and in positive versus negative contexts [12].

1.1.2 Why Do People Use Emojis?

This question can be answered from a twofold perspective, both theoretical and relating to peoples' viewpoints. According to a psychological perspective, emojis would be the analogue for CMC to what nonverbal cues related to facial expressions, gestures, or tone of voice contributing to express emotions are for face-to-face communication [1]. In other words, within text-based messages emojis would function as surrogates of real nonverbal elements to convey emotional meanings [12], supporting the receiver in recognizing the emotions to which the sender is referring to. The ability to recognize others' emotions, based on individuals' physical expressions, has both evolutionary and social origins [13, 15]. Strong evidence supports the existence of universally basic emotions with distinctive expressions, including happiness, sadness, fear, anger, disgust, and surprise [13]. The mentioned ability appears early in childhood, beginning with the ability to identify broad categories such as positive and negative valence which gradually refines with development [35].

According to the media richness theory, the combination of multiple information through texts and emojis would increase richness in communication, facilitating effective CMC [20]. Specifically, media richness would refer to how much channels and goals of a communication enable to increase certainty in exchanging messages [1]. In line with this theoretical approach, some researchers have investigated people's reasons for using emojis, documenting that they result salient for expressing emotions but also for strengthening individuals' messages and expressing humour [12], thus reducing uncertainty in communicative exchanges.

1.1.3 Open Issues in the Use of Emojis

In light of the variety of emojis' uses and functions, it becomes highly relevant to have available sets of emojis which clearly correspond to specific emotional labels. However, notwithstanding the presence of web-based shared labels as those published in reference websites as Emojipedia [14], ambiguity in the meaning of some emojis still persist. It can refer to the attribution, to an emoji, of a positive or negative valence, or of a specific emotional label. Such ambiguity could have negative consequences in terms of the efficacy of messages within CMC. In addition, even in case of no ambiguity in the identification of valence or emotional label, misunderstanding in communication could derive from interpreting differently the level of intensity associated to the same emoji.

1.2 Self-report Instruments to Assess Emotions in Technological Contexts: The Potential of Emojis for Measurement

1.2.1 Assessment of Emotions in Technological and Learning Contexts

For technological contexts, it is currently increasing the need to evaluate the valence or to gather opinions on products, services, issues, policies, and health-care [3]. In other words, it is becoming more relevant to assess the emotional meaning of a variety of objects. The same need characterizes advanced learning technology environments in which emotions arise, and whose assessment gives useful applied suggestions according to the more recent theoretical approaches [19].

Nowadays, self-report instruments still represent the most direct way to have insight on people's emotional world [25]. Among the most frequently used self-report instruments, there are user reviews for technological contexts and self-report questionnaires for learning environments. On the one hand, as regards user reviews, different works are focusing on the analysis of users' sentiments from online forums, in order to classify reviews in terms of valence, i.e., positive, negative, or neutral [3]. However, even when coded with automatic tools, reviews still result expensive methods, given the time they require both for being written and analyzed. On the other hand, self-report questionnaires suffer from limitations typical of self-report instruments such as social desirability bias and memory distortions, and are particularly time-consuming [25]. A possible alternative instrument, preserving the richness of self-report measures and combining it with the speed and visual characteristics of current communication [16], could be developed using emojis as stimuli within a scale assessing intensity of different emotions. Such an instrument would have direct applications generally in technological contexts, but also specifically for advanced learning technology environments in which methods such as the Experience Sampling could be applied [23]. Basing on scales using pictorial stimuli [7, 17, 26, 27, 30], the development of a scale using emojis instead of drawings of faces could also go beyond limitations related to gender and race issues.

1.2.2 Using Emojis as Measurement Tools

A simple and fast way to express evaluations on emotions is using emojis, with or without the corresponding labels. Nowadays emojis are used in a variety of contexts, both for requiring opinions on products and for evaluating people's emotional reactions. In the first case, the common use of emojis to evaluate products, specifically in technological contexts, usually refers to a reduced range of emotions, mainly pleasure or happiness (e.g., as for the American electronic commerce and cloud computing company Amazon), and it does not usually quantify the intensity of the emotion at issue, but only the absence or presence of that emotion (e.g., as for the "I like" judgement requested in social platforms like Facebook), neglecting the salience of intermediate evaluations. Such a use leads to evaluate the object at issue with distortions, in which the participants' responses are influenced by the characteristics of the scale and by the way in which the data are gathered. In particular, it does not guarantee the respect of the basic properties of 'fundamental measurement' (see Sect. 1.2.3). Also for learning contexts, there are some scales including emojis as stimuli, but similar problems arise, for example when scales are formed by emojis ranging from a negative emotion such as sadness to a positive emotion such as happiness, without previously quantifying the intensity of the emotions expressed by the different emojis [23].

1.2.3 The Rasch Models

When it becomes relevant to consider the interaction between different quantitative latent dimensions such as characteristics of persons and items, analyses performed through the Rasch models can be a valuable option to be taken into account [2, 4, 6, 31]. The meaning of the two latent dimensions can vary according to the focus of the assessment. In psychological research, we can for example consider the characteristics of a person in terms of how much positively and/or negatively s/he perceives an item such as an emoji, and the characteristics of an item in terms of the degree with which it expresses an emotion [8, 10]. Compared to measurement traditional approaches, the Rasch models enable to measure latent dimensions respecting the assumptions of the 'fundamental measurement', i.e., linearity, specific objectivity, and stochastic independence [11], usually taken into account within the physical sciences but not within the social sciences. Specifically, through the Rasch models items are calibrated and persons are measured independently from the characteristics of, respectively, the people and the items.

1.3 The Present Study

Our aim was to investigate whether people differentiate emojis of common use according to their valence, positive or negative. In addition, we aimed at quantifying how much positivity or negativity each emoji represented.

2 Method

2.1 Participants

The participants were 110 undergraduate students (*M* = 23.24 years, *SD* = 6.40; 63% female) at the University of Verona, in Northern Italy, with varying socio-economic status. They participated voluntarily, after signing an informed consent form.

2.2 Material and Procedure

The data were gathered through a questionnaire administered with mixed devices (smartphones or laptops). It included the consent form, socio-demographic data like year of birth and gender, and measures on the valence of the presented emojis. The participants were recruited during university lessons, and were asked to answer to the online questionnaire immediately or, if it was not possible, within three days. The mixed-device survey was administered using the Apsym-Survey Software, ApSS [9, 24, 28, 29, 32]. We assessed the valence of a series of emojis selected from the palette of emojis available in WhatsApp cross-platform instant messaging service during January 2018. We selected 81 emojis according to the following criteria: (a) being drawn as stylized faces through a circle, usually yellow (with the exception of the red emoji representing anger and the green emoji expressing disgust); (b) being of the same dimensions (we excluded the emoji representing a bomb and the emoji with the brown hat, which are smaller than the others). The emojis were presented twice, in two blocks. For each emoji in each block, the participants were asked to indicate how much positive and how much negative the expressed emotion was (*How much positive/negative is the expressed emotion?*). The participants could answer moving a slider along a bar ranging from 0 (*not at all*) to 100 (*very much*). The task was adapted from previous ones [16].

3 Results and Discussion

We conducted the Rasch analysis using the Partial Credit Model [22] implemented within the eRm-package [21] of the R-software. To evaluate the fit of the 81 emojis to the Rasch model (examining each emoji for both dimensions, i.e. positive and negative) we considered two criteria. Emojis had a good fit when the infit-t statistics relating to each of them was comprised between −2 and 2 [18], and the *p*-value associated to the Chi square was higher than .05 [5]. Deleting the items with a bad fit made it possible to improve the measurement properties of the whole set of emojis.

Separately by positive and negative dimensions, we report in Tables 1 and 2 the list of the emojis and their scaling values, referred to how much positive and negative valence is expressed by each emoji in an interval logit scale. The infit-t values were included between −2 and 2 for all emojis. We excluded emojis with a Chi square with a $p < .05$, for a total of nine emojis for the positive scale and 31 emojis for the negative scale. To help comparisons between the two dimensions, we transformed the scaling values as values with zero as the minimum value. In addition, we calculated the Person Separation Reliability Index (*PSRI*) [2] for assessing the reliability of the set of emojis, separately for the positive and the negative dimensions. The *PSRI* ranges from 0 to 1, with higher values corresponding to higher reliability. The *PSRI* was good for both the positive (*PSRI* = .927) and the negative dimension (*PSRI* = .923).

Table 1. List of emojis and their scaling values for the positive scale

Emoji	Value	Emoji	Value	Emoji	Value	Emoji	Value	Emoji	Value
	4.854		4.097		3.488		1.626		0.953
	4.839		4.043		3.482		1.598		0.951
	4.824		4.043		3.326		1.353		0.944
	4.816		3.978		3.294		1.308		0.937
	4.754		3.906		3.277		1.299		0.916
	4.746		3.776		3.209		1.294		0.841
	4.730		3.771		3.197		1.281		0.823
	4.730		3.697		3.182		1.266		0.789
	4.452		3.652		3.086		1.178		0.641
	4.339		3.593		3.072		1.153		0.568
	4.304		3.589		2.822		1.111		0.345
	4.150		3.560		2.290		1.089		0.000
	4.117		3.544		2.150		1.029		
	4.100		3.541		2.066		1.022		
	4.100		3.517		1.697		0.962		

Table 2. List of emojis and their scaling values for the negative scale

Emoji	Value	Emoji	Value	Emoji	Value	Emoji	Value	Emoji	Value
	2.451		1.577		0.908		0.632		0.503
	2.299		1.565		0.891		0.612		0.433
	2.265		1.296		0.879		0.608		0.426
	2.225		1.255		0.852		0.587		0.425
	2.223		1.184		0.835		0.584		0.386
	2.065		1.161		0.702		0.577		0.372
	1.946		1.155		0.675		0.575		0.364
	1.782		1.045		0.675		0.545		0.329
	1.644		0.977		0.662		0.522		0.192
	1.582		0.950		0.637		0.511		0.000

4 Conclusions

Clicking on an emoji is nowadays constantly more popular in CMC, but also in a variety of other technological contexts [1, 14, 16, 33]. This study represents a preliminary step within a larger project aiming at reducing the ambiguity in the use of emojis in absence of verbal labels. Through the application of Rasch models [2, 6, 31], we quantified the amount of positive and/or negative valence expressed by a set of emojis of common use. Identifying the valence of emojis with methodologies respecting the properties of the fundamental measurement systems, i.e. being sample-free and item-free [11], is a first phase in the development of an evaluation system formed by emojis. This instrument will include a set of scales formed by emojis representing discrete emotions, to be used in a variety of ways in psychological research as well as in technological learning contexts or for product evaluation purposes.

References

1. Aldunate, N., González-Ibáñez, R.: An integrated review of emoticons in computer-mediated communication. Front. Psychol. **7i**, 1–6 (2017). https://doi.org/10.3389/fpsyg.2016.02061
2. Andrich, D.: Rasch models for measurement. Quantitative applications in the social sciences, vol. 68. Sage Publications, London (1988)
3. Asghar, M.Z., Khan, A., Ahmad, S., Qasim, M., Khan, I.A.: Lexicon-enhanced sentiment analysis framework rule-based classification scheme. PLoS ONE **12**(2), 1–22 (2017). https://doi.org/10.1371/journal.pone.0171649
4. Bianchi, I., Savardi, U., Burro, R.: Perceptual ratings of opposite spatial properties: do they lie on the same dimension? Acta Physiol. **138**, 405–418 (2011). https://doi.org/10.1016/j.actpsy.2011.08.003
5. Bland, J.M., Altman, D.G.: Multiple significance tests: the Bonferroni method. Br. Med. J. **310**(6973), 170 (1995)
6. Bond, T., Fox, C.M.: Applying the Rasch model: fundamental measurement in the human sciences, 3rd edn. Routledge, New York (2015)
7. Brondino, M., Dodero, G., Gennari, R., Melonio, A., Pasini, M., Raccanello, M., Torello, S.: Emotions and inclusion in co-design at school: let's measure them. In: Advances in Intelligent and Soft Computing, vol. 374, pp. 1–8 (2015). https://doi.org/10.1007/978-3-319-19632-9_1
8. Burro, R.: To be objective in experimental phenomenology: a psychophysics application. SpringerPlus **5**(1), 1720 (2016). https://doi.org/10.1186/s40064-016-3418-4
9. Burro, R., Raccanello, D., Pasini, M., Brondino, M.: An estimation of a nonlinear dynamic process using latent class extended mixed models. Affect profiles after terrorist attacks. Nonlinear Dyn. Psychol. Life Sci. **22**(1), 35–52 (2018)
10. Burro, R., Sartori, R., Vidotto, G.: The method of constant stimuli with three rating categories and the use of Rasch models. Qual. Quant. **45**(1), 43–58 (2011). https://doi.org/10.1007/s11135-009-9282-3
11. Campbell, N.R.: Physics: the Eements. Cambridge University Press, Cambridge (2013)
12. Derks, D., Bos, A.E.S., von Grumbkow, J.: Emoticons in computer-mediated communication: social motives and social context. CyberPsychol. Behav. **11**(1), 99–101 (2008). https://doi.org/10.1089/cpb.2007.9926

13. Ekman, P.: An argument for basic emotions. Cogn. Emot. **6**(3–4), 169–200 (1992). https://doi.org/10.1080/02699939208411068
14. Emojipedia (2018). http://emojipedia.org
15. Feldman, R.S., White, J.B., Lobato, D.: Social skills and nonverbal behavior. In: Feldman, R.S. (ed.) Development of nonverbal behavior in children, pp. 259–277. Springer, New York, NY (1982)
16. Gallo, K.E., Swaney-Stueve, M., Chambers, D.H.: A focus group approach to understanding food-related emotions with children using words and emojis. J. Sens. Stud. **32**(e12264), 1–10 (2016). https://doi.org/10.1111/joss.12264
17. Gennari, R., Melonio, A., Raccanello, D., Brondino, M., Dodero, G., Pasini, M., Torello, S.: Children's emotions and quality of products in participatory game design. Int. J. Hum Comput Stud. **101**, 45–61 (2017). https://doi.org/10.1016/j.ijhcs.2017.01.006
18. Giampaglia, G.: How many and which response categories? A Rasch model contribution to develop a good questionnaire [Quante e quali categorie di risposta? Il contributo del modello di Rasch alla costruzione di un buon questionario]. Polena **3**(3), 1000–1034 (2004)
19. Graesser, A.C., D'Mello, S.K., Strain, A.C.: Emotions in advanced learning technologies. In: Pekrun, R., Linnenbrink-Garcia, L. (eds.) International handbook of emotions in education, pp. 473–493. Taylor and Francis, New York (2014)
20. Hsieh, S.H., Tseng, T.H.: Playfulness in mobile instant messaging: examining the influence of emoticons and text messaging on social interaction. Comput. Hum. Behav. **69**, 405–414 (2017). https://doi.org/10.1016/j.chb.2016.12.052
21. Mair, P., Hatzinger, R.: Extended Rasch modeling: the eRm package for the application of IRT models. R. J. Stat. Softw. **20**(9), 1–20 (2007)
22. Masters, G.N.: A Rasch model for partial credit scoring. Psychometrika **47**(2), 149–174 (1982)
23. Meschtscherjakov, A., Weiss, A., Scherndl, T.: Utilizing emoticons on mobile devices within ESM studies to measure emotions in the field. In: Proceedings of the 11th International Conference on Human-Computer Interaction with Mobile Devices and Services, MobileHCI 2009, pp. 1–4 (2009)
24. Pasini, M., Brondino, M., Burro, R., Raccanello, D., Gallo, S.: The use of different multiple devices for an ecological assessment in psychological research: an experience with a daily affect assessment. In: Advances in Intelligent and Soft Computing, vol. 478, pp. 121–129 (2016). https://doi.org/10.1007/978-3-319-40165-2_13
25. Pekrun, R., Bühner, M.: Self-report measures of academic emotions. In: Pekrun, R., Linnenbrink-Garcia, L. (eds.) International handbook of emotions in education, pp. 561–579. Taylor and Francis, New York (2014)
26. Raccanello, D., Bianchetti, C.: Pictorial representations of achievement emotions: preliminary data with primary school children and adults. In: Advances in Intelligent and Soft Computing, vol. 292, pp. 127–134 (2014). https://doi.org/10.1007/978-3-319-07698-0
27. Raccanello, D., Brondino, M., Pasini, M.: Achievement emotions in technology enhanced learning: development and validation of self-report instruments in the Italian context. Interact. Des. Archit. J. **23**, 68–81 (2014)
28. Raccanello, D., Burro, R., Brondino, M., Pasini, M.: Relevance of terrorism for Italian students not directly exposed to it: the affective impact of the 2015 Paris and the 2016 Brussels attacks. Stress Health **34**, 338–343 (2017). https://doi.org/10.1002/smi.2793
29. Raccanello, D., Burro, R., Brondino, M., Pasini, M.: Use of internet and wellbeing: a mixed-device survey. In: Advances in Intelligent and Soft Computing, vol. 617, pp. 65–73 (2017). https://doi.org/10.1007/978-3-319-60819-8_8

30. Raccanello, D., Burro, R., Hall, R.: Children's emotional experience two years after an earthquake: an exploration of knowledge of earthquakes and associated emotions. PLoS ONE **12**(2), 1–21 (2017). https://doi.org/10.1371/journal.pone.0189633
31. Rasch, G.: Probabilistic models for some intelligence and attainment tests. MESA Press, Copenhagen (1993)
32. Schmitz, C.: LimeSurvey: an open source survey tool. LimeSurvey Project Hamburg, Germany (2018). http://www.limesurvey.org
33. Statista (2018). http://www.statista.com/search/?q=WhatsApp
34. Toepoel, V., Lugtig, P.: Online surveys are mixed-device surveys. Issues associated with the use of different (mobile) devices in web surveys. Methods Data Anal. **9**(2), 155–162 (2015). https://doi.org/10.12758/mda.2015.00933
35. Widen, S.C., Russell, J.A.: Children acquire emotion categories gradually. Cogn. Develop. **23**(2), 291–312 (2008). https://doi.org/10.1016/j.cogdev.2008.01.002

The Usability of Multiple Devices for Assessment in Psychological Research: Salience of Reasons Underlying Usability

Daniela Raccanello(✉), Margherita Brondino, Margherita Pasini, Maria Gabriella Landuzzi, Diego Scarpanti, Giada Vicentini, Mara Massaro, and Roberto Burro

Department of Human Sciences, University of Verona, Verona, Italy
{daniela.raccanello,margherita.brondino, margherita.pasini,mariagabriella.landuzzi, diego.scarpanti,roberto.burro}@univr.it, {giada.vicentini,mara.massaro}@studenti.univr.it

Abstract. Focusing on an online survey in psychological research, we evaluated the usability of the device chosen to complete the survey and examined underlying reasons through a usability enquiry method. The participants were 149 undergraduate students who completed a questionnaire for assessing achievement emotions and motivation, with open-ended and closed-ended questions. They also evaluated the usability of the device chosen to complete the survey and reported underlying reasons. We analyzed the data with Generalized Linear Mixed Models. The devices chosen to complete the survey were perceived as highly usable, even if usability was lower for smartphones compared to other devices such as personal computers, notebooks, and tablets. The most relevant reasons regarded characteristics of the tools, followed by those of the tasks and then of users and environments. The findings are discussed taking into account their theoretical and applied relevance for monitoring and improving online psychological assessment.

Keywords: Usability · Multiple devices · Online assessment
Usability enquiry methods · Questionnaires

1 Introduction

Our era provides an amount of new advantages in terms of production and benefits, but on the other hand it offers new challenges in terms of relationships between humans and the last generation of machines [6]. These relationships concern not only having or not having some tools at work, but also taking into account the complexity of these new machines, beyond considering their wide distribution. In addition, because of their powerful and wide range of functions, they have already occupied almost all aspects of the modern life, especially in the western countries, relating to work, communication, leisure, learning, and research contexts. In this scenario, it appears as intriguing the possibility to use these powerful instruments to gather data, elaborate, and share new knowledge for research purposes [13].

T. Di Mascio et al. (Eds.): MIS4TEL 2018, AISC 804, pp. 79–87, 2019.
https://doi.org/10.1007/978-3-319-98872-6_10

The literature on the relationship between humans and machines [8] distinguished between the concept of 'accessibility' and the concept of 'usability'. Accessibility, in terms of the possibility to have access to tools, concerns the encounter between the person's functional capacity and the design of the physical environment, and it is mainly objective in nature [8]. Usability refers to the fact that a person can use what s/he needs, and it's more subjective in nature [8]. Indeed, getting a new tool (e.g., buying it) is different from using it well: Accessibility does not correspond necessarily to usability. In particular, usability refers to human functions, in terms of the feasible and effective possibility to use what people need in everyday life, taking into account all aspects of life tasks and of the surrounding environment, both physical and social [8]. According to the ISO 9241-11, we highlight that usability concerns the effectiveness, the efficiency, and the satisfaction of all kinds of users in the process of achieving their specific goals. On the whole, usability can be thought as depending from a complex range of factors relating to the characteristics of the tool used for a specific task, the user, the task, and the physical and social environment [22].

Usability and usability problems can be assessed with a variety of methods, such as usability testing which concerns peoples' performance in a task; usability inspection in which the focus is on analytic usability-related elements; or usability enquiry methods, in which the users perform a real task in a work context and are asked questions about their feelings and needs [24]. Many studies proposed classification schemes of usability problems regarding human-computer interaction, useful for both formative and summative evaluation [7]. Also models for feeding back the diagnosed problems within the design process have been proposed [7]. However, scarce attention has been paid to the study of the reasons underlying usability when the task at issue is completing an online survey through multiple devices for the purposes of a psychological research.

1.1 The Present Study

In this study usability was evaluated with reference to the completion of an online survey for assessing learning-related motivation and emotions in a psychological research, through usability enquiry methods [7].

We formulated the following research questions and hypotheses:

1. Does the type of device influence the judgement on the usability of the device itself? We investigated whether characteristics such as screen dimension and being touch-enabled were related to usability. Given that the survey included both open-ended questions, in which the participants were requested to write, and closed-ended questions, in which they had to respond filling small dots on Likert-type scales, we hypothesized devices characterized by larger screen dimension and not being touch-enabled to be evaluated as more usable.
2. Which reasons underlying judgements on usability are salient for common people? We explored whether students' representation on the salience of different reasons for the usability of the devices was coherent with what is assumed by theoretical psychological models applied to technological contexts [22] which identify four main factors (user, task, tool, and environment). Specifically, we checked whether

the type of reasons changed according to the type of device and whether people reflect more about easiness or difficulty of completing an online survey; finally, we investigated which reason/s is/are more salient.

2 Method

2.1 Participants

The sample included 149 undergraduate students ($M = 21.38$ years, $SD = 5.69$; 84% female) of the University of Verona, in Northern Italy. This research is part of a longitudinal project on emotions and motivation in the university context. All the students signed an informed consent form for voluntary participation.

2.2 Material and Procedure

The data collection was made through an online questionnaire using the Apsym-Survey Software, ApSS [5, 13, 18, 19], a customized version of the open-source project LimeSurvey [21]. After a short presentation of the research, the participants had to log into the electronic platform with their own personal device and complete it. The questionnaire was administered during normal class; however, the students could choose to complete it also in another moment. It lasted about 15 min.

2.2.1 Achievement Emotions and Motivation

We included eight open-ended questions and 130 closed-ended questions pertaining to different scales (i.e., for a total of 130 items to be evaluated on a 7-point Likert scale), focused on achievement emotions and motivation related to university courses [e.g., 4, 14–17].

2.2.2 Type of Device

We asked to the participants which device they had chosen to complete the survey. Type of devices included smartphones, personal computers, notebooks, and tablets.

2.2.3 Usability

We asked to the participants to rate the usability of the used device (i.e., *How much demanding has been completing the survey with the device you chose?*) on a 7-point Likert scale (1 = *not at all*, 7 = *very much*).

2.2.4 Reasons for Usability

We asked two open-ended questions to investigate the reasons underlying usability, focusing on the easiness or the difficulty in completing the survey through the chosen device (i.e., *Write one or more reason/s that you think has/have facilitated/impeded the completion of this survey from this device*). The coding categories were developed in two ways [20]: inductively [11] and deductively adapting categories from the literature [22]. As a result, we distinguished four categories of types of responses, concerning (a) the user, including reasons related to the individuals, regarding behavioural or psychological elements; (b) the task, including reasons related to the characteristics of the survey; (c) the tool, including reasons related to the characteristics and the functioning of the device; and (d) the environment, including reasons related to the context in which the task was proposed and the tool was used. Each answer could be coded for more than one category. See Table 1 for examples of answers. A first judge coded all the answers and a second judge independently coded 30% of them for reliability: Cohen's k was 0.84, indicating a very good agreement. Disagreement were solved through discussion between judges.

Table 1. Examples of answers for type of reasons (user, task, tool, environment) by response focus (easiness, difficulty).

	Easiness	Difficulty
User	Thinking about my personality Thinking about university handbooks Thinking about university lessons	Thinking about future feelings and emotions I'm not sure about my future emotions Anticipating future moods
Task	Multiple choice structure of the questions Comprehensibility of questions Easy and fast questions	Repetitive questions Open-ended questions Length of the questionnaire
Tool	PC screen dimension enables a good view Touch screen I write answers quickly	Internet connection was slow Reduced screen view Low battery
Environment	Completing it on the train coming home The possibility to complete it anywhere I could complete the questionnaire walking	It's difficult to concentrate being around people Completing it on the train The confusion on the train

2.3 Analysis Procedure

We ran Generalized Linear Mixed Models (GLMM) using the R-software. We used the lmer/glmer functions in the lme4 package [1; for examples of applications, see 2, 3]. We performed Mixed Model ANOVA Tables via likelihood ratio tests (afex package) [23]. We utilized Bonferroni correction for post-hoc tests (lsmeans package) [10]. For rating dependent variables, the GLMM used the Gaussian distribution and the identity link-function; for binary dependent variables, it used the binomial distribution and the logit link-function. For rating variables, we reported the Cohen's d as a measure of effect size, and we calculated the degrees of freedom using the Kenward-Roger correction [9]. For binary variables, we reported the odds ratios (OR) as a measure of effect size [12], considering it modest whether $0.60 \leq OR < 1.00$ or $1.00 < OR \leq 1.70$; moderate whether $0.33 \leq OR < 0.60$ or $1.70 < OR \leq 3.00$; and strong whether $OR < 0.33$ or $OR > 3.00$.

3 Results and Discussion

3.1 Usability and Type of Device

We conducted a GLMM to explore whether the judgment on usability changed according to the type of device. We considered type of device (smartphone, personal computer, notebook, tablet) as the fixed effect, participants as the random effect, and the judgment on usability as the rating dependent variable. There was a significant effect of type of device, $\chi^2(3) = 19.79, p < .001$. Post-hoc tests indicated that scores on usability were higher (meaning lower usability) for smartphones ($M = 1.73$, $SE = 0.08$) compared to the other devices, namely personal computers ($M = 1.00$, $SE = 0.29$; $t(361.4) = 2.51, p = .054, d = 0.27$), notebooks ($M = 1.27, SE = 0.15; t(367.5) = 3.08$, $p = .013$, $d = 0.32$), and tablets ($M = 0.70$, $SE = 0.36$; $t(362.63) = 2.83$, $p = .029$, $d = 0.30$), which did not differ among them. In sum, our findings indicated that judgements on usability were lower for smartphones versus all the other devices used to complete the online survey.

Given that the judgement on usability was different for smartphones compared to all the other devices, for the following analysis we recoded the category related to type of device into a dichotomous one (i.e., 1 = *smartphone*, 2 = *other devices*).

3.2 Reasons for Usability

Concerning the reasons for usability, we ran a GLMM to explore which was the salience of the different types of reasons underlying the judgement on usability, checking also whether such salience differed according to the type of device and to the easiness/difficulty of completing the online survey. We considered type of device (smartphone, other devices), response focus (easiness, difficulty), and type of reason (user, task, tool, environment) as the fixed effects, participants as the random effect, and the frequency of the four reasons as the dependent binary variable. This analysis revealed significant main effects of response focus, $\chi^2(1) = 77.204$, $p < .001$, and type of reason, $\chi^2(3) = 306.722$, $p < .001$. For response focus, post-hoc tests revealed that the mean proportion (expressed in logit scale) of reasons was higher, $z = -4.52$, $p < .001$, $OR = .36$, for responses on easiness ($M = 0.15$, $SE = 0.02$) than for responses on difficulty ($M = 0.06$, $SE = 0.01$). For type of reason, tool-related responses ($M = 0.42$, $SE = 0.03$) were more salient than task-related responses ($M = 0.20$, $SE = 0.02$; $z = 6.82, p < .001, OR = 2.96$). In turn, task-related responses were more salient than both user-related responses ($M = 0.03, SE = 0.01; z = 7.60, p < .001, OR = 9.36$) and environment-related responses ($M = 0.03, SE = 0.01; z = 6.11, p < .001, OR = 8.32$), which did not differ among them.

The effect of type of reason was moderated by two significant two-ways interactions, namely type of device X type of reason, $\chi^2(3) = 17.840, p < .001$, and response focus X type of reason, $\chi^2(3) = 38.771$, $p < .001$. See Figs. 1a and b for mean proportions and 95% confidence intervals. Figure 1a represents the mean scores related to the type of reasons reported by the participants according to the type of device, contrasting smartphones to all the others; Fig. 1b represents the mean scores related to the type of reasons according to the perceived difficulty of the task. On the one hand, post-

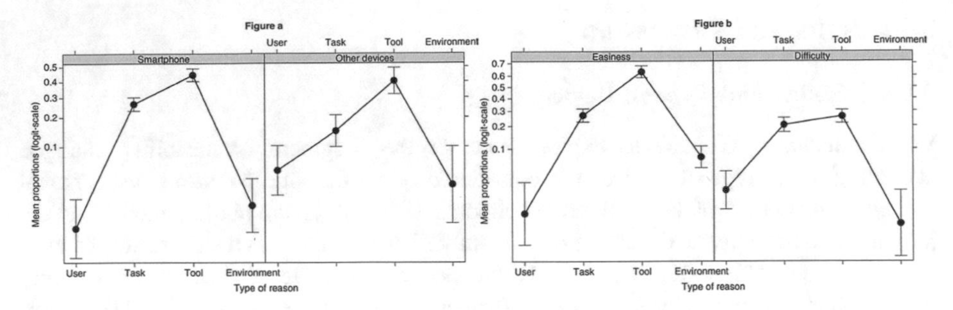

Fig. 1. Mean proportions (95% confidence interval) of type of reasons (user, task, tool, environment) by type of device (smartphone, other devices) for a, and by response focus (difficulty, easiness) for b.

hoc tests revealed that, for smartphone and for responses focused on easiness, tool-related reasons were more salient than task-related reasons (smartphone: $z = 6.18$, $p < .001$, $OR = 2.23$; easiness: $z = 8.70, p < .001, OR = 6.08$), in turn more salient than both user-related reasons (smartphone: $z = 8.07$, $p < .001$, $OR = .29.88$; easiness: $z = 5.82$, $p < .001$, $OR = 13.99$) and environment-related reasons (smartphone: $z = 7.21, p < .001, OR = 15.96$; easiness: $z = 4.28, p < .001, OR = 3.11$), as emerged commenting the main effect of type of reason. On the other hand, for the other devices, tool-related reasons were more salient than all the other reasons (tool vs. task: $z = 4.71$, $p < .001$, $OR = 3.93$; tool vs. user: $z = 6.24$, $p < .001$, $OR = 11.53$; tool vs. environment: $z = 5.03, p < .001, OR = 17.06$). For responses focused on difficulty, tool-related reasons were as salient as task-related reasons, and they were both more salient than user-related reasons (tool vs. user: $z = 6.04$, $p < .001$, $OR = 9.03$; task vs. user: $z = 4.88$, $p < .001$, $OR = 6.27$) and environment-related reasons (tool vs. environment: $z = 5.47$, $p < .001$, $OR = 32.11$; task vs. environment: $z = 4.84$, $p < .001$, $OR = 22.28$).

To sum up, our findings indicated that the reasons underlying usability were more frequent concerning the easiness rather than the difficulties of completing the online survey. In addition, salience was higher for tool-related reasons compared to the others, and for task-related reasons compared to user and environment-related reasons. However, this difference was moderated by the effect of both type of device and judgement on easiness/difficulty of the task.

4 Conclusions

In this study usability was evaluated with reference to the completion of an online survey for assessing learning-related constructs in a psychological research [8, 13, 22]. First, our findings indicate that for this kind of surveys–specifically, including open-ended questions and closed-ended questions with Likert scale–people perceive smartphones as less usable compared to other devices like personal computers, notebooks, and tablets. Keeping in mind how the different devices are characterized in terms of screen dimension and being touch-enabled, our data suggest that these characteristics would be responsible for people's perception of usability. In other words, higher usability was associated to the use of devices with larger screen dimension and not being touch-enabled. In any case, usability was very good for all the types of devices. Future studies could check whether these results generalize to contexts in which, through experimental designs, it is possible to manipulate both characteristics of the devices and characteristics of the surveys, in terms for example of open-ended questions and closed-ended questions with different response scales. Second, we examined the salience of the reasons underlying the judgement on the usability of multiple devices, an issue largely neglected by current literature. On the whole, people reported more reasons reflecting on factors facilitating rather than impeding the completion of the survey, probably mirroring their positive judgement on usability. Regarding the type of reasons, coded according to theoretical models developed within the educational psychology and applied also to technological contexts [e.g., 22], our findings suggest that people represent as more salient those reasons referred to the tool used to complete the survey–namely, smartphones, personal computers, notebooks, and tablets. In addition, task-related reasons were perceived as more salient than user and environment-related reasons. However, this difference did not emerge whether considering only devices different from smartphones, in line with the results related to our first aim. Finally, it is worth noting that, considering only when people reflect on the difficulties of the survey, tool and task-related reasons appear as having the same salience.

At a theoretical level, our study enables to generalize previous literature, taking into account the user's perspective in assessing and justifying the usability of a device, supporting the reliability of key dimensions considered within theoretical learning models [e.g., 22] and offering new knowledge on their relevance for common people. In brief, the devices used to complete the online survey at issue were perceived as highly usable, even if usability was lower for smartphones than for other devices. In addition, the most relevant reasons regarded tools and then tasks. Gathering systematically such information within online surveys can be particularly useful at an applied level. In case of longitudinal surveys, for example, such feedbacks can help to monitor the quality of the usability and to improve progressively possible critical aspects. It is worth noting that, according to our participants, the most salient reasons underlying usability–namely, tool and task-related reasons–were also the factors on which it is easier to intervene, compared to individual and environment-related factors.

References

1. Bates, D., Mächler, M., Bolker, B., Walker, S.: Fitting linear mixed-effects models using lme4. J. Stat. Softw. **67**(1), 1–48 (2015). https://doi.org/10.18637/jss.v067.i01
2. Bianchi, I., Savardi, U., Burro, R., Martelli, M.F.: Doing the opposite to what another person is doing. Acta Psychologica **151**, 117–133 (2014). https://doi.org/10.1016/j.actpsy.2014.06.003
3. Branchini, E., Burro, R., Bianchi, I., Savardi, U.: Contraries as an effective strategy in geometrical problem solving. Think. Reason. **21**(4), 397–430 (2015). https://doi.org/10.1080/13546783.2014.994035

4. Brondino, M., Raccanello, D., Pasini M.: Achievement goals as antecedents of achievement emotions: the 3 X 2 achievement goal model as a framework for learning environments design. In: Advances in Intelligent and Soft Computing, vol. 292, pp. 53–60. https://doi.org/10.1007/978-3-319-07698-0
5. Burro, R., Raccanello, D., Pasini, M., Brondino, M.: An estimation of a nonlinear dynamic process using Latent Class extended Mixed Models. Affect profiles after terrorist attacks. Nonlinear Dyn. Psychol. Life Sci. **22**(1), 35–52 (2018)
6. Dombrowsky, U., Wagner, T.: Mental strain as field of action in the 4th industrial revolution. Procedia CIRP **17**, 100–105 (2014)
7. Ham, D.H.: A model-based framework for classifying and diagnosing usability problems. Cogn. Technol. Work **16**, 373–388 (2014). https://doi.org/10.1007/s10111-013-0267-6
8. Iwarsson, S., Stahl, A.: Accessibility, usability and universal design – positioning and definition of concepts describing person-environment relationships. Disabil. Rehabil. **25**(2), 57–66 (2003)
9. Kenward, M.G., Roger, J.H.: An improved approximation to the precision of fixed effects from restricted maximum likelihood. Comput. Stat. Data Anal. **53**, 2583–2595 (2009)
10. Lenth, R.V.: Least-squares means: the R package lsmeans. J. Stat. Softw. **69**(1), 1–33 (2016). https://doi.org/10.18637/jss.v069.i01
11. Nolen, S.B.: Young children's motivation to read and write: development in social contexts. Cogn. Instr. **25**(2–3), 219–270 (2007). https://doi.org/10.1080/07370000701301174
12. Olivier, J., May, W.L., Bell, M.L.: Relative effect sizes for measures of risk. Commun. Stat. Theory Methods, 1–8 (2017). https://doi.org/10.1080/03610926.2015.1134575
13. Pasini, M., Brondino, M., Burro, R., Raccanello, D., Gallo, S. (2016). The use of different multiple devices for an ecological assessment in psychological research: an experience with a daily affect assessment. In: Advances in Intelligent and Soft Computing, vol. 478, pp. 121–129. https://doi.org/10.1007/978-3-319-40165-2_13
14. Raccanello, D.: Students' expectations about interviewees' and interviewers' achievement emotions in job selection interviews. J. Employ. Couns. **52**(2), 50–64 (2015). https://doi.org/10.1002/joec.12004
15. Raccanello, D., Brondino, M., Pasini, M.: Achievement emotions in technology enhanced learning: development and validation of self-report instruments in the Italian context. Interact. Des. Archit. J. IxD&A **23**, 68–81 (2014)
16. Raccanello, D., Brondino, M., Pasini, M.: On-line assessment of pride and shame: relationships with cognitive dimensions in university students. Adv. Intell. Soft Comput. **374**, 17–24 (2015). https://doi.org/10.1007/978-3-319-19632-9_3
17. Raccanello, D., Brondino, M., Pasini, M.: Two neglected moral emotions in university settings: Some preliminary data on pride and shame. J. Beliefs Values Stud. Relig. Educ. **36**(2), 231–238 (2015). https://doi.org/10.1080/13617672.2015.1031535
18. Raccanello, D., Burro, R., Brondino, M., Pasini, M.: Relevance of terrorism for Italian students not directly exposed to it: the affective impact of the 2015 Paris and the 2016 Brussels attacks. Stress and Health **34**, 338–343 (2017). https://doi.org/10.1002/smi.2793
19. Raccanello, D., Burro, R., Brondino, M., Pasini, M.: Use of internet and wellbeing: a mixed-device survey. In: Advances in Intelligent and Soft Computing, vol. 617, pp. 65–73 (2017). https://doi.org/10.1007/978-3-319-60819-8_8
20. Raccanello, D., Burro, R., Hall, R.: Children's emotional experience two years after an earthquake: an exploration of knowledge of earthquakes and associated emotions. PLoS ONE **12**(2), 1–21 (2017). https://doi.org/10.1371/journal.pone.0189633
21. Schmitz, C.: LimeSurvey: an open source survey tool. LimeSurvey Project Hamburg, Germany (2015). http://www.limesurvey.org

22. Shackel, B.: Usability – Context, framework, definition, design and evaluation. Interact. Comput. **21**, 339–346 (2009). https://doi.org/10.1016/j.intcom.2009.04.007
23. Singmann, H., Bolker, B., Westfall, J., Aust, F.: Afex: analysis of factorial experiments (2016). https://CRAN.R-project.org/package=afex
24. Zhang, Z.: Overview of usability evaluation method (2011). https://www.usabilityhome.com

Collaborative Language Learning Through Computer-Assisted Translation on a Wiki-Based Platform: Operations and Management

Vincenzo Baraniello[1], Luigi Laura[2], and Maurizio Naldi[3(✉)]

[1] Department of Philosophy Literature Art, University of Rome Tor Vergata, Via Columbia 1, 00133 Rome, Italy
baraniello@uniroma2.it

[2] Department of Computer, Control, and Management Engineering "Antonio Ruberti", University of Rome La Sapienza, Via Ariosto 25, 00185 Rome, Italy
laura@dis.uniroma1.it

[3] Department of Computer Science and Civil Engineering, University of Rome Tor Vergata, 00133 Rome, Italy
maurizio.naldi@uniroma2.it

Abstract. An ongoing project is described that aims at developing a computer-assisted translation tool in collaborative language learning, using a wiki approach. The technology stack makes extensive use of Mediawiki and relies on a Ubuntu-powered machine. The GUI closely resembles the familiar Wikipedia environment. Its integration in the syllabus of a university English language course is illustrated. The use of online surveys as a support tool is also described.

Keywords: Computer-assisted translation · Wiki
Language learning · Collaborative learning

1 Introduction

Several methods are employed to teach a second language. The classification proposed in [9] lists the following: audio-lingual; audio-visual; communicative teaching; direct; grammar-translation; task-based. Though translation has often been neglected in English as a foreign language (EFL), it is widely used in foreign language learning process [12,15,19]. Together with text analysis and traditional grammar, translation is at the core of any language learning process, since the link between first and second language is vital to second language learning [9]. In particular, it has been shown that Computer-Assisted Translation (CAT) can be profitably employed in language learning [11]. Computing technologies are progressively being introduced in many language learning methods, shaping the area of Computer-Assisted Language Learning (CALL) [5,14]. A brief history of CALL and its present challenges can be read in [6].

T. Di Mascio et al. (Eds.): MIS4TEL 2018, AISC 804, pp. 88–96, 2019.
https://doi.org/10.1007/978-3-319-98872-6_11

An additional area of interest is Collaborative Learning (CL) [4]. Though CL has an extremely long history and comes in many flavours, and its broad definition as *a situation in which two or more people learn or attempt to learn something together* does not even involve computers [10], it is getting a new impulse from the diffusion of web technologies [8]. In particular, the use of wiki technologies has been mentioned as one of Web 2.0 components that can be used to enhance the learning process [16,17]; its use has been reported in the context of second language learning [20].

Though each of the areas mentioned above has merits by itself, their integration is more than just the sum of the parts, as shown for the mutual benefits of CL and CALL [18], CALL and CAT [11], and CL and CAT [7].

An effort to integrate the three areas of CAT, CL, and CALL is being made with the WALLeT (Wiki Assisted Language Learning and Translation) project at the University of Rome Tor Vergata to develop a non-commercial computer-assisted translation tool, to be used for collaborative language learning. The genesis of the project and its linguistic roots have been described in [1,3].

In this paper, we describe the technological background of the project, and how different technologies are employed to make our tool work in the classroom. We define the CAT tool requirements in Sect. 2 and describe our platform Sect. 3, while its use in the classroom is reported in Sect. 4. Finally, the use of web-based surveys for classroom support is detailed in Sect. 5. Section 6 addresses the technologies employed in the platform.

2 CAT Tools: The Requirements

The core of our tool is the CAT platform. Which must possess specific features, functional to the collaborative learning aim. In this section, we briefly state those desirable features and examine the market offer.

The features we wish to have on our CAT platform are the following:

- web based, so that we can easily access it from any machine, without the need for a specific operating system environment;
- open source, so that we can customize it when needed;
- with support for different translation memories (a baseline requirement);
- highly customizable, since we need to allow teachers to assign tasks to students and monitor their progress;
- free or affordable, without periodic fees attached.

We looked at available CAT tools[1]. A survey of the translation memory (TM) technology in 2006 [13] showed that the most used TM-based tools were commercial (and rather costly), with the first ranked free tool in the list being OmegaT (http://www.omegat.org), however with just a 10% penetration rate.

[1] See, in particular, the Wikipedia page https://en.wikipedia.org/wiki/Computer-assisted_translation#Some_notable_CAT_tools, which lists several widespread CAT tools.

Though things may have changed in the meanwhile, the pie chart reported in Fig. 1, based on the Wikipedia list data, shows us that most tools still need a proprietary license. Unfortunately, no single CAT tool appears to satisfy all the requirements listed above: two - Matecat (http://www.matecat.com) and Memsource (https://www.memsource.com) - are web based, but proprietary.

3 The WALLeT Platform

In this section we describe the WALLeT platform, with a focus on user experience. The technology stack used to build the platform is detailed in Appendix 6.

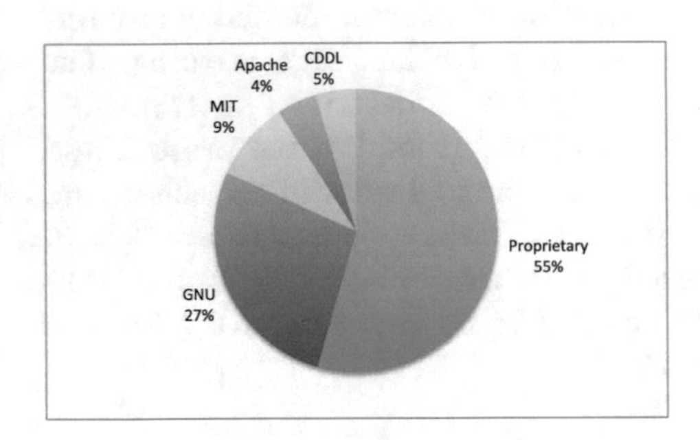

Fig. 1. Distribution of CAT tools by type of license

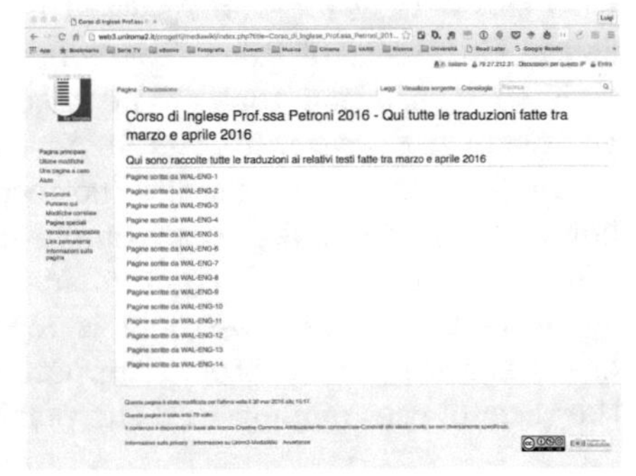

Fig. 2. A page of the WALLeT platform

In Fig. 2 we can see a web page of the current WALLeT platform, reporting all the translation projects carried out on the platform. It closely resembles a Wikipedia page, since both are based on Mediawiki. A wiki is set of structured and interconnected pages; which can be created and edited collaboratively, and provides a versioning system allowing to track the authors of changes and to roll back to the previous version. It is not necessary to use HTML to create/edit a wiki page, since wikis engine like Mediawiki typically use a simplified markup language. In Fig. 3(left) we can see how a page can be edited through a plain text editor, including the text marked using the Mediawiki markup language, also called *wikitext*. For example, a text is shown in bold if it is surrounded by two apostrophes, italic if three apostrophes are used, and bold italic if there are five apostrophes. Bullet items are preceded by an asterisk, and numbered items by an hash. As we will detail in the next section, the need to use wikitext syntax has to be addressed properly in classroom activities.

The lack of a control panel and, most importantly, an explicit command menu may be a source of confusion for inexperienced users of the Mediawiki platform, which requires the use of the *Special Pages* (actual Mediawiki pages but employed to modify the platform functions) in most cases. In Fig. 3(right) we can see the

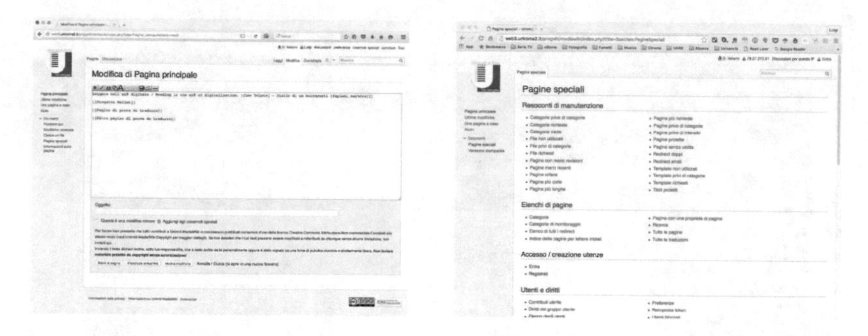

Fig. 3. Left: Editing of a page of the WALLeT platform: note the plain text form, and few commands available above it (in the red box, best viewed in colours). **Right**: the control panel of a Mediawiki platform: its *Special Pages* page.

Special Pages page, that list all the special pages (i.e., commands) available to the current user. For example, if we want to give a user administrative rights, we have to go to the *Special:UserRights* page where we can fill a form with the name of the user we want to promote.

Once a page has been created, it must be marked for translation by an administrator, by selecting the target languages for the translation; a user visiting a page can now see whether the page is available for translation in the selected languages and can start the translation process. The text is divided into fragments, as depicted in Fig. 4. The translation memory is used for suggestions, and the markups in the text are maintained. After the text is translated, the Translate Extension workflow expects a revision of the translation. After the revision, the text is available without warnings to all the users.

4 Integration in Classroom Activities

So far, we have described the general features of WALLeT. The initial effort has been largely devoted to platform development, but we have started experimenting with it rather soon in an actual classroom environment. In this section we report about its inclusion in the syllabus and our first experiences.

The introduction to the platform was carried out by including a module, made of three two-hours sessions, within the *English language* course. The first session included a lecture and a hands-on activity, while the second and the third one were totally devoted to practice, as detailed below:

1. **First session: introduction to the WALLeT platform.** The first hour lecture was devoted to explaining the platform, with a special focus on the wikitext markup language. In the second hour, the students worked in couples, each couple sharing a computer with a single user account. The students were mostly free to experiment with the insertion of the text to be translated, with tutors available to help them on the spot.

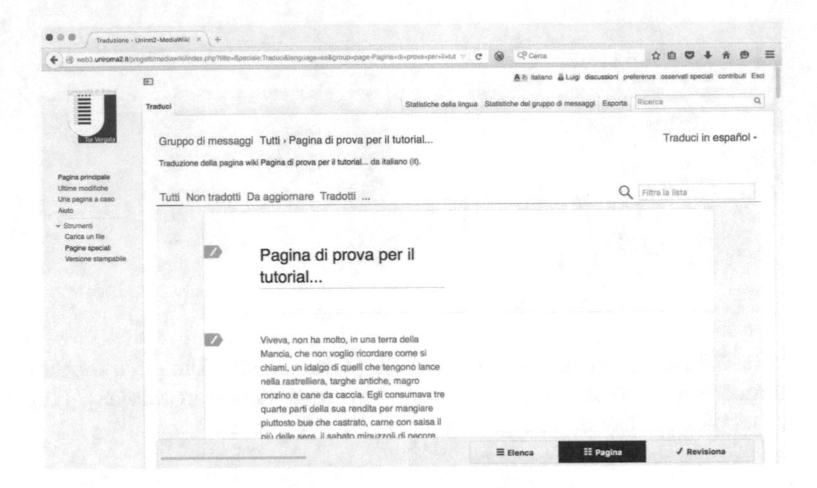

Fig. 4. Sample translation of a page with the original text divided into fragments.

2. **Second session: text insertion and translation.** Each couple of students was given a short piece of English text (in electronic format), to be added in the platform and then translated into Italian. A tutor (with administrative rights) was in charge of marking the corresponding page for translation, to be assigned to a different couple of students.
3. **Third session: translation revision.** In this lesson, each couple was assigned a translated page to be revised, with each text to be *processed* by three different couples of students: one to add the text to the platform, one to translate it, and one more to revise it.

What We Learned in this First Experiment. The students engagement was really high. A major spur to use the platform was that the students had fun with it, though we stressed the collaborative flavour and students were made aware of the advantages their work brought for future student colleagues as well.

However, we also found that the WALLeT platform, due to its wiki nature, can be dispersive for students: if they are free to interact they tend to create dozens of low quality pages, with lots of links in it. We addressed this issue by creating a single page (shown in Fig. 2), and each group was asked to create a link to a single page, to insert their assigned text to be translated.

A problem we had to address was the fact that, in Mediawiki, a newly created page defaults to the language of its creator (Italian in our case), so that all the new users in turn defaulted to be Italian. When they created the pages to include the English text, the page was in Italian for the platform: marking it for the translation into Italian, as planned, led to the same page. We found a tricky workaround, but we suggest starting with the right language to keep it simpler.

The help of tutors was invaluable: first of all, they had administrative rights and were in charge of marking the pages for translation. Such a task was not straightforward, and we believed it could be a source of troubles for students. Their help, indeed, was the key to present a simple workflow to students: first, document editing; second, document translation; third, translation revision. We needed only to teach the students the very basics of the wikitext syntax, and we provided a handy reference sheet for all the most frequent cases. Furthermore, the tutors allowed the students to ask for help in a relaxed way, and thus the overall atmosphere was indeed very positive.

5 The Use of Surveys

After describing the CAT tool that forms the core of the WALLeT platform, we now describe an online survey administration tool that supplements that CAT tool and its role in language learning classes.

We opted to use Google Forms to create and administer online surveys. This tool has been freely available for a while and just requires of the survey administrator to own a Google account. The instructor can invite the whole class to the survey at once by simply providing them with the URL linking the Google Forms page created for the survey (a shortened URL is also available).

Being web-based, the survey is intrinsically multi-platform, bearing no restrictions on the operating system or the device employed by students, and many students take advantage of that, filling in the survey on their smartphone.

The results of the survey are placed on an Excel file, hence easily retrievable by the survey administrator easily processed for statistical analysis.

At present, we are using online surveys for two main purposes: (1) preliminarily assessing students' language skills; (2) getting feedback on the use of the CAT tool. The two tasks are carried out respectively at the beginning of the course and at regular intervals during the course. The preliminary assessment of language skills allows the course instructor to have an overall view of the class level, by including questions concerning the student's language experience (e.g., the number of years of study or the specialized domains s/he deals with) and qualifications, as well as a small test, e.g., asking the student to spot errors in a text or extract keywords from it. An example of such a preliminary survey is reported in [2].

Feedback on the CAT tool is needed to orientate the development of the tool and to introduce remedies for the tool's shortcomings. In particular, we may ask students to express feedback on the GUI, or on the features that the tool offers.

6 The Wiki-Based Platform Technology Stack

In this section, we describe the key components employed in our tool, which builds on technologies that are freely available on the web.

After the evaluation of several options, we decided to base our platform on Mediawiki (https://www.mediawiki.org/), which is the engine powering Wikipedia and complies with our list of requirements, being freely available and open source with a large and active community of developers, in addition to an ever growing list of plug-ins (extensions), which allow to easily add features.

The most important extension of the Mediawiki open source platform for our purposes is undoubtedly the Translate extension (www.mediawiki.org/wiki/Extension:Translate), which supports also translation memories and machine translation with external tools (tmserver, Apertium, Microsoft Translator, Yandex.Translate). The Translate extension is also the core of the translatewiki web-based translation platform (http://translatewiki.net/), supporting the translation of content for open source projects, such as MediaWiki itself and OpenStreetMaps.

Translation Memories Support. The Translate Extension of Mediawiki has several options to support Translation Memories: in particular, it can use as its backend the database of the Mediawiki installation, a remote API service, or a dedicated text search platform such as Apache Solr (http://lucene.apache.org/solr/) or ElasticSearch (www.elastic.co/products/elasticsearch), both based on the search engine Apache Lucene (http://lucene.apache.org). A brief comparison of the available backends is shown in Fig. 5.

The WALLeT Platform Stack. The technology stack employed in the WALLeT platform is shown in Fig. 6. We run our platform on a machine running Ubuntu 16.04 (the last Long Term Support version), to be updated to 18.04 as soon as it is released. The Mediawiki platform needs a LAMP stack, i.e. Linux Apache MySql and PHP, and thus we currently run Apache 2.4, MySql 5.6, and Php 5.6. On top of the LAMP stack there is the Mediawiki platform. The Translate Extension requires Apache Solr (version 5.5) or ElasticSearch (version 5.6) to support different TMs, and both require Oracle Java Virtual Machine (version 1.8.x, note that other JVM are not fully supported).

	Database	Remote API	Solr/ElasticSearch
Enabled by default	Yes	No	No
Multiple sources	No	Yes	Yes
Update with local translations	Yes	No	Yes
Accesses database directly	Yes	No	No
Access to source	Editor	Link	Editor if local or link
Shareable as an API service	Yes	Yes	Yes
Performance	Not scaling well	Unknown	Reasonable

Fig. 5. Backends available to support Translation Memories in Translate Extension

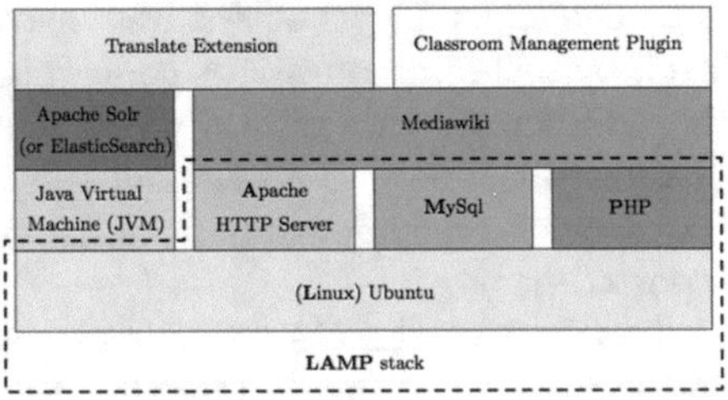

Fig. 6. The technology stack of the WALLeT platform

7 Conclusions

The technology stack behind the experimental WALLeT platform has been illustrated, as well as its use in the classroom. Its features are not found in any single tool among those currently available. The screenshot samples show how the user experience closely resembles the familiar Wikipedia context.

References

1. Baraniello, V., Degano, C., Laura, L., Lozano Zahonero, M., Naldi, M., Petroni, S.: The WALLeT project (Wiki Assisted Language Learning and Translation): bridging the gap between university language teaching and professional communities of discourse. In: The Eighth International Conference in Discourse, Communication and the Enterprise (DICOEN 2015), Naples, 11–13 June 2015
2. Baraniello, V., Degano, C., Laura, L., Lozano-Zahonero, M., Naldi, M., Petroni, S.: Accuracy of second language skills self-assessment through a web-based survey tool. In: Proceedings of the 6th International Conference in Methodologies and Intelligent Systems for Techhnology Enhanced Learning (MIS4TEL 2016), Sevilla (2016)
3. Baraniello, V., Degano, C., Laura, L., Lozano Zahonero, M., Naldi, M., Petroni, S.: A Wiki-based approach to computer-assisted translation for collaborative language learning. In: Li, Y., Chang, M., Kravcik, M., Popescu, E., Huang, R., Kinshuk, Chen, N.-S. (eds.) State-of-the-Art and Future Directions of Smart Learning. Lecture Notes in Educational Technology, pp. 369–379. Springer, Singapore (2016)
4. Barkley, E.F., Cross, K.P., Major, C.H.: Collaborative Learning Techniques: A Handbook For College Faculty. Wiley, Hoboken (2014)
5. Beatty, K.: Teaching & Researching: Computer-Assisted Language Learning. Routledge, Abingdon (2013)
6. Bush, M.D.: Computer-assisted language learning: from vision to reality? CALICO J. **25**(3), 443–470 (2013)
7. Cánovas, M., Samson, R.: Open source software in translator training. Tradumàtica: traducció i tecnologies de la informació i la comunicació **9**, 46–56 (2011)
8. Cecez-Kecmanovic, D., Webb, C.: Towards a communicative model of collaborative web-mediated learning. Australas. J. Educ. Technol. **16**(1), 73–85 (2000)
9. Cook, V.: Second Language Learning and Language Teaching. Routledge, Abingdon (2013)
10. Dillenbourg, P.: What do you mean by collaborative learning. In: Collaborative-Learning: Cognitive and Computational Approaches, vol. 1, pp. 1–15 (1999)
11. Garcia, I., Pena, M.I.: Machine translation-assisted language learning: writing for beginners. Comput. Assist. Lang. Learn. **24**(5), 471–487 (2011)
12. Kim, E.-Y.: Using translation exercises in the communicative EFL writing classroom. ELT J. **65**(2), 154–160 (2011)
13. Lagoudaki, E.: Translation memories survey 2006: users' perceptions around TM use. In: Proceedings of the ASLIB International Conference Translating & the Computer, pp. 1–29 (2006)
14. Levy, M., Stockwell, G.: CALL Dimensions: Options and Issues in Computer-Assisted Language Learning. Routledge, Abingdon (2013)
15. Liao, P.: EFL learners' beliefs about and strategy use of translation in English learning. RELC J. **37**(2), 191–215 (2006)
16. Parker, K., Chao, J.: Wiki as a teaching tool. Interdisc. J. e-learning Learn. Objects **3**(1), 57–72 (2007)
17. Popescu, E.: Using Wikis to support project-based learning: a case study. In: 2014 IEEE 14th International Conference on Advanced Learning Technologies (ICALT), pp. 305–309. IEEE (2014)
18. Popescu, E., Maria, C., Udriştoiu, A.L.: Fostering collaborative learning with Wikis: extending Mediawiki with educational features. In: Advances in Web-Based Learning–ICWL 2014, pp. 22–31. Springer (2014)

19. Ross, N.J.: Interference and intervention: using translation in the EFL classroom. Mod. Engl. Teach. **9**(9), 61–66 (2000)
20. Wang, H.-C., Lu, C.-H., Yang, J.-Y., Hu, H.-W., Chiou, G.-F., Chiang, Y.-T., Hsu, W.-L.: An empirical exploration of using Wiki in an English as a second language course. In: Fifth IEEE International Conference on Advanced Learning Technologies, ICALT 2005, pp. 155–157, July 2005

Learning Scientific Concepts with Text Mining Support

Eliseo Reategui[1](✉), Ana Paula M. Costa[2], Daniel Epstein[1], and Michel Carniato[3]

[1] PPGIE, Federal University of Rio Grande do Sul (UFRGS), Porto Alegre, Brazil
eliseoreategui@gmail.com, daepstein@gmail.com
[2] PPGEDU, Federal University of Rio Grande do Sul (UFRGS), Porto Alegre, Brazil
anapaulametz@gmail.com
[3] Pontifícia Universidade Católica do Rio Grande do Sul (PUCRS), Porto Alegre, Brazil
michelcarniato@hotmail.com

Abstract. This paper evaluates the use of a text mining tool to support learning of science concepts. The tool, called Sobek, extracts relevant information from unstructured data and represents it visually in a graph. Sobek was used here in an experiment with 36 students in 9th grade who had to learn concepts related to the particulate nature of matter. Students were divided in control (16) and experimental group (20). Students in the experimental group interacted with Sobek after reading a few texts, while the students in the control group carried out the activity in a more traditional way (reading/answering questions). Results from the experiment favored students in the experimental group, which led to the conclusion that Sobek did help students in the learning task.

Keywords: Text mining · Refutational texts · Graphic organizers

1 Introduction

A major challenge in Science Education has been the design of effective instructional material to help students learn science concepts. Since the 1970s researchers have been trying to understand how students' conceptions can contribute to learning, and how conceptual change takes place [2]. Another challenge in Science Education is the integration of information and communication technology in current educational practices, especially if we consider how technological innovation should be understood and how science, technology and society should be related [8]. In such context, visual tools may be used to help students represent information such as cause and effect, time sequence, concept comparison, problem structuring and solution planning, relating information to central themes [7]. For instance, concept mapping tools have been used to support learning activities [11], enabling students to identify and represent spatial configuration of elements, relative distance between them, contributing to the development of a network of mental representations [18].

T. Di Mascio et al. (Eds.): MIS4TEL 2018, AISC 804, pp. 97–105, 2019.
https://doi.org/10.1007/978-3-319-98872-6_12

This paper proposes as a main objective to evaluate how a particular text-mining tool with graphical representation can support learning of science concepts. The tool, called Sobek [15], is capable of retrieving important information from texts and representing it graphically. These graphical representations, similar to concept maps, allow students to edit and interact with the information provided, giving them the possibility to further reflect about the texts read. The text mining tool Sobek uses a particular mining algorithm based on the n-simple distance graph model, in which vertices represent terms and edges represent adjacency information [17]. An experiment involving 36 students in 9th grade was carried out to evaluate our approach to science concept learning with the support of text mining. We present in the following sections the text mining tool Sobek, and discuss results of its use in the aforementioned experiment.

2 The Text Mining Tool Sobek

Text mining is a process that aims to retrieve relevant information from semi-structured and unstructured data. It is a research field in constant development, mostly due to the increasing amount of information that is continually produced in our digital society. From books to personal blogs and emails, texts often have the need to be classified and examined. The process is usually statistical, sometimes relying on semantic and syntactic information that highlights important or recurrent concepts from a text [5]. There are different approaches to text mining, such as information retrieval, natural language processing, information extraction, text summarization, supervised and unsupervised learning, probabilistic methods, text streams and social media mining, opinion mining and sentiment analysis [1]. Regarding text mining's fields of application, it has been used in Digital Libraries, in Education and Academic Research, in Life Science, in Social Media and Business Intelligence [21]. In the field of Education, text mining has become more popular especially with the advent of Massive Open Online Courses [20]. Researchers have also evaluated how the mining of student responses from a survey could yield relevant information for management purposes [25], and contrasted the results obtained through the mining of students' opinions about teacher leadership with those of human raters [24].

In this research, we have used and evaluated Sobek, a text mining tool that has been employed in previous studies in the field of Education. It has been used to help students plan and review their own writings [15]; to assist teachers in the evaluation of students' essays [9]; to provide support for the evaluation of student participation in discussion forums [3]. Here, our focus has been different in that we were interested in understanding how Sobek could support reading comprehension and help students understand concepts in the field of Science.

2.1 Sobek Operation

Sobek's operation can be divided into three steps. In the first one, a text T is split into a set of words W, using spaces and punctuation marks as dividers. The set of words W is then condensed in a list of single or compound terms. This process is based on a statistical method that considers the frequency τ with which each word is found in the text. When a subset of words $w_n \in W$ is repeated in W (i.e. w_j, w_{j+1} and $w_{j+2}\}$ with a certain frequency, a compound term is formed (e.g. "Global Warming"). At the end of this process, the system returns a set of terms of size ε. Although the number of terms returned can be manually increased by the user, according to Novak and Cañas [12], no more than 25 terms should be necessary to identify the central idea of a text. Based on this assumption, Sobek's default setting is $\varepsilon = 20$.

Sobek's second step is to identify relationships between terms. This is done by looking for pairs of terms found together in the same sentence along the text, and with no more than z words between them. A relationship between term t_i and t_j implies that the terms are closely related. This relationship may represent cause and consequence, membership, time sequence, or other.

Sobek's final step is related to the arrangement of a visual output for the results of the mining process. This is done by creating a graph $G = (V, E)$ in which terms are represented by vertices (V) and their relationships by edges (E). Vertices are displayed in green boxes that are larger for the representation of terms with higher frequency in the text. Figure 1 shows a graph obtained from a Wikipedia text about Global warming[1].

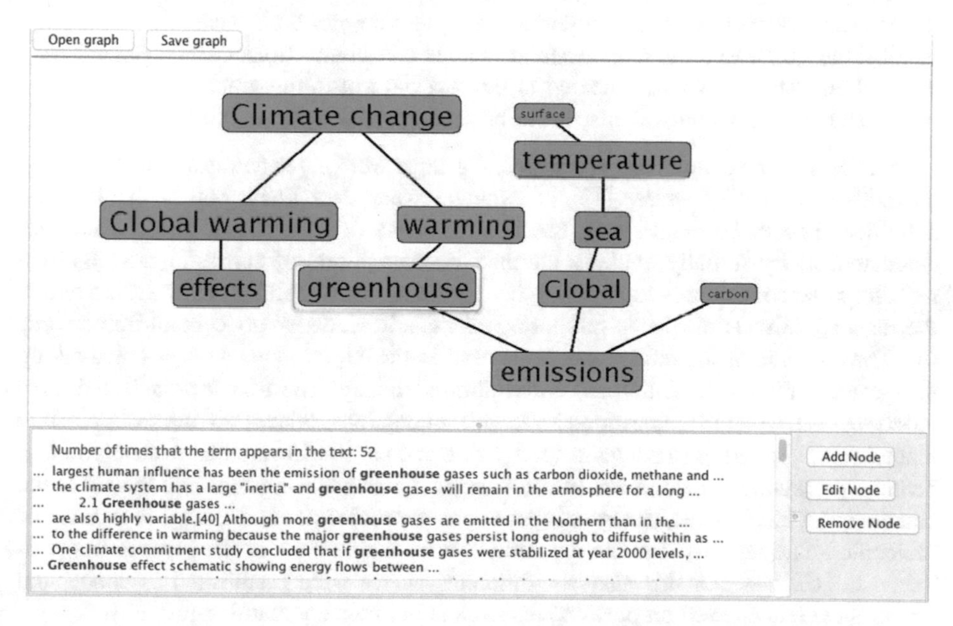

Fig. 1. Graph extracted from Wikipedia text on Global Warming

The user may interact with the graph by moving the position of terms in the screen, adding and removing terms and connections. By clicking on a given term, its frequency is displayed in a text box below the graph, as well as excerpts of the text where the

[1] https://en.wikipedia.org/wiki/Global_warming.

selected term is found. In the example of Fig. 1, the term *greenhouse* was selected. Prefuse API[2] has been used to display the graph in Sobek.

3 Research Methodology

The study presented in this paper is part of a larger set of studies that have been carried out to understand how text mining may be applied to support text comprehension, and in this particular case, learning of concepts related to the *particulate nature of matter*. Thirty-six students in 9th grade participated in the study. Sixteen students formed the control group (10 girls, 6 boys), while twenty other formed the experimental group (6 girls, 14 boys). The control group used a more traditional approach in which students had to read a given text and then answer a questionnaire about it as part of the learning task. As for the experimental group, their learning activity was based on the following steps:

a. Read the text
b. Use Sobek to extract a graph from the text
c. Analyze the graph and highlight the terms shown that are indeed important
d. Analyze connections between terms, trying to interpret their meaning
e. Edit the graph to remove unnecessary nodes and connections
f. Add relevant terms that appeared in the text but not in the graph
g. Make new connections that seem to be relevant and try to interpret their meaning

Students of the control and experimental groups were given a pretest about the topic *particulate nature of matter* [13] to evaluate what they knew before the learning activities. The experiments were based on the use of refutational texts, which are characterized by initially evoking alternative conceptions of scientific models, then refuting these conceptions explicitly. It has been shown that refutational texts are one of the most effective strategies in promoting changes in students' conceptual frameworks [6]. The contents of the refutational texts used in the experiments were selected taking into account their availability and curriculum adequacy. The four texts selected, provided by Özmen [13], questioned the misconceptions related to microscopic and macroscopic properties of matter in solid, liquid and gas due to variables such as cooling, heating, pressure application and phase changes. Concepts analyzed were related to the change of molecule sizes, change of space between molecules and change in number of molecules. The tests used in the experiment were adaptations of the instruments published in [13, 14]. For this analysis, different criteria were established as conceptual learning evidence based on previous research about concept learning [10, 23].

Students' pretest results were evaluated using the t-test to determine whether there was any difference between the scores of the experimental and the control group. No significant difference was found ($p = 0.6010$), which meant that the groups were homogeneous regarding their previous knowledge about the *particulate nature of matter*. During two weeks (2 h per week), students from the control group worked on

[2] http://www.prefuse.org/doc/api/.

the topic using a traditional reading/questionnaire answering approach, while the experimental group worked with Sobek and the given reading strategy. Although students were very autonomous in their use of Sobek in the concept learning activity, their teacher did supervise their work. Two weeks later, the students from both groups were given the exact same test given previously (pretest). Two months later, the students were given a late posttest. Table 1 shows the results obtained in all tests. To compare the performance achieved by students in the tests before and after the intervention, a gain test was used [19]. From the individual scores of the students, three measures were obtained: GAIN1 representing the different between immediate posttest and pretest scores; GAIN2 representing the difference between late posttest and immediate posttest; GAIN3 representing the difference between late posttest and pretest. The gain values were compared using the Mann-Whitney U test. A significantly higher gain was found for students in the experimental group, considering the pretest and the immediate posttest (GAIN1, $U = 223$; $p < 0.05$).

Table 1. Pre-and post-test results

		Average	Standard deviation
Control group (16 students)	Pretest	0.2351	0.0589
	Immediate posttest	0.3051	0.1526
	Late posttest	0.3080	0.1451
Experimental group (20 students)	Pretest	0.2452	0.0558
	Immediate posttest	0.4571	0.2312
	Late posttest	0.4250	0.2527

Table 2 shows the average of scores for each question of the Transformation of Matter Statement Test [14], according to their Alternative Conception.

Table 2. Pre-and post-test results for each alternative conceptions

Alternative conception	Question number	Pretest		Immediate posttest		Late posttest	
		Control (N = 16)	Experim. (N = 20)	Control (N = 16)	Experim. (N = 20)	Control (N = 16)	Experim. (N = 20)
AC1: Change of molecule size	*q1*	0.06	0.3	0.31	0.75	0.13	0.45
	q5	0.56	0.15	0.25	0.5	0.5	0.55
	q9	0.13	0.1	0.19	0.3	0.25	0.35
	q13	0.19	0.25	0.19	0.35	0.38	0.3
	q17	0.19	0.25	0.19	0.55	0.19	0.5
	q21	0.19	0.15	0.06	0.45	0.31	0.55
AC2: Change of space between molecules	*q2*	0.69	0.4	0.44	0.75	0.38	0.7
	q6	0.5	0.55	0.31	0.75	0.25	0.55
	q10	0.5	0.45	0.38	0.75	0.44	0.65
	q14	0.25	0.35	0.31	0.55	0.38	0.6

(*continued*)

Table 2. (*continued*)

Alternative conception	Question number	Pretest		Immediate posttest		Late posttest	
		Control (N = 16)	Experim. (N = 20)	Control (N = 16)	Experim. (N = 20)	Control (N = 16)	Experim. (N = 20)
	q18	0.31	0.5	0.56	0.65	0.5	0.6
	q22	0.31	0.4	0.56	0.5	0.63	0.55
AC3: Change in the number of molecules	*q4*	0.25	0.2	0.56	0.75	0.44	0.5
	q8	0.56	0.5	0.69	0.75	0.56	0.6
	q12	0.31	0.25	0.56	0.45	0.38	0.55
	q16	0.44	0.25	0.56	0.35	0.44	0.4
	q20	0.38	0.05	0.5	0.75	0.5	0.5
	q24	0.38	0.2	0.38	0.6	0.56	0.6

No significant differences were found for the sub-scores computed for each Alternative Conception. Still, when looking for trends regarding the distribution of frequencies for the questions for each alternative conception, some interesting results were identified. For instance, the control group's answers to some questions (*q2, q6* and *q10*) related to the space between particles of the Transformation of Matter Statement Test [14] showed a decrease in the number of correct answers between pretest and posttests:

- "When a solid is melted, the space between the particles…"
- "When a liquid is frozen, the space between the particles…"
- "When a liquid is vaporized, the space between the particles…"

In the case of the experimental group, there was an increase in score for the three questions, reaching 0.70 of correct answers in the late posttest of question *q2*, for example. An open question was also added to the test: "Is there (or is there not) any space between the particles forming a substance when it is solid?" [13]. Results show a certain effectiveness of the experimental intervention. While the control group had an error rate of 37.5% in the immediate posttest and 43.75% in the late posttest, the experimental group showed an error rate of 30% in the immediate posttest and 15% in the late posttest. Although the control group also reduced their error percentage in more than 50% between the pretest and the immediate posttest, this trend was not observed in the delayed posttest. The experimental group, however, was able to maintain a 50% error reduction in the delayed posttest. The hypothesis that supports such results is that the visualization and manipulation of Sobek's graphs by the students was a good instructional strategy to give learners the opportunity to further reflect about the text they had to read. The visual representation provided by Sobek as a result of the mining process is similar to that of concept maps and other graphic organizers. As it has already been shown in conceptual mapping research, getting the students to focus on the understanding of concepts and their relationships is an effective approach to learning [12, 16].

4 Discussion of Results and Limitations

The *particulate nature of matter* is considered as an important topic for the understanding of scientific phenomena. Some researchers argue that introducing this concept in early stages in school life is a powerful strategy to give students the ability to visualize different phenomena at the molecular level [6]. Building, rebuilding and refining concept representations is considered essential for the construction of scientific knowledge [22]. Results in this particular experiment showed that Sobek's capacity to highlight relevant information and allow students to actively interact with mining results (e.g. manipulating visualization features, adding/removing information) is a promising approach to the use of text mining techniques to support learning. However, some limitations are important to be discussed. Some alternative conceptions mentioned in the original study were not present in the pretest, but did appear in the immediate and late the posttests. In future studies it would be important to analyze students' text comprehension after reading the proposed materials, as it is not clear whether sometimes students did understand what was explained in the original texts, which could also explain the low scores obtained in the posttests.

Another aspect important to mention is that, as stated earlier, this study has been part of a larger group of studies to evaluate the use of Sobek in reading comprehension and concept learning tasks. Another similar experiment has been carried out to teach students about the topic *energy*, using the original material provided in [4]. In this case, however, less conclusive results were obtained, possibly because of strong misconceptions that students already carried about differences between *energy, strength* and *matter*. Another limitation refers to the duration of the interventions (2 weeks), which can be considered short because of the complexity of the concepts addressed. Such results demonstrate the need for further research to enable us to identify more clearly in which situations our approach to science concept learning can be more (or less) profitable.

5 Conclusion

This article presented an evaluation of a particular approach to concept learning in science using a strategy based on refutational texts and text mining. Results from an experiment involving 36 students demonstrated the tool's potential to assist students in their learning task, showing a statistically significant gain for students that used the tool in their learning experience in comparison with the ones that did not. The learning strategy proposed here is based on the use of domain knowledge mined automatically from a text provided by the teacher. The learning sequence after this step is focused on getting the student to further reflect about what they read, using the knowledge extracted from the text as a main guideline. A future possibility we envision is to use Sobek together with self-regulated learning strategies to help students in their learning activities. Further studies are also needed to help us determine more precisely in which contexts the use of Sobek can be more advantageous.

Learning from text reading is an extremely important skill for independent and autonomous learning [4]. This research has tried to provide evidence on how a technological tool can support this task, providing some hints on how different graphical representations may help students learn new concepts. For future work, we are also trying to understand how students may benefit from the use of Sobek when tutoring/teaching other students about scientific concepts.

References

1. Allahyari, M., et al.: A brief survey of text mining: classification, clustering and extraction techniques. In: Proceedings of Conference on Knowledge Discovery and Data Mining, KDD, Halifax (2017)
2. Amin, T.G., Smith, C.L., Wiser, M.: Student conceptions and conceptual change: three overlapping phases of research. In: Lederman, N.G., Abell, S.K. (eds.) Handbook of Research on Science Education, vol. 2. Routledge, London (2014)
3. Azevedo, B.F., Reategui, E., Behar, P.A.: Analysis of the relevance of posts in asynchronous discussions. Interdiscip. J. Knowl. Learn. Objects **10**(1), 107–121 (2014)
4. Diakidoy, I., Kendeou, P., Ioannides, C.: Reading about energy: the effects of text structure in science learning and conceptual change. Contemp. Educ. Psychol. **28**(3), 335–356 (2003)
5. Feldman, R., Sanger, J.: Text Mining Handbook: Advanced Approaches in Analyzing Unstructured Data. Cambridge University Press, New York (2006)
6. Guzzetti, B.J., Gynder, T.E., Glass, G.V., Gamas, W.S.: Promoting conceptual change in science: a comparative meta-analysis of instructional interventions from reading education and science education. Read. Res. Q. **28**(2), 116–159 (1993)
7. Hyerle, D.: Visual Tools for Transforming Information into Knowledge. Corwin Press, Thousand Oaks (2009)
8. Kubilay, K., Timurlenk, O.: Challenges for science education. Procedia Soc. Behav. Sci. **51**(1), 763–771 (2012)
9. Macedo, A.L., Reategui, E.B., Lorenzatti, A., Behar, P.A.: Using text mining to support the evaluation of texts produced collaboratively. In: Proceedings of the World Conference on Computers in Education, Bento Gonçalves, pp. 368–377. Springer, Berlin (2009)
10. Mikkil-Erdmann, M.: Improving conceptual change concerning photo-synthesis through text design. Learn. Instr. **11**(3), 241–257 (2001)
11. Novak, J.D.: The origins of the concept mapping tool and the continuing evolution of the tool. Inf. Vis. J. **5**(3), 175–184 (2006)
12. Novak, J.D., Cañas, A.J.: The origins of the concept mapping tool and the continuing evolution of the tool. Inf. Vis. **5**(3), 175–184 (2006)
13. Özmen, H.: Effect of animation enhanced conceptual change texts on 6th grade students' understanding of the particulate nature of matter and transformation during phase changes. Comput. Educ. **57**(1), 1114–1126 (2011)
14. Özmen, H., Kenan, O.: Determination of the Turkish primary students views about the particulate nature of matter. Asia Pac. Forum Sci. Learn. Teach. **1**(8) (2007)
15. Reategui, E., Epstein, D.: Automatic extraction of nonlinguistic representations of texts to support writing. Am. J. Educ. Res. **3**(1), 1592–1596 (2015)
16. Roman, D., Jones, F., Basaraba, D., Hironaka, S.: Helping students bridge inferences in science texts using graphic organizers. J. Adolesc. Adult Lit. **60**(2), 121–130 (2016)
17. Schenker, A.: Graph-theoretic techniques for web content mining, Unpublished Ph.D. thesis, University of South Florida, Tampa (2003)

18. Schroeder, N.L., Nesbit, J.C., Anguiano, C.J., Adescope, O.O.: Studying and constructing concept maps: a meta analysis. Educ. Psycol. Rev. (2017). https://doi.org/10.1007/s10648-017-9403-9
19. Seel, N.M.: Experimental and Quasi-experimental Designs for Research on Learning. Springer, Boston (2012)
20. Shatnawi, S., Gaber, M.M., Cocea, M.: Text stream mining for massive open online courses: review and perspectives. Syst. Sci. Control Eng. **2**(1), 664–676 (2014)
21. Talib, R., Hanif, M.K., Ayesha, S., Fatima, F.: Text mining: techniques, applications and issues. Int. J. Adv. Comput. Sci. Appl. **7**(11), 414–418 (2016)
22. Tytler, R., Peterson, S., Prain, V.: Picturing evaporation: learning science literacy through a particle representation. Teach. Sci. J. Aust. Sci. Teach. Assoc. **1**(52) (2006)
23. Vosniadou, S.: Capturing and modeling the process of conceptual change. Learn. Instr. **4**(1), 45–69 (1994)
24. Xu, Y., Reynolds, N.: Using text mining techniques to analyze students' written responses to a teacher leadership dilemma. Int. J. Comput. Theor. Eng. **4**(4) (2012)
25. Yu, C.H., DiGangi, S.A., Jannasch-Pennell, A.: Using text mining for improving student experience management in higher education. In: Tripathi, P., Mukerji, S. (eds.) Cases on Innovations in Educational Marketing: Transnational and Technological Strategies, pp. 196–213. Information Science Reference, Hershey (2011). https://doi.org/10.4018/978-1-60960-599-5.ch012

Virtual Reality Learning Environments in Materials Engineering: Rockwell Hardness Test

M. P. Rubio[1(✉)], D. Vergara[2], S. Rodríguez[1(✉)], and J. Extremera[2]

[1] University of Salamanca, Salamanca, Spain
{mprc, srg}@usal.es
[2] Catholic University of Ávila, Ávila, Spain

Abstract. The use of advanced Information and Communications Technology (ICT) is becoming really important in teaching-learning activities. This is especially relevant within the field of engineering where many teachers are beginning to use sophisticated virtual laboratories (VL) and computer applications in the classroom. Indeed, results of many teaching experiences validate the usefulness of such virtual tools due to their high efficiency in the teaching-learning process. However, some of the ICT tools and applications used in engineering education are becoming excessively complex and require extensive training to use them, which may be even more difficult than the knowledge they wish to teach. This communication deals with the development of new teaching technologies used in Materials Science and Engineering, specifically a VL based on the step-by-step performance of a Rockwell hardness testing machine. To achieve this goal, a realistic 3D scenario based on non-immersive virtual reality design −similar to the usual videogame environments − is used to increase students' motivation regarding the study of hardness testing of metals. Like any virtual tool which begins to be used, some changes or potential areas of improvement will arise when applied in the classroom during the subsequent years. Any improvement should take into account students' opinions and also consider that a virtual tool must be implemented within an appropriate teaching methodology with an educational aim.

Keywords: Virtual reality · Virtual labs · Materials Science · Interactivity

1 Introduction

The use of virtual laboratories (VL) −normally designed with virtual reality technology − is taking a greater relevance every day in university education. The numerous teaching experiences that have been published in the last decade suggest that the progress of this line of work is spreading incessantly, especially in engineering degrees [1, 2]. This is mainly due to the fact that the machinery used in engineering degrees is usually very delicate and/or expensive [3–5], or that the university itself does not have these facilities and, consequently, the teacher is limited to develop the practical classes of his/her subject.

T. Di Mascio et al. (Eds.): MIS4TEL 2018, AISC 804, pp. 106–113, 2019.
https://doi.org/10.1007/978-3-319-98872-6_13

This way, there are many examples of VL certainly advantageous in the teaching-learning process in engineering because they help instructors to: (i) reduce the costs of practices [3–6]; (ii) make the most of the space available around a machine so that all students receive adequate training in a mass class [7, 48–51]; (iii) avoid problems or accidents, e.g. due to X-ray radiation [8, 9], due to risky chemical experiments [10], etc.; (iv) show the theoretical content in an interactive and effective way before executing the practical content in a real essay [11]; etc. All these advantages justify the use of technology enhanced learning (TEL) tools in engineering education [12, 15–47].

This paper is structured as follows: in the next section, the proposed VL is described, both its design, development and functionalities and user workflow. Sections 3 and 4 include the discussion and conclusions.

2 Proposed Virtual Lab

The main objective of this VL is to familiarize the student with the use of a Rockwell durometer before doing the practical class in the real laboratory of his university. The teaching-learning process try to become as simple and clear as possible, avoiding complex procedures that are difficult to understand.

In the development, the usual workflow is used in the creation of an interactive application and it is summarized in the following diagram (Fig. 1):

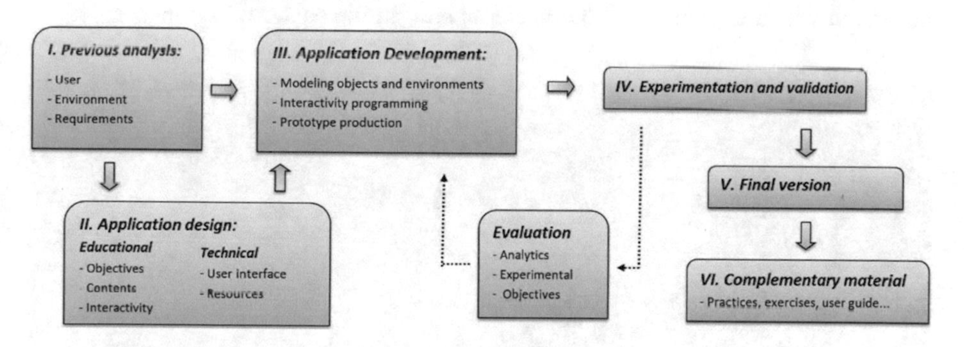

Fig. 1. Workflow in the creation of an interactive application.

In the previous analysis, the behavior of the user of the application as well as the environment and its interactions with it are determined. The design determines the educational and technical needs that must be met.

In this case, it has been designed to meet the following requirements: (i) simulate a realistic room in lighting and materials, similar to a university laboratory, (ii) show a realistic and detailed appearance of the durometer, (iii) explain the step-by-step management of the durometer, requiring real-time student interaction, (iv) program very simple controls and (v) present a final exercise to perform the hardness reading.

The professors of the materials engineering courses do the evaluation and improvement processes until the final application is achieved.

2.1 Development

The VL presented in this article presents a laboratory equipped with a durometer in which the Rockwell hardness measuring process of two specimens is simulated, one in scale RH-B (with ball indenter of diameter 1/16" and 150 kg load) and another in scale RH-C (with diamond indenter and 100 kg load). The durometer used as reference is of the CENTAUR brand and corresponds to the RB2 model.

The durometer and the room that contains it have been modeled using real objects as a reference and with special attention to the detail of the durometer and the elements that make it up. Autodesk 3D Studio Max [13] has been selected for modeling, mainly due to the versatility and the large amount of documentation that exists in this software. In addition, Autodesk 3D Studio Max does not require the payment of licenses if it is used for purely educational purposes.

On the other hand, the programming of the VL has been done using Unreal Engine 4 [14], which allows programming without the need to type code and see in real time what is being done. This programming engine is widely used in the commercial creation of videogames [14] and its license is free if it is not used for profit.

The VL created has the appearance of a video game type "shooter" in first person, which allows to explore with complete freedom the interior of the laboratory through the use of the computer keyboard and mouse. Also, this VL promotes the technical knowledge of this type of durometer and its accessories since it allows to visualize it from any point of view and execute a real-time test. Figure 2 shows a general view of the VL, in which the Rockwell durometer is seen in the center of the image.

Fig. 2. General view of the virtual laboratory.

2.2 User Workflow

The use of this VL involves several stages of action on the part of the student:

- *Free exploration of the laboratory and the durometer*: this helps the student to become familiar with the laboratory and the real machinery, and the VL recreates a reliable academic environment.
- *Start-up of the durometer and selection of the Rockwell scale to execute in the test*: the student must select the scale RH-B or RH-C to execute according to the properties of the material tested.
- *Step by step guide to perform the test*: the VL proposes a sequence of the different phases that must be done to perform a Rockwell test, favoring the student to be aware of each step of the process, e.g. he must press a key to place the specimen, another key to apply the load (Fig. 3), etc. Thus, the VL has been designed to guide the student and ensure a good result in the teaching-learning process.
- *Reading the value of the hardness of the material:* the student must know how to read the quantitative Rockwell hardness measurement from the results shown on the virtual durometer screen (Fig. 4).
- *Answer to the final question*: students must choose, between four possible options defining the hardness of the test executed, which is the correct one (Fig. 4).

Fig. 3. Step-by-step sequence example: instruction to apply the main load.

Fig. 4. Final exercise asking to select the correct hardness value

3 Discussion

From the point of view of the freedom of interaction with the virtual environment created in the VL, there can be multiple levels, ranging from an open world -with total freedom of action and displacement-, to a directed world -restricted to certain actions and movements-. When starting the process of designing a VL it is necessary to decide on the most appropriate grade of interaction to achieve the learning objectives, deciding if an open or directed world is of interest.

In this way, in the case of mechanical tests that have a well-defined use procedure, it is advisable to limit the user's freedom of choice by means of a step-by-step design, i.e. the user is obliged to follow each step of the test execution process. This helps the user (student) is at all times aware of the execution process and quickly and correctly assimilate the steps to follow in the use of a machine. Thus, it can be said that in the directed design the application itself (VL) acts as an instructor in the teaching-learning process. This design is opposed to the open world in which the freedom of action can hinder the concentration of the user in the process that is intended to teach.

Taking into account the step-by-step design, the VL presented in this communication plays short animations that correspond to the different actions that have to be carried out to perform a hardness test, specifically the Rockwell one. Before carrying out each of these actions, the student must read the instructions indicated in the VL. Sometimes the next action is executed at the only possibility of pressing a button (on screen or keyboard), but in others the student must choose between several buttons to execute the next action. This process helps to capture and maintain the student's attention in the test that is running.

4 Conclusions

The VL presented in this paper is a useful tool both for the engineering student, who when using it is familiarized with the process of performing a Rockwell hardness test before seeing it in a real laboratory, as well as for the professor, who you can use in your master classes as a replacement for videos or slides.

The step-by-step design is the most appropriate from an educational point of view in cases similar to the exposed in this communication, i.e. a process consisting of several stages.

Acknowledgments. This work has been supported by project "IOTEC: Development of Technological Capacities around the Industrial Application of Internet of Things (IoT)". 0123_IOTEC_3_E. Project financed with FEDER funds, Interreg Spain-Portugal (PocTep).

References

1. Heradio, R., de la Torre, L., Galán, D., Cabrerizo, F.J., Herrera-Viedma, E., Dormido, S.: Virtual and remote labs in education: a bibliometric analysis. Comput. Educ. **98**, 14–38 (2016)
2. Vergara, D., Lorenzo, M., Rubio, M.P.: On the use of virtual environments in engineering education. Int. J. Qual. Assur. Eng. Technol. Educ. **5**(2), 30–41 (2016)
3. García, J., Entrialgo, J.: Using computer virtualization and software tools to implement a low cost laboratory for the teaching of storage area networks. Comput. Appl. Eng. Educ. **23**, 715–723 (2015)
4. Hilfert, T., König, M.: Low-cost virtual reality environment for engineering and construction. Vis. Eng. **4**, 1–18 (2016)
5. Vergara, D., Rubio, M.P., Prieto, F., Lorenzo, M.: Enhancing the teaching/learning of materials mechanical characterization by using virtual reality. J. Mater. Educ. **38**, 63–74 (2016)
6. Vergara, D., Rodríguez, M., Rubio, M.P., Ferrer, J., Núñez, F.J., Moralejo, L.: Formación de personal técnico en ensayos no destructivos por ultrasonidos mediante realidad virtual. Dyna **93**(2), 150–154 (2018)

7. Vergara, D., Rubio, M.P., Lorenzo, M.: New approach for the teaching of concrete compression tests in large groups of engineering students. J. Prof. Issues Eng. Educ. Pract. **143**(2) (2017). paper 05016009
8. Boukerche, A., Al Hamidi, A., Pazzi, R., Ahmad, L.: Architectural design for the 3D virtual radiology department using virtual reality technology. In: IEEE Workshop on Computational Intelligence in Virtual Environments, pp. 45–52 (2009)
9. Vergara, D., Rubio, M.P.: The application of didactic virtual tools in the instruction of industrial radiography. J. Mater. Educ. **37**(1–2), 17–26 (2015)
10. Xie, Q., Tinker, R.: Molecular dynamics simulations of chemical reactions for use in education. J. Chem. Educ. **83**(1), 77–83 (2006)
11. Dobrzański, L.A., Honysz, R.: On the implementation of virtual machines in computer aided education. J. Mater. Educ. **31**(1–2), 131–140 (2009)
12. Koretsky, M.D., Kelly, Ch.: Industrially situated virtual laboratories. Chem. Eng. Educ. **45**(3), 219–228 (2011)
13. Autodesk 3D Studio MAX: www.autodesk.com/products/3ds-max/overview. Accessed 01 Feb 2018
14. Unreal Engine 4: www.unrealengine.com. Accessed 01 Feb 2018
15. Li, T., Sun, S., Bolić, M., Corchado, J.M.: Algorithm design for parallel implementation of the SMC-PHD filter. Signal Process. **119**, 115–127 (2016). https://doi.org/10.1016/j.sigpro.2015.07.013
16. Lima, A.C.E.S., De Castro, L.N., Corchado, J.M.: A polarity analysis framework for Twitter messages. Appl. Math. Comput. **270**, 756–767 (2015). https://doi.org/10.1016/j.amc.2015.08.059
17. Redondo-Gonzalez, E., De Castro, L.N., Moreno-Sierra, J., De Las Casas, M.L.M., Vera-Gonzalez, V., Ferrari, D.G., Corchado, J.M.: Bladder carcinoma data with clinical risk factors and molecular markers: a cluster analysis. BioMed Res. Int. **2015**, 14 (2015). https://doi.org/10.1155/2015/168682
18. Palomino, C.G., Nunes, C.S., Silveira, R.A., González, S.R., Nakayama, M.K.: Adaptive agent-based environment model to enable the teacher to create an adaptive class. In: Advances in Intelligent Systems and Computing, vol. 617 (2017). https://doi.org/10.1007/978-3-319-60819-8_3
19. Li, T., Sun, S., Corchado, J.M., Siyau, M.F.: A particle dyeing approach for track continuity for the SMC-PHD filter. In: FUSION 2014 - 17th International Conference on Information Fusion (2014)
20. Coria, J.A.G., Castellanos-Garzón, J.A., Corchado, J.M.: Intelligent business processes composition based on multi-agent systems. Expert Syst. Appl. **41**(4 PART 1), 1189–1205 (2014). https://doi.org/10.1016/j.eswa.2013.08.003
21. De La Prieta, F., Navarro, M., García, J.A., González, R., Rodríguez, S.: Multi-agent system for controlling a cloud computing environment. In: LNCS (LNAI and LNBI), (vol. 8154 LNAI). https://doi.org/10.1007/978-3-642-40669-0_2
22. Tapia, D.I., Fraile, J.A., Rodríguez, S., Alonso, R.S., Corchado, J.M.: Integrating hardware agents into an enhanced multi-agent architecture for ambient Intelligence systems. Inf. Sci. **222**, 47–65 (2013). https://doi.org/10.1016/j.ins.2011.05.002
23. Costa, Â., Novais, P., Corchado, J.M., Neves, J.: Increased performance and better patient attendance in an hospital with the use of smart agendas. Logic J. IGPL **20**(4), 689–698 (2012). https://doi.org/10.1093/jigpal/jzr021
24. Rodríguez, S., De La Prieta, F., Tapia, D.I., Corchado, J.M.: Agents and computer vision for processing stereoscopic images. In: LNCS (LNAI and LNBI), (vol. 6077 LNAI) (2010). https://doi.org/10.1007/978-3-642-13803-4_12

25. Rodríguez, S., Gil, O., De La Prieta, F., Zato, C., Corchado, J.M., Vega, P., Francisco, M.: People detection and stereoscopic analysis using MAS. In: Proceedings of the INES 2010 - 14th International Conference on Intelligent Engineering Systems (2010). https://doi.org/10.1109/INES.2010.5483855
26. Baruque, B., Corchado, E., Mata, A., Corchado, J.M.: A forecasting solution to the oil spill problem based on a hybrid intelligent system. Inf. Sci. **180**(10), 2029–2043 (2010). https://doi.org/10.1016/j.ins.2009.12.032
27. Tapia, D.I., Corchado, J.M.: An ambient intelligence based multi-agent system for Alzheimer health care. Int. J. Ambient Comput. Intell. **1**(1), 15–26 (2009). https://doi.org/10.4018/jaci.2009010102
28. Mata, A., Corchado, J.M.: Forecasting the probability of finding oil slicks using a CBR system. Expert Syst. Appl. **36**(4), 8239–8246 (2009). https://doi.org/10.1016/j.eswa.2008.10.003
29. Glez-Peña, D., Díaz, F., Hernández, J.M., Corchado, J.M., Fdez-Riverola, F.: geneCBR: a translational tool for multiple-microarray analysis and integrative information retrieval for aiding diagnosis in cancer research. BMC Bioinform. **10**, 187 (2009). https://doi.org/10.1186/1471-2105-10-187
30. Fernández-Riverola, F., Díaz, F., Corchado, J.M.: Reducing the memory size of a fuzzy case-based reasoning system applying rough set techniques. IEEE Trans. Syst. Man Cybern. Part C Appl. Rev. **37**(1), 138–146 (2007). https://doi.org/10.1109/tsmcc.2006.876058
31. Méndez, J.R., Fdez-Riverola, F., Díaz, F., Iglesias, E.L., Corchado, J.M.: A comparative performance study of feature selection methods for the anti-spam filtering domain. In: LNCS (LNAI and LNBI), (vol. 4065 LNAI), pp. 106–120 (2006)
32. Fdez-Rtverola, F., Corchado, J.M.: FSfRT: Forecasting system for red tides. Appl. Intell. **21**(3), 251–264 (2004). https://doi.org/10.1023/b:apin.0000043558.52701.b1
33. Corchado, J.M., Pavón, J., Corchado, E.S., Castillo, L.F.: Development of CBR-BDI agents: a tourist guide application. In: LNCS (LNAI and LNBI), vol. 3155, pp. 547–559 (2004). https://doi.org/10.1007/978-3-540-28631-8
34. Laza, R., Pavón, R., Corchado, J.M.: A reasoning model for CBR_BDI agents using an adaptable fuzzy inference system. In: LNCS (LNAI and LNBI), vol. 3040, pp. 96–106. Springer, Heidelberg (2004)
35. Corchado, J.A., Aiken, J., Corchado, E.S., Lefevre, N., Smyth, T.: Quantifying the ocean's CO2 budget with a CoHeL-IBR system. In: Proceedings of the Advances in Case-Based Reasoning, vol. 3155, pp. 533–546 (2004)
36. Corchado, J.M., Borrajo, M.L., Pellicer, M.A., Yáñez, J.C.: Neuro-symbolic system for business internal control. In: Industrial Conference on Data Mining, pp. 1–10 (2004). https://doi.org/10.1007/978-3-540-30185-1_1
37. Corchado, J.M., Corchado, E.S., Aiken, J., Fyfe, C., Fernandez, F., Gonzalez, M.: Maximum likelihood Hebbian learning based retrieval method for CBR systems. In: LNCS (LNAI and LNBI), vol. 2689, pp. 107–121 (2003). https://doi.org/10.1007/3-540-45006-8_11
38. Fdez-Riverola, F., Corchado, J.M.: CBR based system for forecasting red tides. Knowl. Based Syst. **16**(5–6 SPEC), 321–328 (2003). https://doi.org/10.1016/S0950-7051(03)00034-0
39. Glez-Bedia, M., Corchado, J.M., Corchado, E.S., Fyfe, C.: Analytical model for constructing deliberative agents. Int. J. Eng. Intell. Syst. Electr. Eng. Commun. **10**(3), 173–185 (2002)
40. Corchado, J.M., Aiken, J.: Hybrid artificial intelligence methods in oceanographic forecast models. IEEE Trans. Syst. Man Cybern. Part C-Appl. Rev. **32**(4), 307–313 (2002). https://doi.org/10.1109/tsmcc.2002.806072
41. Fyfe, C., Corchado, J.M.: Automating the construction of CBR systems using kernel methods. Int. J. Intell. Syst. **16**(4), 571–586 (2001). https://doi.org/10.1002/int.1024

42. Corchado, J.M., Fyfe, C.: Unsupervised neural method for temperature forecasting. Artif. Intell. Eng. **13**(4), 351–357 (1999). https://doi.org/10.1016/s0954-1810(99)00007-2
43. Corchado, J., Fyfe, C., Lees, B.: Unsupervised learning for financial forecasting. In: Proceedings of the IEEE/IAFE/INFORMS 1998 Conference on Computational Intelligence for Financial Engineering (CIFEr), (Cat. No.98TH8367), pp. 259–263 (1998). https://doi.org/10.1109/CIFER.1998.690316
44. Li, T.-C., Su, J.-Y., Liu, W., Corchado, J.M.: Approximate Gaussian conjugacy: parametric recursive filtering under nonlinearity, multimodality, uncertainty, and constraint, and beyond. Front. Inf. Technol. Electron. Eng. **18**(12), 1913–1939 (2017)
45. Wang, X., Li, T., Sun, S., Corchado, J.M.: A survey of recent advances in particle filters and remaining challenges for multitarget tracking. Sensors **17**(12), 2707 (2017)
46. Morente-Molinera, J.A., Kou, G., González-Crespo, R., Corchado, J.M., Herrera-Viedma, E.: Solving multi-criteria group decision making problems under environments with a high number of alternatives using fuzzy ontologies and multi-granular linguistic modelling methods. Knowl Based Syst. **137**, 54–64 (2017)
47. Pinto, T., Gazafroudi, A.S., Prieto-Castrillo, F., Santos, G., Silva, F., Corchado, J.M., Vale, Z.: Reserve costs allocation model for energy and reserve market simulation. In: 2017 19th International Conference on Intelligent System Application to Power Systems, ISAP 2017, Article no. 8071410 (2017)
48. Oliver, M., Molina, J.P., Fernández-Caballero, A., González, P.: Collaborative computer-assisted cognitive rehabilitation system. ADCAIJ Adv. Distrib. Comput. Artif. Intell. J. **6**(3), 57–74 (2017). ISSN 2255-2863
49. Ueno, M., Suenaga, M., Isahara, H.: Classification of two comic books based on convolutional neural networks. ADCAIJ Adv. Distrib. Comput. Artif. Intell. J. **6**(1), 5–12 (2017). ISSN 2255-2863
50. Silveira, R., Da Silva, G.K., Bitencourt, T., Gelaim, A., Marchi, J., De La Prieta, F.: Towards a model of open and reliable cognitive multiagent systems: dealing with trust and emotions. ADCAIJ Adv. Distrib. Comput. Artif. Intell. J. **4**(3), 57–86 (2015). ISSN 2255-2863
51. Chamoso, P., De La Prieta, F.: Simulation environment for algorithms and agents evaluation. ADCAIJ Adv. Distrib. Comput. Artif. Intell. J. **4**(3), 87–96 (2015). ISSN 2255-2863

A Teaching Experience on Information Systems Auditing

Ricardo Pérez-Castillo[1(✉)], Ignacio García Rodríguez de Guzmán[1], Moisés Rodríguez[1,2], and Mario Piattini[1]

[1] Information Technology and Systems Institute, Universidad de Castilla-La Mancha, Paseo de la Universidad 4, 13071 Ciudad Real, Spain
{ricardo.pdelcastillo, ignacio.grodriguez, moises.rodriguez, mario.piattini}@uclm.es
[2] AQCLab, Camino de Moledores, 13051 Ciudad Real, Spain

Abstract. Software has become the keystone in many firms, regardless of the particular sector a company belongs to. Apart from bearing the greater part of the business value of the organisation, software makes it possible to leverage a radical digitalisation of many companies, even for those with a business model that is not IT-intensive. That makes it vital for firms have the capacity to establish mechanisms that enable information system quality to be assured. This has to be done if the systems themselves are to inspire trust and confidence. The auditing of the information systems gathers, groups and assesses evidence with which to judge whether an information system safe-guards assets, maintains the integrity of data, carries out the goals of the organisation effectively, and uses its resources efficiently. This paper presents a teaching experience carried out in the sphere of information system auditing, using a project-based approach. This strategy allows students to have an immersion experience in a fictitious company. They simulate a situation in which they are working for the firm's IT department, establishing internal controls which they later audit. It has been shown that this approach enables students to acquire higher-level competences and skills, while also allowing them to assimilate the contents of the subject more fully.

Keywords: Information system auditing · Project-based learning
Teaching experience

1 Introduction

Over recent decades, software has been growing exponentially. It used to be located in huge computing machines, but now forms part of practically all productive sectors, and it is to be found in any device of our everyday life. In banking, insurance, transport, and the food industry, as well as in education, medicine, sport, leisure, etc.: we don't need much time to think of where software is present. From almost the moment we get out of bed each morning, until we get back into it again at night, we use software, and this inevitably leads us to think about what an important role it plays in our lives. Seeing its significance encourages us to think carefully about how vital it is for the information systems based on that software to be managed and audited in the right way. If we focus

T. Di Mascio et al. (Eds.): MIS4TEL 2018, AISC 804, pp. 114–122, 2019.
https://doi.org/10.1007/978-3-319-98872-6_14

on the business sphere, software has become a keystone, sustaining the main weight of the business value of organizations, and it is a pivotal feature in the support needed for digital transformation. We thus see, for instance, that companies need software to work within the response time demanded by the market, and that they use software to produce higher-quality goods. They also employ software to store all the information about their business and use it to carry out analyses of their data, as well as of the market, making predictions that will enable them to position themselves in the future.

One of the aspects that is vital for organisations at the present time, in the light of all we have just remarked, is that they should be able to trust those software-based information systems (IS) which support all of their work. One of the main tools that monitor all this is auditing. IS auditing is the process of gathering, grouping and assessing assets; it maintains data integrity, carries out the goals of the organisation effectively, and uses resources efficiently [1]. It should be added here that the importance of auditing is such that, according to the latest survey carried out by the ISO [2], a total of 1,106,356 valid certificates were reported for ISO 9001 (including 80,596 issued to the 2015 version) an increase of 7% on last year. ISO and IEC's standard for information security, ISO/IEC 27001 experience the same annual growth of 20% annual increase as last year to 33,290 certificates worldwide. That certification helps organisations to protect and reinforce their IS, allowing them to prove that they have implemented a set of controls that are good enough to assure confidentiality, integrity and availability of their IS.

Taking all of the above into account, IS auditing is currently a field that is attracting a great deal of interest at a business level. This has meant that those university subjects whose task it is to give instruction in this area have become especially significant in degree courses such as Computer Engineering.

This paper presents a teaching experience in the area of IS Auditing; the pedagogical strategy followed was a Project-Based Learning method (PBL) [3]. This methodology created a simulation in which the students were employees of a fictitious company; their task was to specify and establish internal controls, and then audit these at a later point in time.

The remainder of the paper is structured as follows: Sect. 2 presents the most significant international curricula related to IS Auditing. Section 3 provides a description of the teaching experience carried out by the authors in the field under study. Proposals for improvement on teaching in the field of IS auditing are set out in Sect. 4. Lastly, Sect. 5 gives a summary of the conclusions reached.

2 International Curricula

There are international teaching curricula providing guidelines on contents, competences and skills that should be taught on the different Information Technology degrees.

- **ACM.** The ACM (Association for Computing Machinery) [4] provides a series of 5 curricula for the degrees of (i) Computer Engineering, (ii) Computational Science (iii) Information Systems, (iv) Information Technology and (v) Software Engineering. In this series, the teaching curriculum that deals with information system auditing is, in the main, IS [5], although the subject is also tackled to a certain extent in Computational Science [6] and Information Technology [7]. The curriculum that is proposed by the ACM for the degree of Information Systems (IS) consists of 7 compulsory core courses, along with 7 others which are elective courses focusing on different roles or specializations, but only three that address some particular topic of IS auditing: T2 Information and Data Management, T3 Business Architecture, and T4 Project Management. Due to its lack, this curriculum establishes the elective course O5 for auditing and IT controls in IS. Apart from the curriculum established for the degree in IS, the set of ACM curricula deals with auditing with a certain lack of depth when it comes to the curricula of other degree courses. In this sense it is possible to mention the Computational Science [6] or the Information Technology curriculum [7]. The other two remaining curricula in the ACM series do not take in knowledge areas that include IS auditing, but just overlap auditing concepts: the curriculum of Software Engineering [8] deals with the knowledge area of Software Quality, or the curriculum of Computation Engineering [9] defines the knowledge area of Information Security.

- **SWEBOK.** The SWEBOK curriculum [10] is designed for software engineering degrees. That is why IS auditing is not especially relevant in that curriculum. However, IS auditing is addressed in at least 2 of the 15 knowledge areas in SWEBOK: (i) in *Configuration Management* (topic called Software Configuration Auditing); and (ii) the knowledge area of *Software Quality* (topic of Software Quality Management Processes).
- **ISACA.** The ISACA (*Information Systems Audit and Control Association*) is the most important international organization dedicated to IS knowledge, certifications, defence, and education in security and assurance, in addition to corporate governance and management of IT, as well as risk and compliance related to IT. The ISACA curriculum is divided into five knowledge domains: (i) the process of auditing information systems, (ii) governance and management of it, (iii) information systems acquisition, development and implementation, (iv) information systems operations, maintenance and support, and (v) protection of information assets. A complete list of topics is available in [11]. The ISACA curriculum is designed quite specifically for the teaching of information system auditing. This means that it is not likely to have a direct correspondence to an Informatics degree, such as a degree in Information Systems, or in Computational Science. It seems rather to focus on being the curriculum that proposes specific courses on auditing, or that provides a model for subjects within related computer science degrees. This curriculum is the one that most comprehensively brings together all the knowledge areas and subject matter on information system auditing.

3 The Teaching Experience

The teaching experience on which the present approach has been developed takes in a period of 10 years in the subjects of "Informatics System Auditing" and "Auditing of Information Systems" (in the study programme of Computer Engineering and the syllabus of the degree in Computer Engineering) in the Computer Science Faculty of the University of Castilla-La Mancha in Ciudad Real *Escuela Superior de Informática de Ciudad Real* (Universidad de Castilla-La Mancha, UCLM). It also includes a time of teaching in a variety of Masters courses and post-graduate degrees where lectures were (or are) given on subjects about auditing, security and IT Governance. Some of the authors also have experience in rolling out innovative teaching initiatives related to the field of security, which is an area that has much in common with that of auditing [1].

3.1 Precedents

Traditionally, many subjects have been rather theory-based, and it has been difficult to give them a practical focus for the students. They often felt that, far from learning anything, they were trapped in a learning process that had become tedious and routine, since it concentrated mostly on the subject-matter. This view of the subjects on the students' part produces a lack of motivation, damaging the learning mechanisms that enable them to understand and properly take in the competences, skills, and contents. As far as information system auditing is concerned, the first experiences of the authors showed that, although the subjects did include theory-practice assignments, these were just a mere application of the large amount of theoretical content that was being taught. It was thus observed that the simple learning of subject-matter provided no real challenge to the students, apart from their just following the procedures set out for carrying out audits (established by the ISACA). The way of teaching a subject that is essentially more theory-based and difficult to apply in a controlled environment demanded an adaptation of the pedagogical approach. It is not only a matter of adapting the syllabus; the students themselves have to modify their concerns and their learning needs.

3.2 Current Methodology and Teaching Goals

As pointed out above, the learning process is directly related to the ability to "excite" and engage the students, which in turn leads to those individuals being highly-motivated, turning them into the main actor in the development of the subject. The basis of the pedagogical plan that is presented below is "project-based learning" (PBL) [3], where the student, finds out what tools and knowledge they need at any given time, and look for ways to acquire them to solve a problem (project). The current pedagogical programme has been designed for the subject of "Information System Auditing", a subject in the Degree in Computer Engineering in the UCLM. The main goal of this subject is to provide students with the knowledge and the tools necessary for the development of IT audits in IS. This main objective can be broken down into two sub-objectives:

- Understand the context of IT in the organisations, as well as the control mechanisms that these bodies establish for themselves to ensure that they align well with the business goals (i.e., how the organisations create Internal Control – IC from now on).
- Identify and apply IT auditing techniques and tools for the assessment of the control mechanisms which organisations create around their IT (IC auditing of the organisations).

3.3 Sequence of Activities

With this break-down of the main goal, it is obvious that this objective contains an implicit need to have solid knowledge about the structure and development of the IC that must later be audited. Teaching this subject therefore takes place as follows.

1. **Creation of a firm in the IT field.** On the first day of class, the students receive a message telling them that they have been hired as members of the IT department of an organisation. These members of this organisation are divided into sub-groups. The new "professionals" are informed of the working scope of the organisation as well as of how importance they will be as a department that will establish the entire technological infrastructure.
2. **Training period.** The new members of the company go through a time of training, in which the students are given enough instruction in the subject-matter to begin the work on their own. The training period presents concepts such as the basic aspects of auditing and internal control. The pupils are divided into sub-groups of 3–5 members. Each one of these groups is given one or two high-level goals to work with.
3. **Configuration of the technological infrastructure of the company.** After this training period, each group must decide what technological infrastructure their organisation will have. To that end, the groups analyse the business goals and, depending on what these are, they will choose a set of IT technologies and infrastructures that they consider necessary to carry out the business objectives assigned to them. They will thus decide on a final set of IT resources that will work in the organisation.
4. **Development of the internal control.** Once the common technological infrastructure for the organisation has been decided, each group, bearing in mind the business goals of their firm, will develop a part of the IC that enables all those risks that are related to IT and the business goals to be kept under control. This is one of the points where the groups will face the issue of how to identify risks, how to develop mechanisms of control, and how to align this IC with the business objectives. At this point, the teachers will support students by means of brief instructions given in master-classes and "knowledge pills". The students will thus learn to handle COBIT[1] (*Control Objectives for Information and related Technology*), on their own. In this point, students must face the challenge of designing specific IC to lessen the risks. They have not yet received any training on how to do this, and they will have to resolve the problem by using different means, consulting reports and documents, all of which the group will look for on their own. As the IC structure may vary a great deal, the groups will work autonomously to figure out a general IC structure. This will be reported back to the other groups in later feedback, so that a uniform control pattern may be agreed on. Once the IC development of each group has been finished, all the risks related to the technological infrastructure will have been dealt with.

[1] http://www.isaca.org/Knowledge-Center/COBIT/Pages/Overview.aspx - COBIT is an auditing guide *par excellence*, used by IT auditors in their profession.

5. **Development of the audit of the internal control.** Having played the role of IT professionals who have developed IC in a company in the IT field, the groups now radically change their role, to tackle the second objective mentioned at the beginning of this section. This is to acquire the skills and tools of an IT auditor. At this point, each one of the work groups should audit the IC that another group has carried out at an earlier stage. The work groups must learn the IT auditing process. To do so, the students analyse the documentation of the ISACA[2] and especially of the CISA[3] certification, where each particular team learns what the standard process of IT auditing is that they should adapt to the specific IC they have to audit. In addition, the work groups should now become familiar with the "Auditing Standards and Guidelines" (maintain the evidence, audit reports development, etc.). They also learn to use COBIT (or other guides) to identify "those aspects that should be covered" in the IC being studied. This task ends with the development of an audit report which will comply with ISACA's "Auditing Guidelines and Standards".

3.4 Assessment and Comparison of Results

The assessment of the subject establishes three important landmarks: (i) assessment of the IC, which teachers evaluate using a well-known rubric; (ii) assessment of the Auditing Report by using also a rubric, the audits are evaluated by identifying such aspects as what the audit covers, and the rigorous of the process; and (iii) part of the marks corresponds to a final exam which the students must pass. In this there is an assessment of theory and of practical cases where the students must act as auditors and choose the best option. Alternatively, students can pass some smaller exams in a multiple-choice test format that aim for the students to feel motivated to keep content up-to-date, monitoring how the subject is going.

The work method presented has been applied over the last three academic years, so that with that perspective at least 3 conclusions may be drawn following the comparison of results between the current methodology and previous teaching experience:

- The percentage of students passing in this subject is very similar to that obtained in previous editions, before PBL and the simulation of a "real" audit were used.
- There is greater complexity in the tests. We can thus conclude that the students are acquiring not only broader knowledge, but also greater skills in the IS auditing field.
- The questionnaires, filled in anonymously, show a high level of satisfaction about class dynamics. Delegating part of the learning process to the pupil (as per PBL philosophy), and reducing master classes to a minimum, allows students to be the protagonists, with their process of discovery and research also being at the centre.

[2] https://www.isaca.org/pages/default.aspx.

[3] https://www.isaca.org/CHAPTERS7/MADRID/CERTIFICATION/Pages/Page1.aspx.

4 Lessons Learnt and Improvement Proposals

When we talk about teaching, we always have to consider the most important asset we have. We refer to the human factor, namely our students. They matter more than methodologies, learning tools, study programmes and syllabi.

Thanks to the wide experience that our teaching staff has accrued in the subject of Information System Auditing (and other similar subjects), something has been proven. It is that the most important factor when improving the level of acceptance on the part of the students, has been the involvement of the student him/herself in the learning process by simulating a real setting or one that is similar to a current industrial setting.

This simulation made the student feel that the overall direction and efficacy of the IT in "their organization" depended on a proper functioning of the profession (both as a member of an IT department and as a member of the group of auditors). The degree of commitment on the part of the students, along with the greater amount of knowledge acquired, has led to a positive response in the learning process. This meant that not only was the subject-matter learnt; this learning went on to form an integral part of their professional skills, consolidating what was acquired as regards knowledge and skills.

The evolution proposed to address the subject was due to an increasingly demanding experience, rather than being limited to working within the classic paradigm.

As was remarked at the beginning of this article, software and IT quality is one of the main aspects to bear in mind when there is an intention to achieve excellence in organisations. The IT that serves as a "catalyst" in reaching the business goals have developed a host of quality standards related to quality, services, security, etc., all of which are put into effect in the form of quality management systems; these are control mechanisms that in the end go on to form part of the IC of the companies involved. The scope of the development and auditing of the IC, is completely determined by the guidelines and standards of the ISACA. Nevertheless, an auditor who is carrying out his or her professional role has also to face with the compliance with standards such as ISO 20000, ISO 27000, etc. (as an external auditor of the Certification body AENOR). One of the paths to improvement of this present proposal therefore aims to include the certification mechanisms of standards related to IT, giving the students also an even closer view of how the profession of IT auditor works in our present-day world.

It must also be stated that, from the point of view of the international curricula that were analysed, the topic of the technological auditing of software products and projects is dealt with very superficially in the subject. Yet, as the ISACA curriculum demonstrates, these topics are indeed significant. That curriculum contains the topic blocks of "Acquisition, Development and Implementation of Information Systems, along with "Operation, Sustainability and Support of Information systems; these make up 40% of the total curriculum. As a proposal for additional improvement, therefore, there will be the addition of an auditing of a software project, which will be carried out in the context simulated context. In this way, the students will have to carry out the audit on software project management and the underlying software product that is being developed.

5 Conclusions

Over the passing of time, IT has become a catalyst in organisations allowing them to reach their business goals. New technologies, nonetheless, do not only provide benefits. The job of IT auditors is to be able to analyse IS, so that they can assess the performance of the IC in organisations to ensure quality, security and IT governance. It is in this context that the figure of the IT auditor takes on an extra role: their task is also to be an active agent in assessing the proper implementation and certification of the main quality standards related to IT. With all these issues in mind, this paper presents a methodological proposal (underpinned by more than a decade of experience of the teachers) for the teaching of IT auditing, turning students into active subjects in the learning process, discovering the contents, tools and skills, and then putting them into practice. All of these contents and skills, far from being dealt with in a purely theoretical way, are now tackled in a way that is practical and dynamic, simulating a real setting where the student faces decisions similar to those they might find in a real company.

The experience has demonstrated that this method of approaching the subject in question gives the students a competitive edge, not only improving their learning process, but also preparing them for the real world where change is constant, in the ever-evolving realm of IT.

Acknowledgements. This work is part of the ECD project (PTQ-16-08504) (Torres Quevedo Programme of MINECO and part of the SEQUOIA (TIN2015-70259-C2-1-R) (funded by MINECO & FEDER). The work was also supported by the grant Dr. Ricardo Pérez-Castillo has received from JCCM.

References

1. Piattini, M., del Peso, E., del Peso, M.: Auditoría de Tecnologías y Sistemas de Información. RA-MA (2008)
2. ISO: The ISO Survey of Management System Standard Certifications 2016 (2017). https://isotc.iso.org/livelink/livelink?func=ll&objId=18808772&objAction=browse&viewType=1
3. Krajcik, J.S., Blumenfeld, P.C.: Project-based learning. In: Sawyer, R.K. (ed.) The Cambridge Handbook of the Learning Sciences, Cambridge, pp. 317–334 (2006)
4. Curricula Recommendations. Association for Computing Machinery (ACM) (2017). https://www.acm.org/education/curricula-recommendations
5. Curriculum Guidelines for Undergraduate Degree Programs in Information Systems. A volume of the Computing Curricula Series. Association for Computing Machinery (ACM) (2010)
6. Computer Science Curricula 2013. A volume of the Computing Curricula Series. Association for Computing Machinery (ACM) (2013)
7. Information Technology Curricula 2017. A volume of the Computing Curricula Series. Association for Computing Machinery (ACM) (2017)
8. Curriculum Guidelines for Undergraduate Degree Programs in Software Engineering. A volume of the Computing Curricula Series. Association for Computing Machinery (ACM) (2014)

9. Computer Engineering Curricula 2016. A volume of the Computing Curricula Series. Association for Computing Machinery (ACM) (2016)
10. Bourque, P., Fairley, R.E. (eds.): Guide to the Software Engineering Body of Knowledge, Version 3.0. IEEE Computer Society (2014)
11. ISACA: ISACA Model Curriculum for IS Audit and Control, 3rd edn. (2012)

Compliance to eXtreme Apprenticeship in a Programming Course: Performance, Achievement Emotions, and Self-efficacy

Margherita Brondino[1(✉)], Vincenzo Del Fatto[2], Rosella Gennari[2], Margherita Pasini[1], and Daniela Raccanello[1]

[1] Department of Human Sciences, University of Verona, Verona, Italy
{margherita.brondino,margherita.pasini, daniela.raccanello}@univr.it
[2] Faculty of Computer Science, Free University of Bozen-Bolzano, Bozen-Bolzano, Italy
vincenzo.delfatto@unibz.it, gennari@inf.unibz.it

Abstract. We investigated the efficacy of using the eXtreme Apprenticeship (XA) methodology for teaching programming courses at the university by considering an often-neglected aspect: students' achievement emotions in XA tasks. We involved 53 university students who participated in a XA-based programming course. We assessed students' performance in the course, their achievement emotions and self-efficacy. Key results of the study are presented in the paper. Students with a higher compliance towards the course performed better and were characterised by less intense anxiety, anger, and hopelessness compared to those with a lower compliance. Among achievement emotions, only shame mediated the relation between self-efficacy and performance. Such findings are discussed in terms of their theoretical and applied relevance.

Keywords: eXtreme Apprenticeship · Performance · Achievement emotions Self-efficacy

1 Introduction

The methodology considered in this paper is eXtreme Apprenticeship (XA). It was introduced at the University of Helsinki as a new educational approach in introductory programming courses at the Bachelor level [22]. XA itself is based on the Cognitive Apprenticeship (CA) approach, that is, on learning a task as apprentices, by observing how a master performs it and gaining feedback. XA puts even stronger emphasis on learners and the entire process of learning: learners acquire a new cognitive skill, such as programming, by doing many small exercises, under the guidance of 'masters' available to give students on-demand feedback. More specifically, XA is based on the two following principles: (a) learning by doing, that is learning through many weekly mandatory exercises, chunked and organised in coherent sets of progressive difficulty, each coming with clear achievement goals; and (b) formative assessment, carried out through continuous bidirectional feedback between teachers and students.

T. Di Mascio et al. (Eds.): MIS4TEL 2018, AISC 804, pp. 123–130, 2019.
https://doi.org/10.1007/978-3-319-98872-6_15

Whereas the literature has investigated different aspects associated to students' performances and motivations with XA in programming courses, it has not considered emotional aspects associated to learning of programming with XA. Therefore, the general aim of the study presented in this paper is to investigate the efficacy of using the XA methodology for teaching programming courses at the Bachelor level at university, in terms of both students' performance and so-called achievement emotions, taking into account the compliance with which the students adhered to the methodology itself [12].

1.1 Background

Activities on XA have been conducted in Bolzano in different high schools but mainly at the Free University of Bozen-Bolzano (Unibz) in the Operating Systems (OS) course, dealing with Bash programming at the Bachelor level. With respect to environments used in teaching other programming languages, e.g., Java, the Bash environment is considered more difficult to master for nowadays students. However it is still a skill taught in courses at an intermediate level at university, which is further deepened only by professionals working as system administrators. XA has already shown positive effects on students of OS or similar programming courses. In [3, 4] pre and post surveys were used to investigate university student perceptions of XA, showing their positive perception of the approach. Exercises are relevant features of XA. Therefore students' performances in XA exercises have also been measured [5]. Such measures enabled teachers to organise exercises in progressive levels of difficulty for university students, as recommended by XA. Moreover, a semi-automated tool for the assessment of XA exercises was also started, given that it is a demanding activity for teachers in XA-based teaching [6].

In spite of the relevance of XA and its several applications to teaching OS and similar programming courses, scarce attention has been paid until now to the study of affective and motivational aspects associated to students' performances with XA exercises, particularly from a psychological perspective. Such aspects are all instead relevant and worth investigations, as explained in the following.

The collaborative nature of the learning context of the XA should give rise to those conditions typical for example of cooperative learning methodologies based on constructivism [19, 20] and responsible for improvements in students' performance. However, performance could be better for those students who adhere most strictly to the ways of working that are proposed in the XA context. In other terms, whether the collaborative nature of the XA is assumed to promote learning and associated performance, it is plausible that this effect is more relevant for those students who are collaborative themselves, for example being in time in executing and giving back to the masters the assignments.

Beyond performance, another dimension that could be associated with a high compliance to the XA methodologies could be the affective dimension, in terms for example of achievement emotions as reactions to learning activities or outcomes [12–14]. Achievement emotions can be distinguished on the basis of at least two underlying dimensions, valence and activation [8]. Examples of positive activating emotions are enjoyment, hope, and pride; of positive deactivating emotions are relief and relaxation; of negative activating dimensions are anxiety, anger, and shame; and of negative deactivating dimensions are boredom and hopelessness [11]. Achievement emotions focus on learning activities or outcomes. However, in a context highly collaborative in nature, we could presume that achievement emotions with a particularly marked social component, such as pride and shame, could play a more marked role [17, 18]. Such emotions have been defined as moral emotions, 'linked to the interests or welfare either of society as a whole or at least of persons other than the judge or agent' [9, p. 853], or self-conscious emotions, implying the ability to think about the self and having a consequent immediate punishment or reinforcement to people's behaviours [22]. They are highly relevant for human adaptation to the social context, influencing the relation between individuals' moral standards and decisions on behaviours [22].

One theoretical framework which has recently given a large impetus to research in the field of achievement emotions is Pekrun's control-value theory [12–14]. The theory considers antecedents and outcomes of emotions, and simultaneously recognizes the cyclic nature of the relations. Two important proximal antecedents are perceived control and value. Perceived control refers to "appraisals of control over actions and outcomes" [14, p. 124], while perceived value to the "perceived degree of importance for oneself" [14, p. 125].

Concerning outcomes, achievement emotions would deeply influence learning, also by functional mechanisms such as working memory, information processing, and self-regulation [12–14]. Links with achievement have been documented to be positive for positive activating emotions. They would be frequently variable for positive deactivating emotions and for negative activating emotions. Finally, they would be negative for negative deactivating emotions.

Finally, it is worth noting that the control-value theory assumes that achievement emotions are domain-specific [12–14]. In other words, both achievement emotions and their motivational antecedents are organized in specific ways, that differ according to subject domains such as mathematics, science education, reading, or writing. However, only recently researchers have begun to investigate achievement emotions related to informatics, and specifically to university programming courses.

1.2 Aims and Hypotheses

Focusing on the use of the XA methodology for teaching programming courses at the Bachelor level at university, the main aim was to study students' performance and achievement emotions, considering the compliance with which the students adhered to the methodology itself [12]. We hypothesised that students with a higher compliance to the XA methodology had a better course-related performance (Hypothesis 1). We also expected them to be characterised by more intense positive emotions and less intense negative emotions related to studying for the programming course (Hypothesis 2).

In addition, we explored whether assumptions concerning emotions from the control-value theory [12–14] could be generalized to the specific context of the XA methodologies, focusing on the relations between appraisals of emotions, achievement emotions, and performance. In particular, we investigated the possible mediational role of achievement emotions in the relation between self-efficacy as a control appraisal and performance. We expected that the mediational role could be more salient for those emotions implying a social component, such as pride and shame (Hypothesis 3), in light of the specific characteristics of the XA context. The paper reports the main results of the study and discusses its implications for future work concerning XA and teaching of programming courses.

2 Method

2.1 Participants

The participants were 53 students (42 males and 11 females) who participated in the eXtreme Apprenticeship OS course, in the first year of the Bachelor in Computer Science.

2.2 Measures and Procedure

The students completed a questionnaire including measures on self-efficacy before the course. *Self-efficacy* was measured using three ad hoc items, with which the students were asked to assess their mastery of shell scripting and Linux OS in general, using a Likert scale (1 = *totally disagree* and 5 = *totally agree*).

Achievement emotions in relation to the course exercises were measured as follows. The OS course and its many small exercises were organised into progressive levels per topic and difficulty, in line with the results in [5] and following XA principles. A total of 45 mandatory exercises, focused on Bash programming, were organized into weekly thematic sets and assigned during 6 weeks. An actual achievement goal was established to each exercise in order to promote a student's sense of competence and sense of autonomy over the learning progress. Students had circa 2 weeks to complete each set of Bash exercises. Each exercise was assessed as either passed or not passed by the course teacher. Students could resubmit the same exercise as many times as they wished in the given time frame, and receive, following the XA approach, formative assessment feedback from the course teacher. Teachers randomly selected one exercise per level, for a total of six; after each exercise, students completed a short self-report instrument concerning achievement emotions, developed on the basis of Pekrun's control-value theory [12–14], the Achievement Emotions Adjective List, AEAL [2, 16–18]. The questionnaire includes 30 adjectives related to three positive activating emotions (enjoyment, hope, pride), two positive deactivating emotions (relief, relaxation), three negative activating emotions (anxiety, shame, anger), and two negative deactivating emotions (boredom, hopelessness). The students were asked to indicate how they felt in relation to the exercise, evaluating how much each word described their feelings on a 7-point Likert scale (1 = *not at all* and 7 = *completely*). The order of the words was randomised and kept constant. The mean value of students' reports was used in this study.

Performance was assessed considering the evaluation on the aforementioned six exercises. Mean performance was computed, considering the six exercises. Performance scores were standardised to be in the interval from 0 to 1.

Compliance to the XA learning approach was also ex post evaluated, considering the number of the six exercises delivered in time at the end of the program. Students were divided into two groups on the basis of the number of delivered exercises: We chose to select the High Compliance group with a threshold over or equal 5 delivered exercises to be sure to select students who had delivered more than 80% of exercises. The High Compliance (HC) group, 5–6 delivered exercises, comprised 36 students, and the Low Compliance (LC) group, with from 0 to 4 delivered exercises, comprised 17 students.

2.3 Analysis Procedure

We calculated descriptive statistics, intercorrelations, and *t*-tests using SPSS version 21.0 for Windows. Level of significance was set at $p < .05$. We ran mediation analyses using the PROCESS macro for SPSS [10; www.processmacro.org]. See Table 1 for descriptive statistics (mean values and standard deviations).

Table 1. Mean values, standard deviations, t-tests, and p-values for the considered dependent variables (significant tests are reported in bold)

	LC (N = 17)		HC (N = 36)		Independent sample *t*-test	
	M	*SD*	*M*	*SD*	$t(df = 51)$	*p*-value
Performance	**0.72**	**0.29**	**0.95**	**0.10**	**−4.30**	**.001**
Enjoyment	2.69	0.95	3.07	0.82	−1.48	.146
Hope	3.27	0.84	3.38	0.90	0.31	.755
Pride	3.14	1.12	3.06	0.78	0.59	.558
Relief	2.88	0.82	2.75	0.78	−0.41	.683
Relaxation	3.43	1.06	3.55	0.82	−0.47	.639
Anxiety	**2.15**	**0.82**	**1.64**	**0.80**	**2.15**	**.036**
Shame	1.57	0.74	1.33	0.51	1.40	.168
Anger	**2.40**	**0.86**	**1.93**	**0.73**	**2.05**	**.045**
Boredom	2.34	0.97	2.17	0.82	0.68	.499
Hopelessness	**2.01**	**0.82**	**1.47**	**0.56**	**2.83**	**.007**

3 Results and Discussion

3.1 The Effect of XA Compliance on Performance and Achievement Emotions

A first set of analyses concerns the amount of compliance to the XA methodology, measured considering the number of exercises delivered by the students. The size of the Low Compliance group was smaller than the size of the High Compliance group (17 vs. 36) maybe suggesting a quite good level of effect of this methodology on learners' engagement. We conducted a series of independent sample *t*-tests, with group (Low Compliance, High Compliance) as the independent variable, and performance and the ten achievement emotions as dependent variables. See Table 1 for t-tests and p-values for the considered dependent variables.

Performance was higher for the High Compliance group than for the Low Compliance group. Concerning the different level of achievement emotions associated with the test experience, no significant differences were found for the five positive emotions. Considering the negative achievement emotions, three out of five were different in the two groups: anxiety, anger, and hopelessness. For all these negative achievement emotions, the High Compliance group showed a significant lower level than the Low Compliance group.

3.2 The Mediation Role of Achievement Emotions

Regression analysis was used to investigate the hypothesis that achievement emotions mediates the effect of self-efficacy on performance. Only one negative emotion was found to be a mediator in the relationship between self-efficacy and performance. Results indicated that self-efficacy was a significant and negative predictor of shame, $b = -.38$, $t(45) = -3.05$, $p = .004$, and that shame was a significant and negative predictor of performance, $b = -.14$, $t(44) = -2.65$, $p = .011$. The higher the level of self-efficacy before starting the course, the lower the level of shame in performing the task. At the same time, the lower the shame level, the higher the performance. These results support the mediational hypothesis. Self-efficacy was no longer a significant predictor of performance after controlling for the mediator, shame, $b = .06$, $t(45) = 1.15$, $p = .255$ (not significant) consistent with full mediation. Approximately 23% of the variance in satisfaction was accounted for by the predictors. High level of self-efficacy was associated with approximately .23 points higher performance as mediated by shame.

4 Conclusions

This study focused on affective, motivational, and performance dimensions associated with the XA methodology applied within an introductory OS programming course. Only recently the advantages of this methodology for students' learning have been documented, also beginning to focus on psychological constructs [e.g., 22].

The main aim regarded the efficacy of the XA methodology focusing on students' compliance about it. Our findings confirmed Hypothesis 1: performance was better for those students for whom compliance was higher. At a theoretical level, this result enables to extend the knowledge on the factors responsible of the efficacy of the XA methodology. In addition, there can be several implications at an applied level. For example, the gathered evidence can support all those actions that can help students to be more compliant during the courses, from teachers' interventions for raising students' awareness of the relevance of being compliant, to the adaptation of the exercises in relation to students' performances and reported achievement emotions, e.g., [1].

In addition, our data documented that, on the whole, more compliant students were characterized by better indicators relating to their achievement emotions [12–14]. Partially confirming Hypothesis 2, our findings indicated that levels of negative activating emotions such as anxiety and anger, and levels of negative deactivating emotions such as hopelessness were lower for the high compliance group compared to the low compliance group. If we consider emotions as one of the component of wellbeing [7], such a result further suggests the efficacy of the XA methodology.

Secondarily, we focused on the generalizability of the relations between control appraisals, achievement emotions, and performance emotions [12–14] as related to the XA methodology applied in the programming course. The mediational role of achievement emotions between self-efficacy, as a control appraisal, and performance was documented only for shame (Hypothesis 3). It is worth noting that shame is one of the emotions that could be presumed to play a salient role in a context in which the collaborative nature of learning is particularly stressed, for its social connotation. Future research could examine how the compliance to the XA methodology could moderate the emerged relation. However, it is worth noting that these findings are only a preliminary step in exploring this issue, given limitations of this study such as, for example, the reduced sample size.

On the whole, our findings enabled to support the efficacy of the XA methodology in a programming course examining psychological components, extending current knowledge and giving some hints at the applied level.

References

1. Alrifai, M., Gennari, R., Vittorini, P.: Adapting with evidence: the adaptive model and the stimulation plan of TERENCE. Adv. Intell. Soft Comput. **152**, 75–82 (2012). https://doi.org/10.1007/978-3-642-28801-2_9
2. Brondino, M., Raccanello, D., Pasini, M.: Achievement goals as antecedents of achievement emotions: the 3 X 2 achievement goal model as a framework for learning environments design. Adv. Intell. Soft Comput. **292**, 53–60 (2014). https://doi.org/10.1007/978-3-319-07698-0
3. Del Fatto, V., Dodero, G., Gennari, R.: Assessing student perception of extreme apprenticeship for operating systems. In: Proceedings of ICALT 2014, pp. 459–460 (2014a). https://doi.org/10.1109/icalt.2014.137
4. Del Fatto, V., Dodero, G., Gennari, R.: Operating systems with blended extreme apprenticeship: what are students' perceptions? Interact. Des. Archit. **23**(1), 24–37 (2014b)
5. Del Fatto, V., Dodero, G., Gennari, R.: How measuring student performances allows for measuring blended extreme apprenticeship for learning Bash programming. Comput. Hum. Behav. **55**, 1231–1240 (2016). https://doi.org/10.1109/ICALT.2014.137
6. Del Fatto, V., Dodero, G., Gennari, R., Gruber, B., Helmer, S., Raimato, G.: Automating assessment of exercises as means to decrease MOOC teachers' efforts. Smart Innov. Syst. Technol. **80**, 201–208 (2018). https://doi.org/10.1109/ICALT.2014.137
7. Diener, E.: Subjective wellbeing: the science of happiness and a proposal for a national index. Am. Psychol. **55**(1), 56–67 (2000). https://doi.org/10.1037/0003-066X.55.1.34
8. Feldman Barrett, L., Russell, J.A.: Independence and bipolarity in the structure of current affect. J. Pers. Soc. Psychol. **74**(4), 967–984 (1998). https://doi.org/10.1037/0022-3514.74.4.967
9. Haidt, J.: The moral emotions. In: Davidson, R.J., Scherer, K.R., Goldsmith, H.H. (eds.) Handbook of Affective Sciences, pp. 852–870. Oxford University Press, Oxford (2003)
10. Hayes, A.F.: Introduction to mediation, moderation, and conditional process analysis. a regression-based approach, 2nd edn. The Guilford Press, New York (2018)

11. Kleine, M., Goetz, T., Pekrun, R., Hall, N.: The structure of students' emotions experienced during a mathematical achievement test. ZDM Int. J. Math. Educ. **37**, 221–225 (2005)
12. Pekrun, R.: The control-value theory of achievement emotions: assumptions, corollaries, and implications for educational research and practice. Educ. Psychol. Rev. **18**(4), 315–341 (2006). https://doi.org/10.1007/s10648-006-9029-9
13. Pekrun, R., Muis, K., Frenzel, A.C., Goetz, T.: Emotions at school. Routledge, New York (2018)
14. Pekrun, R., Perry, R.P.: Control-value theory of achievement emotions. In: Pekrun, R., Linnenbrick-Garcia, L. (eds.) International handbook of emotions in education, pp. 120–141. Taylor and Francis, New York (2014)
15. Raccanello, D.: Students' expectations about interviewees' and interviewers' achievement emotions in job selection interviews. J. Employ. Couns. **52**(2), 50–64 (2015). https://doi.org/10.1002/joec.12004
16. Raccanello, D., Brondino, M., Pasini, M.: Achievement emotions in technology enhanced learning: development and validation of self-report instruments in the Italian context. Interact. Des. Archit. J. IxD&A **23**, 68–81 (2014)
17. Raccanello, D., Brondino, M., Pasini, M.: On-line assessment of pride and shame: relationships with cognitive dimensions in university students. Adv. Intell. Soft Comput. **374**, 17–24 (2015). https://doi.org/10.1007/978-3-319-19632-9_3
18. Raccanello, D., Brondino, M., Pasini, M.: Two neglected moral emotions in university settings: some preliminary data on pride and shame. J. Beliefs Values: Stud. Relig. Educ. **36**(2), 231–238 (2015). https://doi.org/10.1080/13617672.2015.1031535
19. Sharan, S., Sharan, Y.: Small group teaching. Englewood Cliffs, NJ (1992)
20. Slavin, R.E.: Student team learning: A practical guide to cooperative learning. National Education Association of the United States, DC (1991)
21. Tangney, J.P., Stuewig, J., Mashek, D.J.: Moral emotions and moral behavior. Annu. Rev. Psychol. **58**, 345–372 (2007). https://doi.org/10.1146/annurev.psych.56.091103.070145
22. Vihavainen, A., Paksula, M., Luukkainen, M.: Extreme apprenticeship method in teaching programming for beginners. In: Proceedings of SIGCSE 2011, pp. 93–98. ACM (2011)

Learning Analytics in the Classroom: Comparing Self-assessment, Teacher Assessment and Tests

Michael D. Kickmeier-Rust[1,2](✉) and Lenka Firtova[3]

[1] Graz University of Technology, Graz, Austria
michael.kickmeier@phsg.ch
[2] University of Teacher Education, St. Gallen, Switzerland
[3] SCIO s.r.o, Prague, Czech Republic
lfirtova@scio.cz

Abstract. Learning Analytics is an important trend in education. In conventional classroom settings, however, a sound basis of digital data for analytics is lacking. Therefore, it is important to develop the methodologies and technologies to utilize the scattered and heterogeneous bits of available data as effective as possible. It is also important to deploy simple and usable tools to teachers that could help them within the context conditions and constraints of their daily work. In this paper, we introduce a prototypical approach for learning Analytics in the classroom and a simple data collection tool named Flower Tool. This tool enables collecting and comparing students' self-assessments with teacher-lead assessments and the results of external tests. In a first field study, we gathered feedback from students and teachers about the approach, indicating a strong acceptance and a number of potential advantages for the assessment and reflection processes in the classroom.

Keywords: Learning Analytics · Self-assessment · Formative assessment

1 Learning Analytics in the Classroom

The history of Learning Analytics is remarkably short; the influential work of George Siemens, that almost all work on Learning Analytics is citing, was published in 2011 [1]. In the same year, the first Learning Analytics and Knowledge (LAK) conference was organized. A year or two before, SOLAR was founded and Ryan Baker published articles about data mining techniques in education [2, 3]. Ever since, a very strong and lively community pushed the research activities and the development of solutions and products. Today, Learning Analytics has become an integral element of most learning management systems (LMS) and educators across all education levels are using features of Learning Analytics on a daily basis to improve their teaching and to give students feedback about learning processes. A significant part of the success of Learning Analytics is due to European initiatives and projects. The LACE project (www.laceproject.eu), for example, carried out groundbreaking work in reducing fragmentation in the field and bringing research and application communities together.

T. Di Mascio et al. (Eds.): MIS4TEL 2018, AISC 804, pp. 131–138, 2019.
https://doi.org/10.1007/978-3-319-98872-6_16

Thus, using Learning Analytics and educational data mining are more than recent buzz words in educational research: they signify one of the most promising developments in improving teaching and learning. While many attempts to enhance learning with mere technology failed in the past, making sense of a large amount of data collected over a long period of time and conveying it to teachers in a suitable form is indeed the area where computers and technology can add value for future classrooms. However, reasoning about data, and in particular learning-related data, is not trivial and requires a robust foundation of well-elaborated psycho-pedagogical theories.

The fundamental idea of Learning Analytics is not new. In essence, the aim is using as much information about learners as possible to understand the meaning of the data in terms of the learners' strengths, abilities, knowledge, weakness, learning progress, attitudes, and social networks with the final goal of providing the best and most appropriate personalized support. Thus, the concept of Learning Analytics is quite similar to the idea of formative assessment. "Good" teachers of all time have strived to achieve exactly this goal. However, collecting, storing, interpreting, and aggregating information about learners requires a solid set of available data – ideally complete and rather homogenous data. While such basis is available with digital learning solutions such as online courses and MOOCs, the reality in typical K18 classrooms is different. The basis of digitally available data is rather weak; the few bits of data available are rather heterogeneous and incomplete and come from a variety of different sources (ranging from learning apps on mobile phones to homework in Google Docs).

Lea's Box (www.leas-box.eu) is a project that focuses exactly on this problem. In the center are methodologies and technological solutions to aggregate various heterogeneous data from all kinds of sources over a long period of time and collected by several different persons (teachers, students, peers).

The project utilizes so-called *Competence-based Knowledge Space Theory* [4, 5] and Open Learner Models [6] as anchor point to bring the bits of data together. In essence, data is collected from all possible sources via the xAPI interface (https://experienceapi.com/overview/) and serve as evidence for a central competence model. This model distinguishes the observable performance and evidence from the underlying aptitude of a learner, which is not directly observable [7]. Competence model and individual evidences feed a learner model, which is structured and visualized in an open, transparent and intuitive way. The project manifest in form of a Web platform for teachers and learners provide links to the existing components and interfaces to a broad range of educational data sources. Teachers will be able to link the various tools and methods that they are already using in their daily practice and that provide software APIs (e.g., Moodle courses, electronic tests, Google Docs, etc.) in one central location. More importantly, the platform hosts the newly developed LA/EDM services, empowering educators to conduct competence-based analysis of rich data sets.

A key focus of the platform is on enabling teachers not only to combine existing bits of data but also to allow them to "generate" and collect data in very simple forms. In the past, tools have been developed to support teachers making electronic records in a simple, quick and device independent way. Studies have shown that such simple features are much appreciated by teachers and related outputs allowed teachers to gain deeper insights into classroom process and individual learning [8].

2 Objective vs Subjective Assessment: Lea's Flower Tool

Recent research in the context of assessment emphasize the learning potential of assessment [9] and the importance of formative assessment. Self and peer assessments are being increasingly used in higher education to help students learn more efficiently. However, there are few papers discussing these methods [10]. Self-assessments enables the individual to reflect own learning achievements [11] while in peer assessment apply standards to the work of their peers in order to judge that work [12]. When using a correct assessment system, students perceive assessment as a motivating and productive part of their education because this procedure informs them if they are good at learning and are able achieve proposed goals [13]. Both self and peer assessment can complement the assessment of teachers and put it into a more general light. All types of assessments (self, peer, teacher) are subjective assessments. Specifically, the assessment of teachers might be considered subjective, incorrect or even unfair by individual students. Comparing and juxtaposing different sources of assessment can provide new views and ignite deeper reflection processes about achievements and the assessment process itself – by learners and teachers. When the subjective assessment can be complemented with the results of an (more or less) objective external test, this effect can even be increased. In the context of the Lea's Box project, we developed a tool that enables the comparison of self-assessment, teacher-lead assessment and an external test.

The fundamental idea of the tool is to provide students with a nice graphical interface for conducting self-assessments in a specific topic (Fig. 1a–c). A student can select from several domains (e.g., mathematics, language, etc.). The interface for each domain is designed in form of a flower (Fig. 1b). The leafs of the flower indicate subtopics; when clicking on a leaf, the student can access a questionnaire for this specific topic and can assess herself (Fig. 1c). Depending on the value of self-assessment, the leafs fill with color (a very positive assessment leads to a fully filled and colored leaf). From the same interface, the students can take an external test about the same competencies and aspects. Teachers have a similar interface and can rate their students along the same criteria. Finally, students and teachers can compare the different assessment and enter a negotiation process. To investigate the acceptance of the tool as well as interaction dynamics, we conducted a field study with schools in the Czech Republic.

3 Field Study

3.1 Study Setup

In total, we deployed the tool to 18 teachers; each teacher used the tool in at least one class, so overall 598 students participated in the study and used the flower tool. As domains we used math, Czech, English and general academic achievements and prepared the self-assessment questionnaires via the flower's leafs. These included not only ratings about the academic achievements but also aspects such as motivation, easiness, effort, and satisfaction. After students have completed the self-assessment

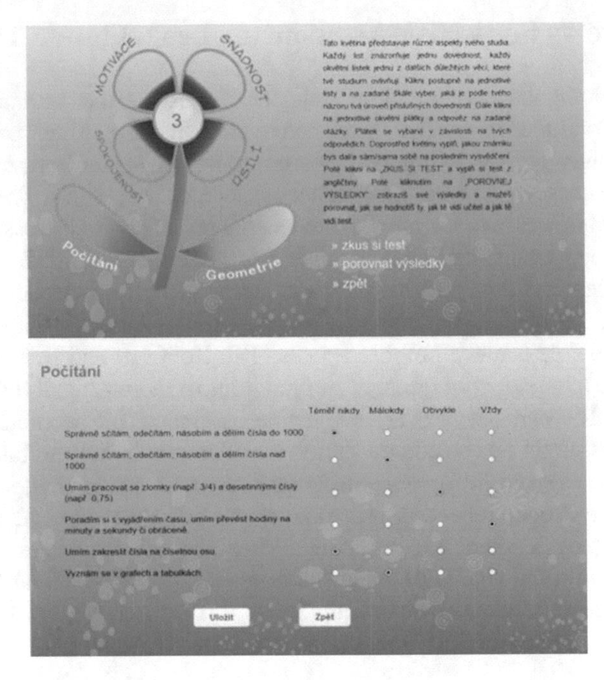

Fig. 1. Screen captures of the Lea's Box Flower Tool. The top panel (a) shows the domain selection screen, the middle panel (b) shows the flower visualization for a chosen domain, and the bottom panel (c) shows a student self-assessment questionnaire.

questionnaires, they were redirected to the corresponding external test. These tests are based on the Czech standard assessment of academic achievements, which are developed and deployed by the Czech company SCIO (https://www.scio.cz/english/). Teachers evaluated each student separately. Finally, students and teachers were asked to compare self-assessment, teacher assessment and external assessment (Fig. 2).

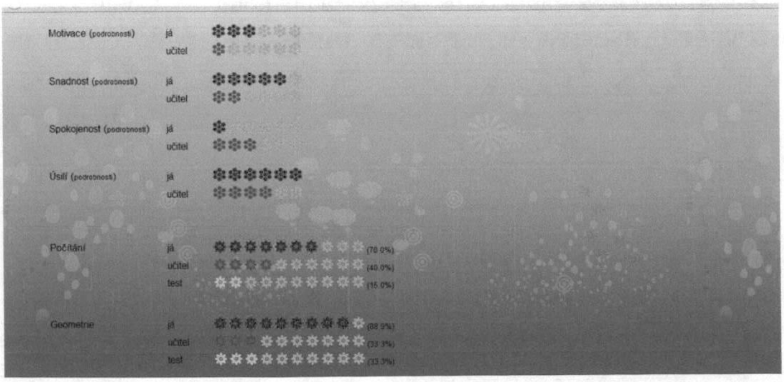

Fig. 2. Comparison of self-assessment, teacher assessment, and external test

3.2 Results

Aim of the field study was to explore the possibilities, to identify the acceptance, and to receive feedback for future developments. As initial step, we asked teachers and students about the usefulness of the tool as such and the comparisons of different assessments as a methodology for the classroom. As shown in Fig. 3, most students highlighted the usefulness of the approach by rating it mainly with 3 "useful" or 4 "very useful". Equally positive were the results for the trust in the approach and the results, as shown in Fig. 4.

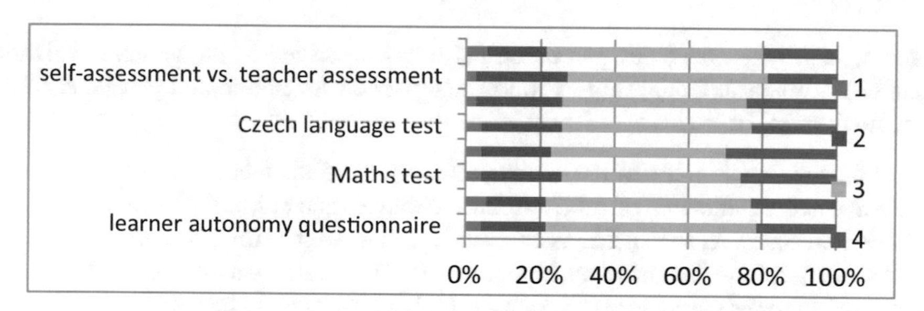

Fig. 3. Usefulness of the tools and the approach (1 .... not useful at all, 4 ... very useful).

Additional results of our survey in schools were as follows:

- 37% of students would quite recommend the system to students from a different classroom and 44% would highly recommend it, while 33% of teachers would probably not recommend the system to their colleagues and 67% would quite recommend it;
- most students would quite like (39%) or very much like (20%) to use the system on a regular basis; 83% of the teachers said they would quite like to use the system on a regular basis;
- 72% of the students would recommend their teacher to buy the Flower Tool;
- 80% of the students would prefer the Flower Tool to just taking the tests it contains (however, interestingly, only two thirds of the students think their teacher would choose this option as well);
- in comparison, half of the teachers would prefer the Flower Tool to just making students take the tests (however, 67% of the teachers believe students themselves would prefer the Flower Tool);
- 30% of the students would quite like to use the Flower Tool in other subjects as well and 34% of the students would very much like that; in comparison, 67% of teachers said they would quite like to use the Flower Tool in other subjects but the rest said they would not like to do that.

A second part of the field study was realized in form of focus groups and phone interviews with the representatives of the schools and institutions where the field studies had taken place. We used structured interviews to get more context-specific and in-depth feedback related to the suitability of the system in that particular educational

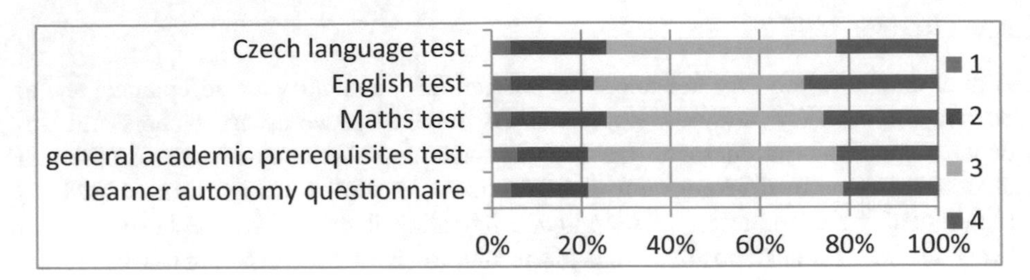

Fig. 4. Trust in the results (1 …. very low, 4 … very high).

context, with a special focus placed on the areas described in the general evaluation matrix. The main findings were as follows (separated for public and private schools):

Public schools:

- As for the functionality of the system, teachers liked the fact it included self assessment and the evaluation of subject-related behavior and attitudes (motivation, learner autonomy etc.). As for teacher assessment, teachers were at times uncertain about how to assess pupils in domains such as motivation. More training and experience would be needed to make them more experienced in this type of assessment.
- Teachers said the average time pupils spent using the system before their results were displayed was adequate and acceptable. However, they said the stability and speed could still be improved. The system was, at times, quite slow, which was probably caused by slow wireless connection combined with high demands on the internet connection placed by the system.
- The results presented by the system seemed to be valid and teachers trusted them, with the exception of occasional errors in the database, in which case the system displayed erroneous results.
- When asked whether they would prefer the Lea's Box system to what they normally use, all teachers unanimously opted for the Lea's Box system. As the school where the evaluation took place is one of our established partners, we are preparing several follow-up activities we could offer to this school as a part of the exploitation of the system.

Private schools:

- As for the functionality of the system, teachers especially appreciated the possibility to combine different source of information, such as self-assessment and teacher assessment.
- The stability and speed of the system was good, the system did not crash, the only problem teachers mentioned was the fact that some pupils had to log in repeatedly. Teachers said the system could be easily used on a regular basis without any major obstacles.
- Teachers doubted the results presented by the system a bit, because the tasks and evaluation materials presented by the system were rather lengthy and demanding and they believed some pupils may have lost focus.
- They said they could imagine using the system on a regular basis.

4 Conclusions

The scientific depth of the field study described in this paper is obviously limited. The reason for this lies in the fact that the Flower Tool is applied in real school settings and in the first instance the technical robustness had to be proved (these procedures are out of the scope of this paper) as well as the general acceptance and conceptual applicability. In this study in the Czech Republic, we found that learners and teachers do appreciate the approach of having a direct visual comparison of different assessment sources. Specifically juxtaposing the teacher's assessment with the external test in certain cases lead to interesting dynamics. Students (and teachers) frequently reported that in case of a better test results as opposed to the teacher's assessment, students could use this comparison as a starting point for discussions with the teacher and a solid background to negotiate assessments and even grading. A common statement from students was "*look, I was way better than you thought I am*", referencing a better test results than the teacher's assessment. This is a great benefit since it strengthens the students' position in the classroom and their self-confidence. This is specifically important for the weaker students. In terms of Open Learner Modelling, the approach adds another level of transparency and trust and elements for "persuasion". This concept refers to the possibility to negotiate assessment and grading results based on integrating further evidences (cf. [14] for an overview).

Overall, such simple tools like the Flower Tool, not only have positive main effects, they illustrate the importance of making Learning Analytics are more integral part of school education. Even such simple tools can contribute to collect data on a more frequent basis and compare different resource. More importantly, in the light of the general system, as described initially, all sources (self-assessment, teacher-assessment, external tests) can service as evidence for a joint and more valid and credible appraisal of the learner. Lately, this is the main idea of the Lea's Box project. In future steps, we will explore the effects of transparent assessments and comparisons over a longer period.

Acknowledgements. This work was conducted in the context of the LEA's BOX project, contracted under number 619762, of the 7th Framework Programme of the European Commission. This document does not represent the opinion of the EC and the EC is not responsible for any use that might be made of its content.

References

1. Siemens, G., Long, P.: Penetrating the fog: analytics in learning and education. EDUCAUSE Rev. **46**(4), 30 (2011)
2. Baker, R.S.J.D.: Data mining for education. In: McGaw, B., Peterson, P., Baker, E. (eds.) International Encyclopedia of Education, vol. 7, 3rd edn., pp. 112–118. Elsevier, Oxford (2010)
3. Baker, R.S.J.D., Yacef, K.: The state of educational data mining in 2009: a review and future visions. J. Educ. Data Min. **1**(1), 3–17 (2009)
4. Doignon, J., Falmagne, J.: Knowledge Spaces. Springer, Berlin (1999)

5. Albert, D., Lukas, J.: Knowledge Spaces: Theories, Empirical Research, and Applications. Lawrence Erlbaum Associates, Mahwah (1999)
6. Bull, S., Ginon, B., Boscolo, C.: Introducing learning visualisations and metacognitive support in a persuadable open learner model. In: Proceedings of LAK 2016, 25–29 April 2016, Edinburgh, UK (2016)
7. Reimann, P., Kickmeier-Rust, M.D., Albert, D.: Problem solving learning environments and assessment: a knowledge space theory approach. Comput. Educ. **64**, 183–193 (2013)
8. Kickmeier-Rust, M.D., Albert, D.: Competence-based knowledge space theory: options for the 21st century classrooms. In: Reimann, P., Bull, S., Kickmeier-Rust, M.D., Vatrapu, R., Wasson, B. (eds.) Measuring and Visualizing Learning in the Information-Rich Classroom, pp. 109–120. Routledge, New York (2015)
9. Taras, M.: Summative and formative assessment: perceptions and realities. Act. Learn. High Educ. **9**(2), 172–192 (2008)
10. Amo, E., Jareño, F.: Self, peer and teacher assessment as active learning methods. Res. J. Int. Stud. **18**, 41–47 (2011)
11. Kayler, M., Weller, K.: Pedagogy, self-assessment, and online discussion groups. Educ. Technol. Soc. **10**(1), 136–147 (2007)
12. Falchikov, N.: Improving Assessment Through Student Involvement: Practical Solutions for Aiding Learning in Higher and Further Education. Routledge, New York (2005)
13. Munns, G., Woodward, H.: Student engagement and student self-assessment: the REAL framework. Assess. Educ.: Princ. Policy Pract. **13**(2), 193–213 (2006)
14. Ginon, B., Boscolo, C., Johnson, M.D., Bull, S.: Persuading an open learner model in the context of a university course: an exploratory study. In: Proceedings of ITS, 7–10 June 2016, Zagreb, Croatia (2016)

Towards a Model of Early Entrepreneurial Education: Appreciation, Facilitation and Evaluation

Elisabeth Unterfrauner[1](✉), Christian Voigt[1], and Sandra Schön[2]

[1] Centre for Social Innovation, Linke Wienzeile 246, 1150 Vienna, Austria
{unterfrauner, voigt}@zsi.at
[2] Salzburg Research Forschungsgesellschaft m.b.H., Jakob Haringer Strasse 5/III, 5020 Salzburg, Austria
Sandra.schoen@salzburgresearch.at

Abstract. This paper introduces the Maker movement as a bottom-up movement, placing digital fabrication technologies on people's desks to produce "almost anything". It explores further the pedagogical value of making in education in general and in early entrepreneurial education in particular. Making as a pedagogical approach is analysed referencing established pedagogical concepts as well as a qualitative study including makers and managers of maker spaces. Although maker education has so far only rarely been introduced in formal education, there are many initiatives that bring making and formal education together. According to maker experts, formal education would benefit from making because it is well suited to develop practical skills such as prototyping, supporting creativity and promoting critical reflection. In conclusion we describe a model of introducing making in early entrepreneurial education and conclude with a proposed assessment framework for measuring its impact, which will be tested in an on-going project funded by the European Commission.

Keywords: Making · Early entrepreneurial education · Maker pedagogy

1 Introduction

Thanks to the availability of digital technologies such as 3D printers, laser cutters and CNC (Computer Numerical Control) machines, digital fabrication and prototyping have become widely accessible for anyone and are no longer limited exclusively to industries. The number of maker spaces and Fab Labs (fabrication laboratories) that make their facilities and digital fabrication tools available to their members are constantly growing in the recent years. Currently there are around 1,200 Fab Labs globally[1]. In these workshops makers design and fabricate their own prototypes, some meant for personal use, others for commercialisation. The inter-changeability of bits and atoms, from design to physical artefacts, is being called the Maker movement [1]. It could be said that the Maker movement represents a return of interest to the physical side of innovation following the almost complete shift to the digital side with the

[1] http://www.fabfoundation.org/index.php/fab-labs/index.html.

T. Di Mascio et al. (Eds.): MIS4TEL 2018, AISC 804, pp. 139–146, 2019.
https://doi.org/10.1007/978-3-319-98872-6_17

dot-com bubble, the rise of the participatory Web 2.0 and the diffusion of Open Source Software. Neil Gershenfeld [2] called the Maker movement the next digital revolution as it placed the means of fabrication on people's desks.

It started as a community-based, socially driven bottom-up movement, but today its potential to impact on society is manifold, in terms of environmental, economic and social impact.

Most maker spaces and Fab Labs offer educational activities for children and adults, from kindergartens up to university students [3], recognising the pedagogical value of making. However, there are few examples where making has been introduced to school settings [4], for instance as a school subject, and where the impact and value of making have been scientifically analysed. This paper seeks to close this gap in regard to early entrepreneurial education (EEE) proposing an assessment framework to diagnose the impact of maker activities on attitudes, knowledge and skills that favour an entrepreneurial "spirit". We depart with exploring the pedagogical value of making in general in reference to established pedagogical concepts as well as a qualitative study with makers and maker managers [5]. We will further describe the DOIT (Entrepreneurial skills for young social innovators in an open digital world) approach, a model of entrepreneurial education that is put into practice in the framework of a EU (European Union) funded project[2]. Finally we will deliver an assessment framework for analysing the impact of maker activities in EEE.

2 The Pedagogy of Making

The pedagogy of making builds on several pedagogical pioneers, from reform pedagogues to constructivists, from Montessori [6] to Piaget and Papert [7], who all support self-regulated learning [8], where learners decide on their learning goals and on the when and the how. In this learning setting teachers acquire the role of tutors assisting the learner in their learning paths replacing the traditional teacher-learner relation.

Making is hands-on learning, where makers learn from others, from trial and error, often in interdisciplinary and collaborative teams [9–11]. In this, making is similar to problem solving and project based learning approaches. Making includes a desire to produce things more collaboratively by improving design suggestions of others or by simply copying, mashing or personalising existing design elements. Making is thus theoretically and historically founded on "learning by doing" principles [12, 13]. As a pedagogical approach, it is learner-centred and project oriented, while allowing learners to follow their individual goals [14]. According to the Horizon report, which anticipates technological trends having an impact on educational settings, maker education will have an increasing impact on education in the following years [15].

[2] https://www.doit-europe.net/.

2.1 What Makers Say - Empirical Results from Interviews with Makers

In order to understand the value and impact of making, we have conducted 40 interviews with experts in making, i.e. makers and managers of maker initiatives across Europe and asked them, among other questions, how they perceive the educational potential of making [3].

In qualitative analysis of the interviews, it became clear that makers themselves believe that the Maker movement already has an impact on education, as there are numerous examples of collaborations between maker spaces and educational institutions. They either invited school classes to the maker space or installed pop-up maker spaces at schools or rented out some machines to trained teachers, although no integration in the school curriculum is known to date. In respect to the educational potential of making, the interviewees named entrepreneurial education, STEAM (Science technology engineering arts and mathematics) education, and as pedagogical approach collaborative and interdisciplinary learning in particular. For instance, a maker at Fab Lab Barcelona said: *"...this could be a way to introduce them to a new way of production as well as also teaching the young people who are interested in technology...(...) get things moving on a different level when you bring two skilled individuals together by combining these skills"*(maker, Spain). Children are taught to be creative themselves, "*which might lead to further growth of the DIY (Do-it-Yourself) community, which then in turn could have a considerable impact on production processes*" (maker, Denmark). Also, local job creation was named as an argument for developing 21st century skills through making and keeping a well trained work force in the region if making found its ways into formal education.

The makers see maker pedagogy as preparing children better for real life situations: "*I think that they are having this traditional education that is not preparing them for the real world, to be competitive (...). When one day they have to start working and they are being educated like you just sit and listen and here are 10 pages and then (...) they do not know what to do*" (maker, Croatia). Kids' interest in 3D printing and other digital manufacture technologies can be easily triggered and many makers and maker initiative managers claim that the incorporation in formal education would be a necessary step to prepare children for the skills that are needed today to compete on the labour market. Maker initiatives can also provide room for education for disadvantaged kids and young adults by empowering them and thus maker initiatives would have the potential to break barriers and give access to people from different social backgrounds: "*Part of the task, which we set ourselves is of course to try to break barriers, especially for pupils who would never get the idea to study because they grow up in a social environment where they have no contact at all to universities. (...) social origin determines the educational career a lot here (Düsseldorf, Germany). And one of our tasks, which we set out to do, is to provide a bit of support there. (...) When I say we were successful here, even though we have no proper measuring tool for it*" (maker manager, Germany).

After all, making contributes to a paradigm shift in its anti-consumerist perspective that leads from pure consumption to producing and further to prosuming, which is the merge of producing and consuming: *"I don't want to have a kid who thinks 'OK, I want this and where can I buy it', instead of 'I want this and how can I make it'"* (maker,

Croatia). Maker education is about opening black boxes, giving first hand experiences how artefacts are produced and know-how regarding the production cycle: *"One of the things could be that people stop being consumers, but instead become more creators. (...) Today we are in a society of consumerism, so people buy things (...). I really believe that will change the way society works"* (maker, Denmark). Maker objects are believed to create a different awareness of products in general and many interviewed makers showed an anti-consumerism attitude. Makers want to know how products are made, what the product consists of, and open these "black boxes". Some argue that they would like to be in control of the production phase as buying off the shelf means missing out on the different production steps and losing that knowledge.

2.2 Maker Education in Social Media

Previous studies analysed the 'making' related discussions in social media [16]. After analysing a total of 50,097 tweets with #makerspaces (12,180 occurrences), #makerEd was one of the more prominent hashtags for indicating the discussion of making in education (4,370 occurrences). However, more informative than the absolute numbers are the co-occurrences of keywords characterising these two data sets. Hence, #makerspaces are most frequently mentioned in conjunction with libraries, schools and STEAM showing how much 'making' is already connected with traditional places for learning, at least as far as the informal debate on Twitter is concerned.

Based on a qualitative screening of these tweets, these words indicate the contextual constraints of introducing making into formal education, where 'classes' are the standard unit for teacher – learner interactions and where 'rubrics' are critical instruments to assess learning (see Fig. 1). 'Free' refers to the availability of freely available materials supporting 'maker education' that are promoted over the Twitter network. This is not necessarily an indication that only free materials are thought after.

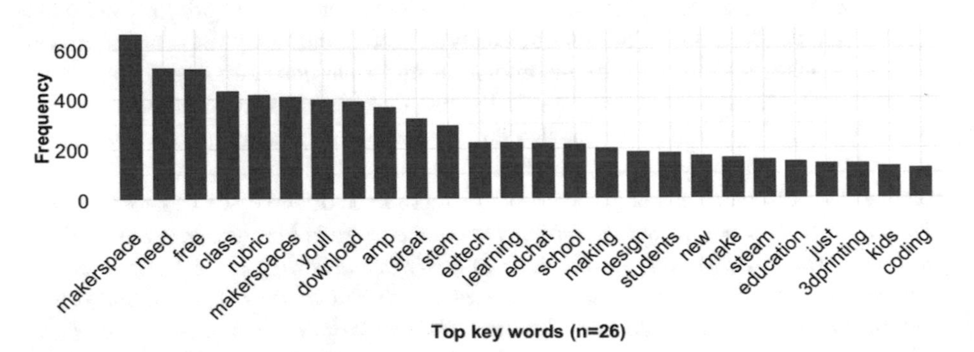

Fig. 1. Co-word analysis of 4,370 tweets containing #MakerEd (Nov 2015).

3 Making in Early Entrepreneurial Education

At first glance, making supports many skills relevant to entrepreneurship: setting goals and devising paths to achieve them, the ability to integrate the skills of others (collaboration) and it is highly interdisciplinary, since making proceeds iteratively and supports primarily problem-based learning. Nevertheless, despite the promising nature of introducing making in education, there are only rare examples where making has been integrated with formal education, apart from short term arrangements such as project or maker days [e.g. 17] at schools or teachers who incorporated making in their classes without the curricular support other forms of learning have.

3.1 Objectives of Early Entrepreneurial Education

As stated in the 2016 Eurydice report, defining goals and learning methods for entrepreneurship education is still an underdeveloped area in most European countries [18]. First and foremost, we do not propose a narrow, commercial definition of entrepreneurship education, but work on the basis of the more comprehensive definition provided by the EC thematic working group: "Entrepreneurship education is about learners developing the skills and mind-set to be able to turn creative ideas into entrepreneurial action. This is a key competence for all learners, supporting personal development, active citizenship, social inclusion and employability" [18].

Putting skills and attitudes at the core of the definition, means that some broader objectives, e.g. creativity, planning or teamwork, are already addressed in other programmes, run by schools. A first overview is provided by Lackéus [19]: *Entrepreneurial attitudes*: self-confidence, self-efficacy, sense of initiative, ambiguity tolerance, perseverance; *Entrepreneurial skills*: creativity, planning, financial literacy, managing resources, managing uncertainty/risk, teamwork; *Entrepreneurial knowledge*: assessment of opportunities, identifying with the role of an entrepreneurs – self-reflection, how-to knowledge (accounting, finance, marketing and communications)

3.2 Facilitation and Evaluation of Early Entrepreneurial Education

The EEE approach as developed in the DOIT project builds on the advantages of maker pedagogy since its effect on most of the above mentioned attitudes, skills and knowledge is promising. However, it is not meant to replace existing EEE such as entrepreneurial games or companies [18] but adding to these more traditional settings. Specifically, the focus of DOIT is on social entrepreneurship, assisting children between the age of 6 and 16, in their path from the ideation phase with the identification of a "problem" from their life worlds to calibrating options for businesses based on their inventions. The programme will be tested in 10 different pilots across Europe with 50 children each, thus involving 500 children in total. The following table (Table 1) gives an overview of the programme elements.

The evaluation method is based on a mixed method approach, with quantitative and qualitative measures. It follows a pre-post design, comparing the baseline data before and after the programme, where possible. For some of the above listed dimensions standardised psychological assessment tools are available, for instance, for measuring

Table 1. DOIT programme elements of early entrepreneurial education (EEE)

EEE elements	Description of possible activities	How the activity could be supported	Evaluation dimenson
1. Motivation (Do it because you can)	Students get motivated by early successes or by envisioning the scope of their possibilities	Presenting/telling success stories that motivate, e.g. by peers	Self-confidence Self-efficacy Sense of initiative
2. Co-design (Do what matters)	Students are asked to collect and select potential ideas for innovations, this includes methodologies and approaches to identify the true roots of a problem, e.g. talking with relevant stakeholders	Methods and Materials to detect true roots of a problem, Creativity tools	Self-confidence Sense of initiative Creativity
3. Co-creation (Do it together)	Students will make the project a reality collaboratively – including more knowledgeable others (entrepreneurs, makers)	Planning methods Interdisciplinary group working	Creativity, sense of initiative, planning, managing resources, managing uncertainty/risk/teamwork
4. Iterate (Start it now)	The development of projects is focusing on concrete prototypes and their continuous improvement	Lean prototyping methods using different materials, understanding the decomposition of design challenges	Teamwork, creativity, managing resources
5. Reflection (Do it better)	Within and after the development of the projects, students are asked to reflect their work and to get and give feedback for better (future) results.	Moderation skills; Reflection and feedback phase; sharing of failure experiences	Assessment of opportunities, managing resources
6. Scaling (Do more of it)	Depending on students' age, project results are brought to a bigger group of users	Developing plans for scaling. Testing the robustness of a solution if replicated multiple times	Assessment of opportunities, financial literacy, managing resources, managing uncertainty/risk
7. Reaching out (Do inspire others)	Students are asked to share their ideas and projects to a wider public	Public presentation and sharing of the idea and the (success) story	Role of entrepreneurs, entrepreneurial career options

creativity (e.g. TSD-Z) and self-confidence (e.g. CFSEI-3) [20, 21]. For others, a self-rating survey will be developed to cover dimensions such as planning capacity or the perceived role of entrepreneurs. Not for all dimensions and EEE elements a pre-post comparison is feasible, e.g. for the assessment of opportunities or teamwork. For these, qualitative instruments will be used along the path accompanying the different programme elements. Interviews with facilitators and children will be carried out to understand if and how children identify with the role of entrepreneurs, how the deal with uncertainties or think of entrepreneurial career options. Semi-structured interview guidelines based on critical incidence technique (CIT) steering self-reflection and self-evaluation will be developed for interviewing two (randomly selected) children per pilot.

Furthermore, an artefact analysis of the developed prototypes will be carried out.

Facilitators will be asked to fill in a researcher diary at various occasions and will be interviewed after the programme reflecting based on their observations throughout the programme and on their dairy entries. With qualitative content analysis software the qualitative data will be analysed and complemented with the quantitative analysis constitute a rigorous evaluation framework.

4 Conclusions and Outlook

The argument that formal education in general and early entrepreneurial education in particular would benefit from making and the maker pedagogy is not without foundation but empirical data for grounding these claims are lacking. Setting goals and devising paths to achieve them, collaborating with others, in project based learning environments are typical characteristics of maker work as well as entrepreneurial activities. However, making as subject has so far not been introduced to formal education settings - with a few exceptions. The project DOIT constitutes an attempt to bring making and social entrepreneurship education together and to analyse in sound and rigorous manner its effects on the development of entrepreneurial skills, attitudes and knowledge. Thus, we will systematically evaluate the DOIT programme based on mixed method approach combining qualitative and quantitative measures and contribute to the science base as empirical insights into the effect of maker pedagogy in reference to opportunities and constraints are currently lacking.

Acknowledgment. DOIT has received funding from the European Union's Horizon 2020 research and innovation programme under grant agreement No 770063.

References

1. Millard, J., Unterfrauner, E., Voigt, C., Katsikis, O.K., Sorivelle, M.N.: The maker movement in Europe: empirical and practitioner insights. In: Presented at the 4TH ICT for Sustainability Conference (ICT4S), Toronto (accepted)
2. Gershenfeld, N.: How to make almost anything: the digital fabrication revolution. Foreign Aff. **91**(6), 43–57 (2012). Council on Foreign Affairs

3. Unterfrauner, E., Schrammel, M., Hofer, M., Fabian, C.M., Voigt, C., Deljanin, S.R., Sorivelle, M.N., Devoldere, B., Haga, H.: Final case study report focusing on cross-case analysis (2017)
4. Invent To Learn – Making, Tinkering, and Engineering in the Classroom. http://inventtolearn.com/
5. Unterfrauner, E., Voigt, C.: Makers' ambitions to do socially valuable things. In: Di Lucchio, L., Imbesi, L., Atkinson, P. (eds.) The Design Journal, vol. 20(Sup1), pp. 3317–3325. Taylor & Francis Online (2017)
6. Montessori, M.: The Montessori Method. Transaction Publishers, Chicago (2013)
7. Ackermann, E.: Piaget's constructivism, Papert's constructionism: what's the difference. Future Learn. Group Publ. **5**, 438 (2001)
8. van Hout-Wolters, B., Simons, R.-J., Volet, S.: Active learning: self-directed learning and independent work. In: New Learning. pp. 21–36. Springer, Heidelberg (2000)
9. Bell, S.: Project-based learning for the 21st century: skills for the future. Clear. House **83**, 39–43 (2010)
10. Kaltman, G.S.: Hands-on Learning. Corwin, Thousand Oaks (2009)
11. Bruffee, K.A.: Collaborative Learning: Higher Education, Interdependence, and the Authority of Knowledge, 2nd edn. John Hopkins Press, Baltimore (1999)
12. Papert, S.: Situating constructionism. In: Harel, I., Papert, S. (eds.) Constructionism. Ablex Publishing, Norwood (1991)
13. Papert, S.: The Children's Machine: Rethinking School in the Age of the Computer. Basic Books, New York (1994)
14. Schön, S., Boy, H., Brombach, G., Ebner, M., Kleeberger, J., Narr, K., Rösch, E., Schreiber, B., Zorn, I.: Einführung zu Making-Aktivitäten mit Kindern und Jugendlichen. Book on Demand, Norderstedt (2016)
15. Becker, S.A., Cummins, M., Davis, A., Freeman, A., Hall, C.G., Ananthanarayanan, V.: NMC Horizon Report: 2017 Higher, Education edn. The New Media Consortium, Austin (2017)
16. Voigt, C., Montero, C.S., Menichinelli, M.: An empirically informed taxonomy for the Maker Movement. In: International Conference on Internet Science, pp. 189–204. Springer, Florence (2016)
17. Craddock, I.L.: Makers on the move: a mobile makerspace at a comprehensive public high school. Libr. Hi Tech. **33**, 497–504 (2015)
18. Eurydice: Entrepreneurship Education at School in Europe: Eurydice Report. Publications Office of the European Union (2016)
19. Lackéus, M.: Entrepreneurship in education: What, why, when, how. Backgr. Pap. (2015)
20. Urban, K.K., Jellen, H.G.: Test zum Schöpferischen Denken-Zeichnerisch (TSD-Z). Swets (1995)
21. Battle, J.: Culture-Free Self-Esteem Inventories, Third Edition (CFSEI-3). Pro.Ed, Austin (2002)

Awareness of School Learning Environments

Margarida Figueiredo[1], Henrique Vicente[2,3], Jorge Ribeiro[4], and José Neves[3](✉)

[1] Departamento de Química, Escola de Ciências e Tecnologia, Centro de Investigação em Educação e Psicologia, Universidade de Évora, Évora, Portugal
mtf@uevora.pt

[2] Departamento de Química, Escola de Ciências e Tecnologia, Centro de Química de Évora, Universidade de Évora, Évora, Portugal
hvicente@uevora.pt

[3] Centro Algoritmi, Universidade do Minho, Braga, Portugal
jneves@di.uminho.pt

[4] Escola Superior de Tecnologia e Gestão, ARC4DigiT – Applied Research Center for Digital Transformation, Instituto Politécnico de Viana do Castelo, Viana do Castelo, Portugal
jribeiro@estg.ipvc.pt

Abstract. Now, and in the times that follow, student education should focus on developing inclusive skills such as problem-solving and decision-making, where the role of the learning environment plays a crucial part, i.e., it is a process where the screen of the universe of discourse is accomplished in order to consider not only the complex relationships that flow among the objects that populate it, but also its inner structure, co-existing incomplete/unknown or even self-contradictory information or knowledge. As a result, we will focus on the development of an *Intelligent Social Machine* to assess *Learning Environments* in high schools, based on factors like *School* and *Disciplinary Climates* as well as *Parental Involvement*. The formal background will be to use *Logic Programming* to define its architecture based on a *Deep Learning-Big Data* approach to *Knowledge Representation* and *Reasoning*, complemented by an *Evolutionary* approach to *Computing* grounded on *Virtual Intellects*.

Keywords: Artificial Intelligence · Intelligent Learning Environments Logic Programming · Knowledge Representation and Reasoning Evolutionary Computation · Intelligent Social Machine

1 Introduction

Intelligent Learning Environments (*ILEs*) can generally be defined as computer-based educational systems based on various *Artificial Intelligence* (*AI*) techniques to enhance students' learning experience and help them achieve their learning goals. In this context, a key issue for education systems in general and for *ILEs* in particular, is the ability to harness these new paradigms to create, maintain, and share the knowledge that these systems embed. This will enable *ILEs* to benefit from shared information

T. Di Mascio et al. (Eds.): MIS4TEL 2018, AISC 804, pp. 147–155, 2019.
https://doi.org/10.1007/978-3-319-98872-6_18

from distinct systems related to learning content and student activities, thereby reducing the complexity of system development and maintenance while enhancing personalization and contextualization and interaction. In fact, some studies suggest that schools can help improve students' success by enhancing their *ILEs* [1–3]. Under this setting, qualities such as resilience, perseverance, ethics, conscience and leadership should be developed not only at home but also in schools. The development of global skills such as problem-solving and decision-making requires the involvement of all educators (i.e., schools, teachers and parents) [4–6]. The report of the *International Student Assessment Program* (*PISA*) 2015 shows that the proportion of students who skipped at least once a full day of school in the two weeks prior to the *PISA* test increased by about 5 percentage points between 2012 and 2012 in *OECD* countries and 2015 [7]. This finding is worrying, as truancy is one of the factors influencing the school climate, influencing teacher work and adversely affecting the performance of all students in the class. In addition, factors such as security and order, teacher relations and cooperation, academic expectations, leadership and career development have been identified as key to improving the school environment. Indeed, the analysis and development of a new approach to *ILEs*, able to attend the need to predict the failure according to technical and non-technical criteria in a school environment, is a hard task, namely in terms of the huge number of possible scenarios. Indeed, the current state-of-the-art for service-life teaching is at empiric and empiric–mechanistic levels, and does not provide any suitable answer even for a single failure criterion. Consequently, it is imperative to achieve qualified models and qualitative reasoning methods, in particular due to the need to have first-class environments at our disposal where defective information is at hand. The foregoing shows that a proactive strategy is needed to solve the problems associated with improving the *ILEs*. In fact, different conditions with complex relationships between them must be taken into account where the available data may be incomplete/unknown (e.g., missing answers to some questions in the questionnaire) and/or self-contradictory (e.g., questions related with the same problem with inappropriate answers). To overcome these problems, this work presents a *Mathematical Logic* based computational machine that is featured as an *Intellect* and labelled as an *Intelligent Social Machine* (*ISM*) that is put at the forefront to help decision makers to further development of *ILEs*. Therefore, the next section gives a brief description of an pioneering *Deep Learning* (*DL*) approach to *Knowledge Representation* and *Reasoning* (*KRR*) followed by the presentation of a case study to screen the factors affecting *ILEs* and the perspectives of the *ISM*. Finally, conclusions are drawn and directions for future work are outlined.

2 A Deep Learning Route to Knowledge Representation and Reasoning

Knowledge Representation and Reasoning (*KRR*) aims to understand the complexity in information. Automated reasoning capabilities enables a system to "fill in the blanks" since in the real world incomplete information or data with gaps are common. Although *KRR* has been grounded on a symbolic logic in vector spaces, the fundamentals and attributes of the logical functions described in the present study go from discrete to continuous, allowing for the representation and handling of unknown, incomplete, or even self-contradictory information/knowledge. Indeed, such fact denotes the key distinction of the presented approach, otherwise it would be only symbolic logic in vector space, where the data items remain essentially discrete, and therefore no added value would be attained. A data item is understood as find something smaller inside when ones taking anything apart, and it is mostly formed from different elements, namely the *Interval Ends* where its value may be situated, the *Quality-of-Information* (*QoI*) it carries, and the *Degree-of-Confidence* (*DoC*) put on the fact that its value is inside the interval ends just referred to above. These are just three of over an endless element's number. Indeed, one can make virtually anything one may think of by joining different elements together. In other words, viz.

- What happens when one splits a data item? The broken pieces become data item for another element, a process that may be endless; and
- Can a data item be broken down? Basically, it is the smallest possible part of an element that still remains the element.

This makes one's route from *Deep Learning* (*DL*) to *KRR*. It is based on the fact that the data items (i.e., the fundamentals and attribute's values of the logical functions that make the universe of discourse). Therefore, the proposed approach to this issue, put in terms of the logical programs that elicit the universe of discourse, will be set as productions of the type, viz.

$$predicate_{1 \leq i \leq n} - \bigcap_{1 \leq j \leq m} clause_j(([A_{x_1}, B_{x_1}](QoI_{x_1}, DoC_{x_1})), \cdots$$
$$\cdots, ([A_{x_m}, B_{x_m}](QoI_{x_m}, DoC_{x_m}))) :: QoI_j :: DoC_j$$

that engender one's view to *DL*. n, $\cap$, m and A_{x_m}, B_{x_m} stand for, respectively, for the cardinality of the predicates' set, conjunction, predicate's extension, and the interval ends where the predicates attributes values may be situated. The metrics *QoI* and *DoC* show the way to data item dissection [8, 9], i.e., a data item is to be understood as the data's atomic structure. It consists of identifying not only all the sub items that are thought to make up an data item, but also to investigate the rules that oversee them, i.e., how $[A_{x_m}, B_{x_m}]$, QoI_{x_m}, and DoC_{x_m} are kept together and how much added value is created.

3 Case Study

3.1 Data Collection

To achieve the goals of this study, a questionnaire was developed and applied to a sample of 291 students aged 15 to 19 years old with an average of 17 ± 2 years. The gender distribution was 47.6% and 52.4% for men and women, respectively. The questions in the questionnaire were divided into three sections, the first of which contains the questions about the *School Climate* (Table *School Climate*, Fig. 1). The second includes the questions relating to the students' opinion on *Parental Involvement*

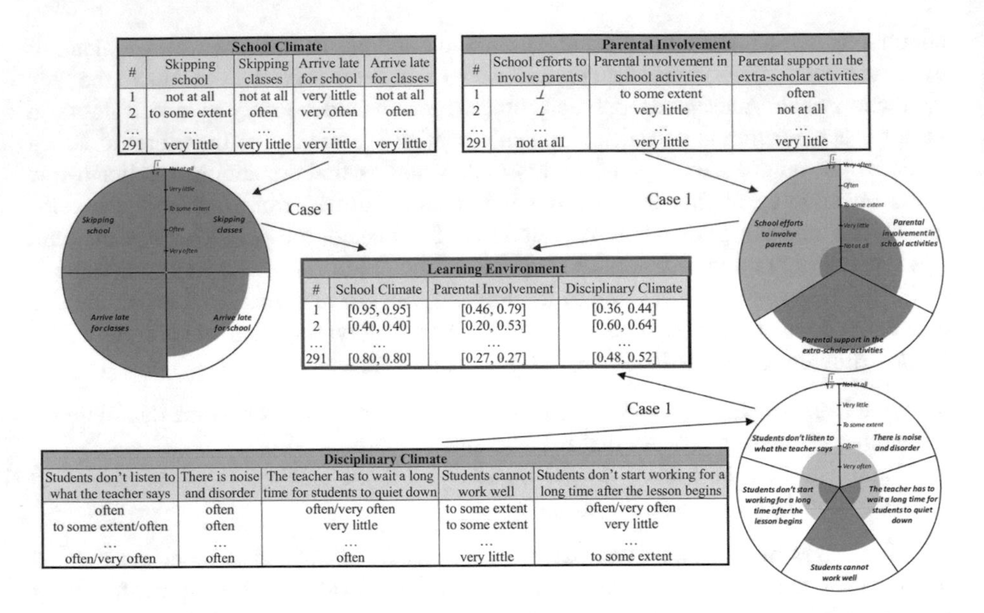

School Climate				
#	Skipping school	Skipping classes	Arrive late for school	Arrive late for classes
1	not at all	not at all	very little	not at all
2	to some extent	often	very often	often
...	...	...	...	...
291	very little	very little	very little	very little

Parental Involvement			
#	School efforts to involve parents	Parental involvement in school activities	Parental support in the extra-scholar activities
1	$\perp$	to some extent	often
2	$\perp$	very little	not at all
...	...	...	...
291	not at all	very little	very little

Learning Environment			
#	School Climate	Parental Involvement	Disciplinary Climate
1	[0.95, 0.95]	[0.46, 0.79]	[0.36, 0.44]
2	[0.40, 0.40]	[0.20, 0.53]	[0.60, 0.64]
...	...	...	...
291	[0.80, 0.80]	[0.27, 0.27]	[0.48, 0.52]

Disciplinary Climate				
Students don't listen to what the teacher says	There is noise and disorder	The teacher has to wait a long time for students to quiet down	Students cannot work well	Students don't start working for a long time after the lesson begins
often	often	often/very often	to some extent	often/very often
to some extent/often	often	very little	to some extent	very little
...	...	...	...	...
often/very often	often	often	very little	to some extent

Fig. 1. A knowledge-based fragment of an extension of the relational database for the different issues that characterize the *Learning Environment*.

(Table *Parental Involvement*, Fig. 1). The last one contains questions related to the students' opinion on the *Disciplinary Climate* (Table *Discipline Climate*, Fig. 1).

3.2 Feature Extraction

The feature extraction's process focused on the more relevant issues involved in each topic affecting the *LE*, namely *School Climate*, *Disciplinary Climate* and *Parental Involvement* [1, 4–7]. Each was included in the above-mentioned questionnaire and the answers stored in the respective table as shown in Fig. 1. Qualitative values are also used to classify the different issues, given in terms of the scale *not at all*, *very little*, *to some extent*, *often* and *very often*, and evaluated according to the method described in Fernandes et al. [10].

4 Evolving Systems

The *Intelligent Social Machine's* architecture is structured as an *Intellect* and set as an ensemble of entities designated as symbolic neurons. To each neuron is associated a logic program or theory, given by the extensions of the predicates that make their corpus. The genome is given in terms of an ensemble of neurons, being each one coded with two types of genes [11], viz.

- Processing genes that specify how each neuron will assess its output; and
- A set of connection ones which specify the potential connections to other neurons, built in terms of the extensions of the predicates that model their inner universe of discourse.

According to the knowledge representation and reasoning formalism presented in Sect. 2 and above, the processing genes are structured by the label $gpn_n(t)$ from the ordered theory $OT = (T, <, (S, \prec))$, where T, $<$, S and $\prec$ correspond, respectively, to the knowledge base of the gene in a clausal form, a non-circular order relation over the clauses, a set of priority rules and one non-circular relation order over those rules. The non-circular order is necessary by two reasons, i.e., by the relative importance of the rules and by the operational usability in which a logic program written (e.g., PROLOG) needs to set some concrete order over the set of rules. In this sense a processing gene can be described as follows, viz.

$$gpn(t) = <T, M, Q, C, Intervals\ Ends, QoI, DoC>$$

where T corresponds to the logic theory that make up the inner neuron's universe of discourse, M the inference mechanism, Q the question (or sub-problem) to solve, and C the scenarios under which Q is to be addressed. QoI and DoC stand for themselves (Fig. 2). It is now possible to give a schematic view of how to model the universe of discourse in a dynamic or evolutionary environment, where the extensions of two or

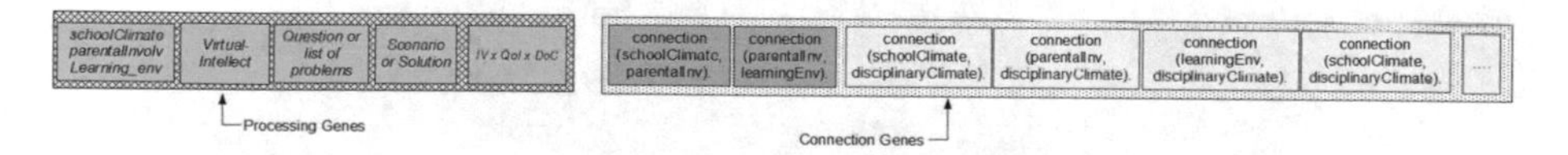

Fig. 2. The genome at its initial state.

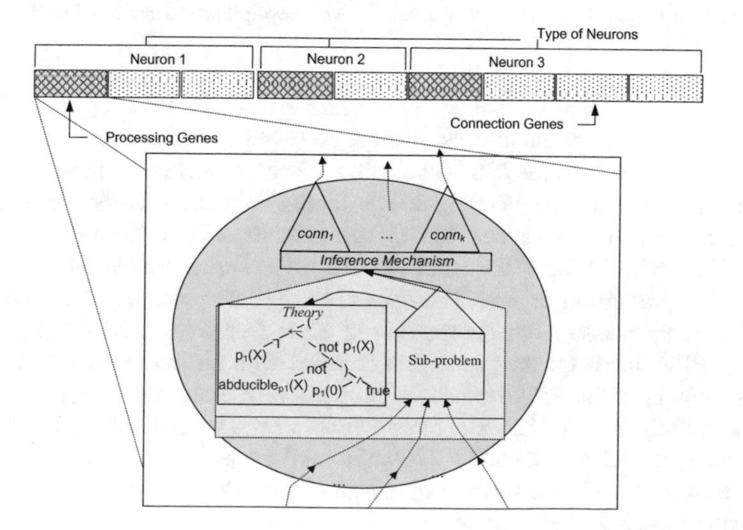

Fig. 3. The genome's scheme and its inner inference mechanism.

more predicates are projected into a fusion space that inherits a partial structure from their inputs and emerge with a structure of its own (Figs. 2 and 3).

The input of each neuron is given as a list of sub-problems to be solved according to the diverse scenarios referred to above (Fig. 1). Indeed, the evolutionary process to set the *Intelligent Social Machine* (*ISM*) starts with an estimate representation of the universe of discourse in terms of its predicates' extensions and proceeds in order to optimize their attributes' metrics, i.e., the interval ends $[A_{x_j}, B_{x_j}]$, QoI_j, and DoC_j, once a problem to be solved is set as a theorem to be proved. This evolutionary process is depicted in Figs. 4, 5, 6 and 7, i.e., the *Intellect* evolution is grounded on a theorem proving course.

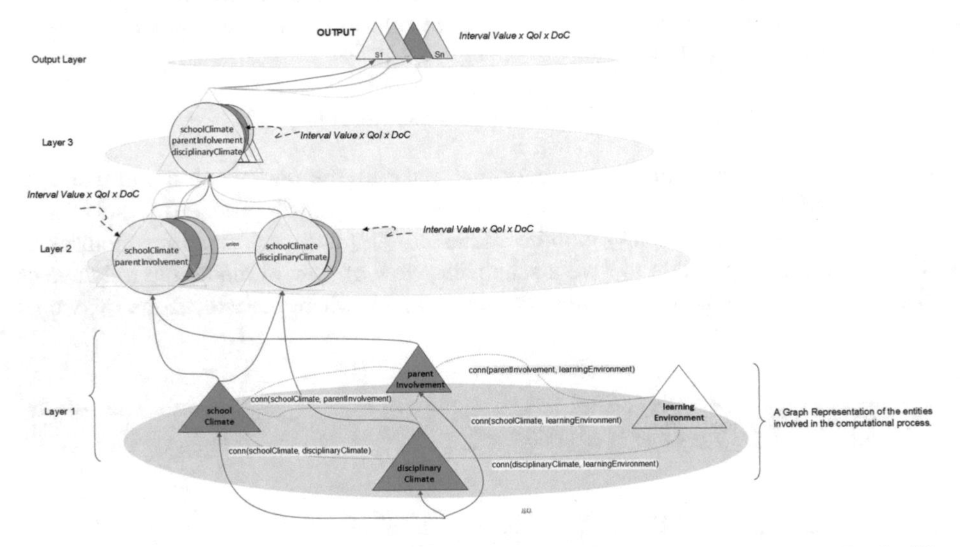

Fig. 4. Input: ? (schoolClimate (ID, A, B, C, D), disciplinaryClimate (ID, E, F, G, H), parentInvolvement (ID, I, J, K)).

A set with 291 records was used to the *ISM's* analysis. In order to guarantee the statistical significance of the attained results, 20 (twenty) experiments were applied in all tests. Table 1 presents the *ISM's* confusion matrix. A perusal to Table 1 shows that the model accuracy was 91.4% (i.e., 263 instances correctly classified in 291). Based on coincidence matrix it is possible to compute the *ISM's Sensitivity*, *Specificity*, *Positive Predictive Value* (*PPV*) and *Negative Predictive Value* (*NPV*). *Sensitivity* measures the proportion of *True Positives* (*TP*) that are correctly identified as such, while *Specificity* measures the proportion of *True Negatives* (*TN*) that are correctly identified. *PPV* stands for the proportion of cases with positive values that were correctly diagnosed, while *NPV* denotes the proportion of cases with negative values that were successfully labeled [12]. The *Sensitivity* is 92.9% while the *Specificity* is 89.6%. *PPV*, in turns, is 91.2% whereas *NPV* is 91.7%. All the performance metrics mentioned above are close to 90% and seem to suggest that the *ISM* exhibits a good performance in assessing the *LE* in high schools.

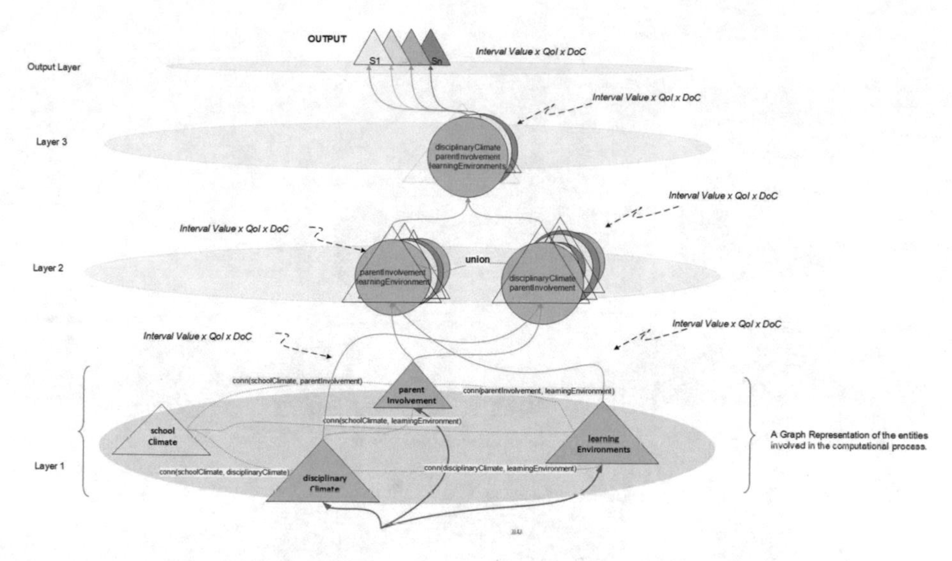

Fig. 5. Input: ? (disciplinaryClimate (ID, A, B, C, D), parentInvolvement (ID, E, F, G), learningEnvironment (ID, H, I, J)).

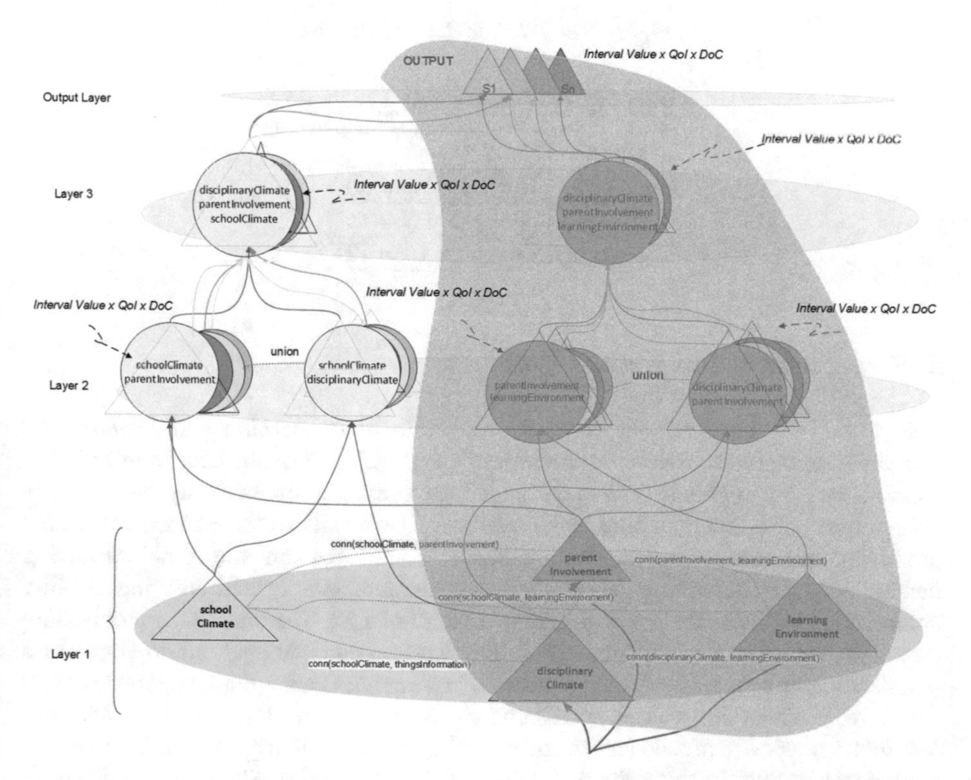

Fig. 6. Input: ? (disciplinaryClimate (ID, A, B, C, D), parentInvolvement (ID, E, F, G), learningEnvironment (ID, H, I, J)).

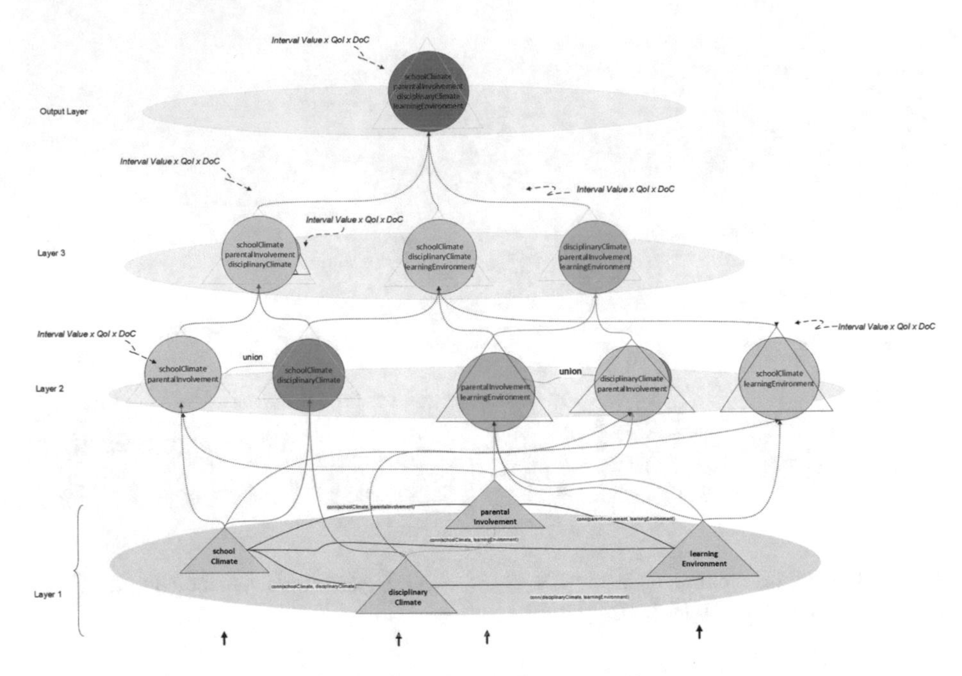

Fig. 7. The *ISM's* at the present time.

Table 1. The *ISM's* confusion matrix.

Output	Model Output	
	True (1)	False (0)
True (1)	TP = 145	FN = 11
False (0)	FP = 14	TN = 121

5 Conclusions and Future Work

This work may have set one route to apply *Machine Learning* and *Evolutionary Computation* methods to define suitable ways to develop *ILEs*. Indeed, in this work a methodology for problem solving grounded on symbolic, evolutionary and connectionist approaches to computing was presented. The results so far attained show how promising an *ISM* is, strengthening one's confidence on the problem-solving methodology just referred to above, reducing unpredictability and ensuring stability among all actors contributing to the improvement of *LEs*. The underlying architecture turns the *ISM* into a versatile, creative and powerful computational tool to engender a practically infinite variety of data processing and analysis capabilities, adaptable to any conceivable situation. Future work encompasses the development of *Conscious Machines*, under a symbolic and mathematical approach to computing, presenting also a good opportunity to study the real nature of the *Artificial* or *Synthetic Intelligence*.

Acknowledgments. This work has been supported by COMPETE: POCI-01-0145-FEDER-007043 and FCT – Fundação para a Ciência e Tecnologia within the Project Scope: UID/CEC/00319/2013.

References

1. Kraft, M.A., Marinell, W.H., Yee, D.: School Organizational Contexts, Teacher Turnover, and Student Achievement: Evidence from Panel Data. March 2016, https://steinhardt.nyu.edu/scmsAdmin/media/users/sg158/PDFs/schools_as_organizations/SchoolOrganizationalContexts_WorkingPaper.pdf. Accessed 23 Jan 2018
2. Richardson, C., Mishra, P.: Learning environments that support student creativity: developing the SCALE. Think. Skills Creat. **27**, 45–54 (2018)
3. Shernoff, D.J., et al.: Student engagement as a function of environmental complexity in high school classrooms. Learn. Instr. **43**, 52–60 (2016)
4. Poveya, J., et al.: The impact of parental involvement, parental support and family education on pupil achievements and adjustment: a literature review. Int. J. Educ. Res. **79**, 128–141 (2016)
5. Castro, M., et al.: Parental involvement on student academic achievement: a meta-analysis. Educ. Res. Rev. **14**, 33–46 (2015)
6. Thapa, A., Cohen, J., Guffey, S., Higgins-D'Alessandro, A.: A review of school climate research. Rev. Educ. Res. **83**, 357–385 (2013)
7. OECD: PISA 2015 Results (Volume II): Policies and Practices for Successful Schools, PISA. OECD Publishing, Paris (2016)
8. Fernandes, F., et al.: Artificial neural networks in diabetes control. In: Proceedings of the 2015 Science and Information Conference (SAI 2015), pp. 362–370. IEEE Edition (2015)
9. Silva, A., et al.: Length of stay in intensive care units – a case base evaluation. In: Fujita, H., Papadopoulos, G.A. (eds.) New Trends in Software Methodologies, Tools and Techniques, Frontiers in Artificial Intelligence and Applications, vol. 286, pp. 191–202. IOS Press, Amsterdam (2016)
10. Fernandes, A., Vicente, H., Figueiredo, M., Neves, M., Neves, J.: An adaptive and evolutionary model to assess the organizational efficiency in training corporations. In: Dang, T.K., et al. (eds.) Future Data and Security Engineering. LNCS, vol. 10018, pp. 415–428. Springer, Cham (2016)
11. Neves, J., et al.: Evolutionary intelligence in asphalt pavement modelling and quality-of-information. Prog. Artif. Intell. J. Springer (2011). https://doi.org/10.1007/s13748-011-0003-5
12. Florkowski, C.M.: Sensitivity, Specificity, Receiver-Operating Characteristic (ROC) curves and likelihood ratios: communicating the performance of diagnostic tests. Clin. Biochemist. Rev. **29**(Suppl. 1), S83–S87 (2008)

Gamification and Learning Analytics to Improve Engagement in University Courses

Fabio Cassano(✉), Antonio Piccinno, Teresa Roselli, and Veronica Rossano

Department of Computer Science, University of Bari, Bari, Italy
{fabio.cassano1, antonio.piccinno, teresa.roselli, veronica.rossano}@uniba.it

Abstract. Gamification is one of the most used techniques to improve active participation and engagement in different kinds of contexts. The use of game techniques is effective in pushing subjects to be involved in an activity. Since the early childhood, indeed, the promises of rewards are useful to affect specific behaviors. On the other hands, the learning analytics have been largely implemented in education in order to improve the assessment and the self-assessment of students, above all in e-learning settings. The research presented in this work aims at combining gamification techniques and learning analytics to improve the engagement in University courses. The paper describes a model of gamification and a learning dashboard defined based on data in Moodle e-learning platform. A pilot test of an app android in which both the solutions have been implemented pointed out promising results.

Keywords: Gamification · Learning analytics · E-learning engagement Learning dashboard

1 Introduction

Mobile devices (such as smartphones and tablets) allow people to be connected and communicate all over the world. Thanks to the increasing Internet communication speed and the more powerful mobile Central Processing Unit (CPU), mobile devices can be used for a wide variety of tasks. According to the age of the user, the device is prevalently used: to play mobile games, to use email or messaging, to play videos etc. For the university students, the mobile phone is a critical device. As a matter of facts, it is used by boys and girls attending the university, not only to check the exam dates and share notes, but also to message information about lessons.

E-learning is a modern way to allow university students to attend lessons virtually, using the Internet connection. It is common for many universities to offer online courses, the MOOCs (Massive Open Online Courses) phenomenon is a proof of this trend [1, 2]. Many online platforms, such as, Coursera, EdX, Iversity deliver online contents to all students who have either a Personal Computer (PC) or a mobile device. This strategy has been so successful, that the main universities all over the world, both public and private, publish on those platforms their courses.

T. Di Mascio et al. (Eds.): MIS4TEL 2018, AISC 804, pp. 156–163, 2019.
https://doi.org/10.1007/978-3-319-98872-6_19

The MOOCs have been a big revolution in Education, since the university courses can be attended by all people that need to acquire specific knowledge or competences without to necessary being physical in the site where the lesson is given. This means that even non-university students can access to high education without physical limitations. Moreover, these courses (and universities) release attending certificates, once the course has been completed, and qualification certificates if the student successfully passes the related exam. These certifications can be used to improve the curriculum vitae and then the personal job. Unfortunately, one of the main problems in using MOOCs and all kind of online courses is the engagement and motivation [3]. The flexibility and the freedom to attend the e-learning courses often translate into a high dropout rate [4]. Many problems can distract the student from the aim and, after a failure, the motivation without the teacher support dramatically drops [5]. This students' failure is a problem also for the educational system [6]. In this context, in order to mitigate this problem of online courses and activities, we propose to apply the gamification approach to improve the student's active participation and engagement in online university course. In order to measure the impact of gamification, a mobile application was developed to let the student be more engaged in attending online activities in her/his university course and, in general, in the university life.

This paper is organized as follows: in the following section some related works are reported; Sect. 3 describes the adopted gamification approach for monitoring engagement and Sect. 4 how it is implemented in a mobile application; in Sect. 5 some preliminary results of the user testing is reported, and Sect. 6 concludes the paper.

2 Related Works

E-learning systems are widely used in both the university and the work domain. This form of instruction has grown up during the last twenty years thanks to the greater Internet speed and the powerful devices that can play videos. There are many guidelines on how the e-learning systems should be designed [7]. Those fall into the HCI field, where the users need to be considered during all the stages of the development process. For example, more and more solutions use the same framework to deliver different contents [8]. The study of how people really learn from this new way to teach is constantly monitored by the companies that deliver e-learning contents [9]. The e-learning is thus a recent and evolving topic. More and more techniques are used in order to engage people and let them be more proficient in following courses and learn. One of them is the so called "Gamification".

Gamification is defined as "the use of game design elements in non-game contexts" [10]. The term "gamification" is sometimes controversial, but the definition given above and the survey provided in [10] clarify that "gamified" applications are different from (video) games, serious games or just software applications that provide a playful interaction, like those considered in [11].

Gamification has been proved that improve participation and engagement in e-learning activities [12, 13], in fact, suggests that gamification strategies, aligned with instructional objectives and user context, are effective in improving student participation and encouraging extracurricular learning. Moreover, game elements such as points, badges, and leaderboards, are useful strategy in Massive Open Online Courses as [14–16]. On the other hands, once that the online environment stimulates student motivation and engagement, it is necessary to measure it. In e-learning settings, characterized by the distance both in terms of time and places, it is necessary to monitor and track the student activities in online environments. The Learning Analytics (LAs), i.e. the measurement, collection, analysis and reporting of data about learners and their contexts [17] are very useful to meet this objective. There are some research works that investigate on the relationships among the LAs and engagement [18, 19]. The main novelty of the proposal described in this paper is to combine the use of game elements in order to foster student engagement in online activities in academic contexts and the LAs in order to measure and visualize the level of engagement for each single student.

3 The Gamification Approach for Monitoring Engagement

In e-learning platforms, keeping track of the user's learning activities is very important to make effective and reliable assessment. To improve the quality of assessment in online courses, even with a large population (as happens in MOOCs), in literature different solutions have been implemented [20–22]. An interesting solution is the use of a Learning Dashboards (LD) in the e-learning environments to visualize student's engagement in e-learning paths. Usually, LDs allow to visualize the Learning Analytics. The LAs can be automatically or manually collected by the system. In this research, LAs and LDs have been used in order to keep high student's engagement and motivation. To address this challenge, VeeU2.0, a learning dashboard for Moodle, has been designed and developed to support assessment by both teachers and learners in e-learning courses [3, 21]. The defined model has been conceived to monitor the engagement in an e-learning course through measures of the student's participation, in terms of user's actions in wikis and social posts. Following the game mechanics, eXperience Points (XP) were defined. In order to classify the kind of activities performed in the e-learning platform, the XP were subdivided in Degree Course XP (DC) and Single Course XP (SC). The student can gain DC performing general activities (Table 1) in the e-learning platform and in all online courses published for her/his degree course. The SC points are gained performing general activities in any specific course that student is attending (Table 1). In other words, if the student accesses (Activity A1) to the e-learning platforms to browse all the online courses of her/his degree course s/he gain 10 DC points. If the student creates a wiki page in the "Programming course" s/he gains 5 SC points.

When the user reaches a certain amount of XP, it gains a new level. Every new level (for a maximum of 100 levels) allow the student to gain a higher reputation in both virtual and real class. In order to make visible this reputation, badges can be collected according to both the points gained and the levels reached. There are three type of badges Gold, Silver and Bronze and they can be achieved both for the degree course or for a specific course. Moreover, the badges can be achieved for each kind of activities performed by the student, thus a student can have a Gold badge for the DC access points (Activity A1) but any badge for SC creating wiki pages (A6) this means that the student mainly surfs in the e-learning platforms only to download the learning resources or to read news, but s/he is not an active participant to online activities.

Table 1. The XP points gained by the user according to the action performed

Activity type	XP gained
(A1) Access	10
(A2) Read a wiki page	2
(A3) Put a “like” to a post	2
(A4) Publish a post on the dashboard	2
(A5) Comment a post on the dashboard	3
(A6) Create a wiki page	5
(A7) Edit a wiki page	5

4 A Mobile App for VeeU2.0

VeeU2.0 is a learning dashboard developed as plugin for Moodle, the e-learning platform in use in our University. The main goal of VeeU2.0 is to make students and teachers aware of their engagement in e-learning environment. As a matter of fact, teachers and students need to be aware of what kinds of interactions are occurring in the virtual space and how the building up knowledge process happens. The dashboard offers two points of views, one addressed to the teacher, who can visualise the trend of the entire class or of a specific student, and one addressed to the student, who can visualise her/his rate of participation in each activity and can compare her/his data with those of the other students. From the teachers' points of view, the information visualised are useful in order to monitor the level of students' participation and interest in the subject. This information can lead the teacher to change the teaching strategies in order to improve the teaching effectiveness. From the student point of view, the visualised data can help the student's self-assessment that could be pushed to improve her/his efforts in the learning process.

In order to improve the efficacy of VeeU2.0, a mobile app has been developed and the gamification model was applied. The mobile version (the language is Italian) reported here, as well as the web app, offers two points of view. One addressed to the teacher, who can view how many accesses students have done for each course (Fig. 1a), which resources have been downloaded, how many quizzes have been completed and so on. Moreover, the teacher could visualize information about each single student in order to verify how her/his learning process is going on. For each course, the number of access of the student is represented using a histogram together with a line indicating the mean of accesses of the class in the same period (see Fig. 1b).

Gamification techniques has been implemented in the app for the student point of view (Fig. 2). Once the user has selected the degree course, the app shows his/her progress (Fig. 2a): on the top of the screen the course name is shown and below the level reached and the overall XP gained are given (“Livello 3” and “99/600” respectively in the figure). For each course the student can visualise the list of the resources

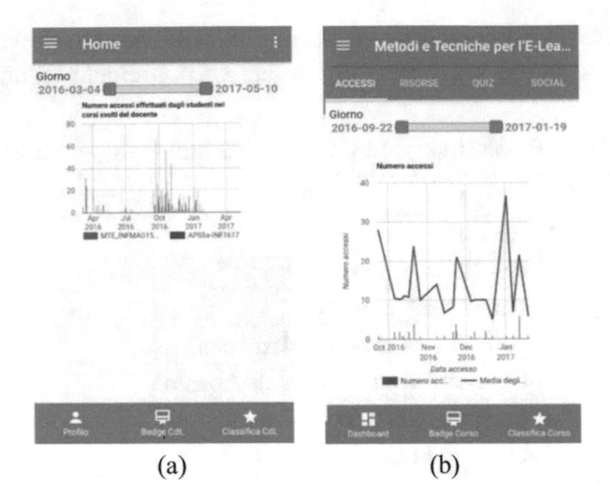

Fig. 1. The mobile app from the teacher's point of view showing students' accesses (a) for all courses and (b) for a specific course

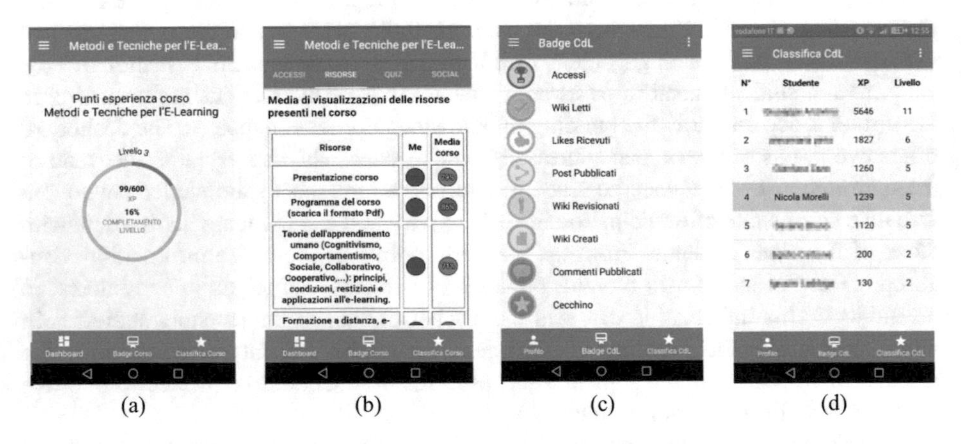

Fig. 2. The mobile app from the students' point of view showing (a) progresses, (b) available resources, (c) badges achieved and (d) the leader board

available in the course and the mean of the students that have visualised each resource (Fig. 2b). The green colour is used for the visualised resources, the red for non-visualised ones.

In the badge section, the user can see how many (and which type) of badges s/he has achieved (Fig. 2c). Every time a new badge has reached, a popup message is shown with details of the badge and how it is possible to increase the number (the value can be shown by clicking on it). Finally, the leader board of the users shows all the users, for the selected university course, with the relative amount of XP gained and the level (Fig. 2d).

5 User Testing

In order to evaluate the usability of the application, we performed a user testing with real end users. The user testing aims at analysing the user behaviour during the interaction with the system. To this aim, we defined a list of 9 tasks that users have had to accomplish. The "Thinking aloud" technique was used in order to better understand the interaction problems. For lack of space, we cannot report further details on the user test performed. The sample was composed of 15 students attending one of the Computer Science degree courses at the University of Bari. All students use regularly the LMS to access the content of the different courses.

Each student used the system alone. A facilitator gave the instructions to the student and an observer annotated all the significant information about student's behavior. During the test the success rate was used as objective measure. The success rate has been calculated as follows: *Success rate = (S + (P * 0,5))/N*. Where: *S* is the number of tasks successfully completed; *P* is the number of tasks partially completed; *N* is the sum of all tasks. The results of the test are shown in Table 2.

Table 2. Results from the user tests with 15 users and 9 tasks (S: success, P: partial, F: failure)

	Task1	Task2	Task3	Task4	Task5	Task6	Task7	Task8	Task9
S1	S	S	S	S	S	S	S	S	S
S2	S	P	S	S	S	S	S	P	S
S3	S	S	S	S	S	S	S	S	S
S4	S	S	S	P	F	S	S	S	S
S5	S	S	S	S	S	S	S	S	S
S6	S	S	S	S	S	S	S	P	S
S7	S	P	S	S	S	S	S	S	S
S8	S	S	S	S	S	S	S	S	S
S9	S	S	S	S	S	S	S	S	S
S10	S	S	S	S	P	S	S	S	S
S11	S	S	S	S	S	S	S	S	S
S12	S	S	S	S	S	S	S	S	S
S13	S	S	S	S	S	S	S	S	S
S14	S	S	S	S	P	S	S	S	S
S15	S	P	S	S	S	S	S	P	S

The success rate of 95% could be considered a positive indication about the usability of the system. Moreover, in order to have a qualitative evaluation of the system, the students were asked to answer to a 5-Likert scale questionnaire about the perceived usefulness of the app during the learning process.

Analysing the results, the system has reached a good level of acceptance. The 85% of the students have appreciated the use of the system and hope that it would be used in the future. Moreover, the student's appreciation is visible also in the Fig. 3.

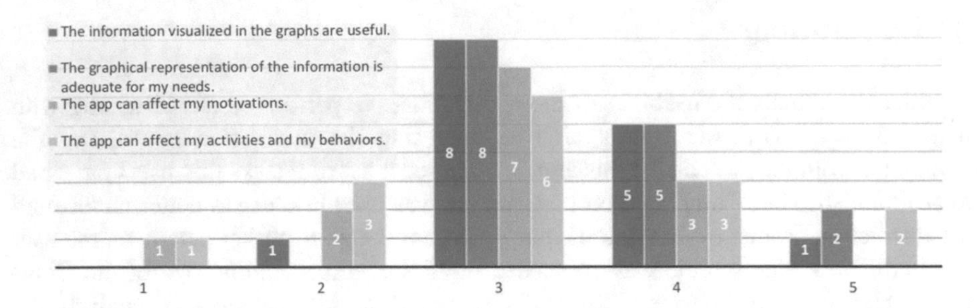

Fig. 3. Results of the students'appreciation: from 5-Likert scale questionnaires administered to the 15 participants of the user test, where "1" means "not at all" and 5 is "definitely"

In particular, to the questions about motivation 10 students out 15 give an evaluation between 3 and 4. For what concerning the engagement in e-learning activities, 5 students out 15 give an evaluation higher than 3 (neutral value). The user testing has revealed also a number of weakness that should be addressed in the next future, but the results are promising, and this can lead the research to further developments.

6 Conclusions

In this paper we have proposed a mobile application, based on the VeeU2.0 Learning Dashboard, that allow university student and teachers to be aware about their engagement, using the gamification technique. The students, using the app and interacting with the system, gain XP points that are used to rise their own level. The application presents for each DC and SC, a leader board about the most proactive students. The teacher can evaluate the ongoing student's engagement through a specific view.

We performed a preliminary evaluation test with 15 users to perform some tasks and then a 5-Likert questionnaire has been administered. The preliminary results show that the proposed app is promising with a good usability and acceptance rate.

References

1. Kennedy, J.: Characteristics of massive open online courses (MOOCs): a research review, 2009-2012. J. Interact. Online Learn. **13**(1), 1–16 (2014)
2. Liyanagunawardena, T.R., Adams, A.A., Williams, S.A.: MOOCs: a systematic study of the published literature 2008-2012. **14**(3), p. 26 (2013)
3. Pesare, E., Roselli, T., Rossano, V.: Visualizing student engagement in e-learning environment. In: 22th International Conference on Distributed Multimedia Systems (DMS), pp. 26–33. Knowledge Systems Institute, Skokie, IL 60076, USA (2016)
4. Chakor, Y.A., El Faddouli, N.-e: Abandonment of learners MOOC problematic analysis and proposed solutions. Int. J. Comput. Appl. **153**(2), 35–37 (2016)
5. Bates, A.W.T.: Technology, E-learning and Distance Education. Routledge, London (2005)

6. Bennett, R.: Determinants of undergraduate student drop out rates in a university business studies department. J. Further High. Educ. **27**(2), 123–141 (2003)
7. Clark, R.C., Mayer, R.E. (eds.): E-learning and the Science of Instruction: Proven Guidelines for Consumers and Designers of Multimedia Learning, 4th edn. Wiley, Hoboken, NJ, USA (2016)
8. Garrison, D.R.: E-Learning in the 21st Century: A Framework for Research and Practice. Psychology Press, London (2003)
9. Sun, P.-C., Tsai, R.J., Finger, G., Chen, Y.-Y., Yeh, D.: What drives a successful e-Learning? An empirical investigation of the critical factors influencing learner satisfaction. Comput. Educ. **50**(4), 1183–1202 (2008)
10. Deterding, S., Dixon, D., Khaled, R., Nacke, L.: From game design elements to gamefulness: defining "gamification". In: 15th International Academic MindTrek Conference: Envisioning Future Media Environments, pp. 9–15. ACM, New York, NY, USA (2011)
11. Salah, A.A., Schouten, B.A.M., Göbel, S., Arnrich, B.: Playful Interactions and Serious Games. IOS Press (2014)
12. Darina, D., Christo, D., Gennady, A., Galia, A.: Gamification in education: a systematic mapping study. J. Educ. Technol. Soc. **18**(3), 75–88 (2015)
13. de-Marcos, L., Domínguez, A., Saenz-de-Navarrete, J., Pagés, C.: An empirical study comparing gamification and social networking on e-learning. Comput. Educ. **75**, 82–91 (2014)
14. Klemke, R., Eradze, M., Antonaci, A.: The flipped MOOC: using gamification and learning analytics in MOOC design—a conceptual approach. Educ. Sci. **8**(1), 25 (2018)
15. Chang, J.-W., Wei, H.-Y.: Exploring engaging gamification mechanics in massive online open courses. J. Educ. Technol. Soc. **19**(2), 177–203 (2016)
16. Mazarakis, A.: Using gamification for technology enhanced learning: the case of feedback mechanisms. Bull. IEEE Tech. Comm. Learn. Technol. **17**(4), 6–9 (2015)
17. Long, P., Siemens, G.: Penetrating the fog: analytics in learning and education. EducausE Rev. **46**(5), 31–40 (2011)
18. Phillips, R., Preston, G., Roberts, P., Cumming-Potvin, W., Herrington, J., Maor, D., Gosper, M.: Using academic analytic tools to investigate studying behaviours in technology-supported learning environments (2010)
19. Pesare, E., Roselli, T., Rossano, V.: Engagement in Social Learning: Detecting Engagement in Online Communities of Practice, pp. 151–158. Springer International Publishing (2017)
20. Muñoz-Merino, P.J., Ruipérez-Valiente, J.A., Alario-Hoyos, C., Pérez-Sanagustín, M., Delgado Kloos, C.: Precise effectiveness strategy for analyzing the effectiveness of students with educational resources and activities in MOOCs. Comput. Hum. Behav. **47**, 108–118 (2015)
21. Pesare, E., Roselli, T., Rossano, V., Di Bitonto, P.: Digitally enhanced assessment in virtual learning environments. J. Vis. Lang. Comput. **31**, 252–259 (2015)
22. Siemens, G.: Learning analytics: envisioning a research discipline and a domain of practice. In: 2nd International Conference on Learning Analytics and Knowledge (LAK), pp. 4–8. ACM, New York, NY, USA (2012)

The Online Self-assessment *Mygoalin* Supporting Self-regulated Learning: User-Feedback for Optimization

Rebecca Fill Giordano(✉)

Alpen Adria University Klagenfurt, Interdisciplinary Research and Education (Klagenfurt, Graz, Vienna), Klagenfurt, Austria
rfillgio@edu.aau.at

Abstract. As self-regulated learning (SRL) is the prerequisite and the basis for *lifelong learning*, students should be supported in their way of organizing their learning processes. Therefore, an online self-assessment called *mygoalin* was developed based on state-of-the-art theoretical frameworks of educational psychology. This paper deals with the analysis of user-feedback to assess user-friendliness and usability of *mygoalin* with the aim to optimize the self-assessment tool for students. The qualitative content analysis comprises $N = 67$ students, who completed *mygoalin* and then answered different questions regarding usability and critical issues. The original qualitative answers were paraphrased based on the qualitative content analysis according to Mayring. Two categories (I *elements* and II *review type*) were defined and a total of 988 paraphrased statements were recorded. Already the number of 604 (60.5%) positive statements and the frequent mention of the usefulness and user-friendliness, suggest that *mygoalin* seems to be useful for self-reflection. Of these, 262 (26.3%) are critical statements and 132 (13.2%) suggest improvements. Qualitative analysis led to certain insights for optimizing *mygoalin*. The needs for a revision to improve the usability and an extension of the tool were discussed. Thus, *mygoalin revised* has become an online learning assistant supporting SRL by giving the opportunity for a standardized but still individual way.

Keywords: Self-regulated learning · Qualitative content analyses
Online self-assessment

1 Introduction

In the era of *lifelong learning*, personal development and the promotion of self-regulated learning (SRL) is a central goal for individuals as well as for society [1–3]. This is reflected both in the EU Guidelines for secondary and higher education and in the various *lifelong learning* programs for adolescents and adults [4]. However, as SRL is the prerequisite and the basis for *lifelong learning*, students need to be supported in organizing their learning processes. Therefore, this paper deals with the optimization of the new online self-assessment *mygoalin* for the analysis and support of self-regulated learning for students (www.mygoalin.com).

T. Di Mascio et al. (Eds.): MIS4TEL 2018, AISC 804, pp. 164–171, 2019.
https://doi.org/10.1007/978-3-319-98872-6_20

1.1 Theoretical Framework: SRL

There is no clear definition of the term SRL in literature but there are several definitions and approaches. Different authors [e.g. 5–9]) try to give an adequate definition. First, there is conformity at a descriptive level. Like Zimmerman [10] also other authors emphasize that students can be described as self-regulated to the degree that they are motivationally, metacognitively and behaviorally active participants in their personal learning process. Second, all authors claim that self-regulated learners are able to autonomously set goals, select appropriate techniques and strategies, and continually evaluate and correct their learning process [e.g. 11, 12]. Third, all authors describe the importance of different components in SRL [e.g. 13]. These include cognitive, metacognitive, volitional, emotional, motivational aspects. These theoretical frameworks are included in a new SRL model and base for the development of the online self-assessment *mygoalin* for students.

1.2 Practical Support: SRL

Therefore, the practical support consists in helping students to be active participants in their personal learning process selecting appropriate strategies to reach personal goals. Various training programs [6, 14] indicate that in general the following steps can be conducive for learners to achieve their goals:

Step 1. Activate SRL-awareness and self-reflection to understand SRL
Step 2. Acquire know-how by trying SRL, and by getting to know and use SRL-methods and techniques to understand in which context certain strategies are productive
Step 3. Consolidate SRL-strategies using them daily to achieve individual goals for understanding the usefulness of different methods and techniques

The online self-assessment *mygoalin* was developed to subsequently operationalize these steps. The first module analyses and promotes SRL in theory and practice especially regarding steps 1 and 2. For evaluating, whether *mygoalin* does this in an appealing way, this study analyses the user-feedback using a qualitative content analysis approach. The results lead to the conclusion which specific improvements are needed.

2 Research Questions

This study evaluates the user-feedback conducting a qualitative content analysis for answering the following key questions:

1. Quantitative: How many statements are positive, negative or suggest optimization?
2. Qualitative: Which aspects are described as positive or negative and what is their content? Which suggestions for optimization are frequently made? What should be improved and changed in *mygoalin* and how?

Then, the results obtained from the empirical qualitative content analysis are interpreted to determine the satisfaction and usability of potential users and for quality assurance. The main goal is to improve *mygoalin* for giving an optimal learning assistant.

3 Methods

The qualitative content analysis according to Mayring [15] is intended as a satisfaction analysis to identify user-friendliness and suggestions for optimization. For this, a group of students is asked to perform *mygoalin* and to critically evaluate it afterwards.

3.1 The Online Self-assessment *Mygoalin*

Mygoalin is based on theories of cognitive and humanistic psychology as well as established models of self-regulated learning [16]. The new theoretical cyclical 4-layered model of *mygoalin* includes two state-like layers the *cognitive* and *metacognitive strategies*, as well as two trait-like layers *motivation & resource management* as well as *personal styles & characteristics*. In addition, it demonstrates a processual character with the pre-actional, actional and post-actional phases. *Mygoalin* corresponds to a online self-assessment, which functions as a questionnaire with nine scales (*memorization, elaboration, organizational strategies, planning & control, monitoring & evaluation, time management, interest, accuracy* and *discipline*). In total, the online tool counts 81 concise items measuring nine scales. The response format corresponds to an analogue scale (continuum with two poles). After having accomplished *mygoalin*, the testee receives personal feedback, which includes improvement-strategies if the score is under average.

Quality Criteria. Different quantitative data analyses reveal that *mygoalin* meets the main quality criteria objectivity, reliability and validity [17]. *Mygoalin* is objective because it corresponds an instructor-free online-environment and the automatic individual feedback is based on norms. Moreover, reliability (Cronbach's alpha) vary between $r = .621$ and $r = .878$ for the main scales. The results of external validity-studies show as expected, that strong students give less variation in their performance than weaker students. Moreover, successful students report an increased use of certain strategies, greater interest, more discipline and stronger internal success attribution, as well as more effort and motivation than the less successful.

A stepwise regression analysis ($N = 203$) reveals that the three predictors *discipline, internal success attribution*, and *planning & control* explain 26% of the variance of the students' grades ($r = .511$) [17]. This indicates a successful analysis of external criterion validity leading to similar results as a previous study [18].

3.2 The Procedure for Qualitative Content Analysis in Theory

The central idea of the qualitative content analysis according to Mayring [15, 19] is to start from the methodological basis of a Quantitative Content Analysis in a classic way but to conceptualize the process of assigning categories to text passages as a qualitative-interpretive act, following content-analytical rules. The reduction occurs by developing a category system, which preserves the essential content and gives an image of the basic material. This is carried out with the help of a content-analytic process step by step model. Thus, the rule-based approach makes the method of qualitative content analysis systematically and intersubjective comprehensible [20], whereby different reviewers should arrive at the same result.

3.3 Description of the Sample and the Procedure in Practice

The source material for the qualitative content analysis was collected during an e-learning course *Business Psychology Test Procedures* at the Ferdinand Porsche FernFH in Wiener Neustadt in summer semester 2016. $N = 67$ students participated in this study. Of these, 42 were female and 25 males. The assignment consisted in filling out the online self-assessment *mygoalin*, reading the feedback and answering the two following questionnaires:

Q-A: What did you find appealing and why? How helpful did you find the feedback? Do you want to change your learning methods because of *mygoalin* and if so, what?

Q-B: What did you find less satisfactory and why? What did *mygoalin* not fulfill or what is missing in your opinion? Describe your suggestions for optimization.

The qualitative content analysis of the user-feedback includes the following steps:

Definition of the Analysis Technique. A combination of structuring by forming categories and summarizing the content is used by comparing and collecting main individual statements. Since the source text material (original comments) corresponds to a continuous text, this is checked for relevant text passages or contents. Particularly interesting phrases are defined as coding units (first reduction). Afterwards a binding of similar coding units with similar statements (second reduction) lead to the output of a certain amount of paraphrased main statements.

Inductive Definition of Categories. After having analyzed 20% of the source text material an inductive category formation occurs, and the two main content categories I *elements,* and II *review type* were defined. Afterwards all original comments are additionally examined for the content categories I *elements* (MY = *mygoalin* in general, IN = Instruction, DD = Demographic Data, IT = Items, RS = Response Scale, SC = Scales, FB = Feedback, IS = Improvement-Strategies) and II *review type* (POS = Positive, NEG = Constructive Criticism, OP = Suggestion for Optimization).

Quantitative Analysis of the Contents (Frequencies). For each category I and II, the number of main statements (frequencies) were counted (see - Table 1).

Procedure for Qualitative Analysis and Conclusions. Since all basic material relate the same categories I and II, the paraphrasing of main statements is done in sub-steps:

The main statements are checked for redundancies and, if necessary, one of two identical statements mentioned by the same person is deleted (e.g. *mygoalin is helpful* and *mygoalin was a support*).

All main statements that do not say more in substance than the category itself (e.g., *the response scale was good,* RS POS) are not suitable for a differentiated content analysis and are therefore excluded.

The binding of similar coding units (assigned to the same categories I and II) works as follows: e.g. the individual statement IN POS *The instruction is clear and comprehensible*, as well as IN POS *The instruction was understandable and left no questions open* becomes the main statement IN POS *The instruction is clearly understandable and comprehensible*.

The paraphrased statements were counted and are presented in the following chapter (the frequency shown in brackets).

Those paraphrased statements, which were frequently made by more than five students, are evaluated and considered in the optimization and development of *mygoalin*.

4 Results

The user-feedback of $N = 67$ students were analyzed and a total of 988 statements were recorded. Table 1 gives a summary of all counted paraphrased statements for questionnaire A and B (QA and QB). Percentages always refer to the same category (except in the Total column).

Table 1. Number of counted paraphrased statements of user-feedback on *mygoalin* using a qualitative content analysis ($N = 67$).

II Review	Positive	Negative	Optimization	Total
I *Element*				
MY	195	61	28	284
	68.66%	21.48%	9.85%	28.5%
IN	15	–	1	16
	93.75%	–	6.25%	0.02%
DD	1	6	7	14
	7.14%	42.86%	50.00%	1.4%
SC	12	11	8	31
	38.71%	35.48%	25.80%	3.1%
IT	70	43	14	127
	55.12%	33.86%	11.02%	12.7%
RS	65	13	8	86
	75.58%	15.12%	9.30%	8.6%
FB	218	87	27	332
	65.66%	26.26%	8.13%	33.3%

(continued)

Table 1. (*continued*)

II Review	Positive	Negative	Optimization	Total
IS	28	41	39	108
	25.93%	37.96%	36.11%	10.8%
Frequencies	604	262	132	998
Total	60.5%	26.3%	13.2%	100%

MY = *mygoalin* in general, IN = Instruction, DD = Demographic Data, SC = Main Scales, IT = Items, RS = Response Scale, FB = feedback, IS = Improvement-Strategies

More than half of the statements (604) are positive comments (60.5%), which suggest that *mygoalin* seems to be useful for self-reflection. 262 (26.3%) correspond to critical statements and 132 (13.2%) statements suggest optimization for *mygoalin*.

Also, *mygoalin* in general (MY) counts more positive (195, 68.66%), than negative statements (61, 21.48%). In comparison, rather few concrete suggestions for optimization are given (28, 9.85%). The critical remarks regarding *mygoalin* apply to the slight adaptability to individual framework conditions (9) and the perceived superficiality (6). This could be remedied by a modular design which relates to the demographic data, thus age and learning experience require certain items and feedback variants including also suitable reference samples.

While the instruction (IN) almost exclusively gets positive comments (15, 93.75%), the needed demographic data (DD) should be referenced to the route guidance of the data entered, because there are a few critical statements (6, 42.86%).

Also, the main scales (SC) show a small amount of positive (12, 38.71%) and negative statements (11, 35.5%) thus no optimization seems needed.

The items (IT) become rather positive (70, 55.12%) than negative (43, 33.86%) statements. A group (21) describe the items explicitly as understandable, clear, precise, accurate, comprehensible and easily answerable. However, they are also criticized as falsifiable (8). The falsifiability of questionnaire data is a fundamental dilemma [21]. However, since the questionnaire is for self-reflection and not for personnel selection, this critical point does not seem particularly relevant. Nevertheless, at the beginning of *mygoalin*, the *goal of self-reflection* and the subsequent *target agreement* should be underlined. Another criticism of some students (8) concerns repetitive and similar questions. During the construction process, great care was taken to ensure that specific questions were formulated. Moreover, content validity involving experts was conducted.

The response scale (RS) is described almost exclusively positive (65, 75.58%), as it allows free, accurate answers while being pleasant and intuitive to handle. Since this answer format does not correspond to a classical Likert scale but to an innovative approach, the user-feedback is more pleasing to this answer format than expected.

Most of the statements concern the personal feedback (FB) to the students (332, 33.3%). Of these, 218 (65.66%) the majority is positive and 87 (26.26%) are critical comments, but only 27 (8.13%) of these contain suggestions for optimization. The feedback is helpful as self-appraisal for many students (26), convincing (16), helpful and valuable (16) graphically appealing, easy to grasp (10), detailed (9), instructive (9), understandable (9), great (7), clear (7), appealing in content (6) and helpful because it shows strengths and weaknesses (6). However, for a small number of students (12), the feedback is too unspecific or insufficiently detailed and therefore the weaknesses cannot be understood (12). Therefore, in the presence of weaknesses, *mygoalin* should describe why this may be and how one can counteract this problem. *Mygoalin* is unlikely to offer interesting news to already mature students with a deep self-reflextion (9). As a concrete reminder for all students, *mygoalin* should summarize the *top 3 strategies* for the *ideal learning strategist*.

Concerning the improvement-strategies (IS) only those students who reached an under average score in a certain scale, received a list of relevant strategies. However, 28 (25.93%) students show positive and 41 (37.96%) negative statements. Some people (8) describe it as helpful and meaningful. The criticism primarily refers to the fact that these strategies and techniques are not displayed, when the participant reaches an average or even better score (14). In addition, they do not contain any concrete instructions (5) or examples (5) of how to do it better. Therefore, these learning methods should be accessible to every user. Beyond that, concrete instructions and examples of techniques should be given in *mygoalin*.

5 Conclusion

Qualitative content analysis leads to certain insights for an optimization of *mygoalin*. The new online tool should maintain a modular structure: in the user profile (*module profile*) the demographic data (age, success and learning experience) should determine question groups and the selection of reference tables. After the processing of the self-assessment *mygoalin*, there should follow a *goal agreement* (*module goal*), in which the students can set their personal targets. In addition, to improve, all students should be given access to the *module strategies*, which comprise different methods and learning techniques together with concrete instructions and examples. Moreover, all participants should get information about the top 3 strategies, which influence success. *Mygoalin* is the result of technological innovative times, corresponding to the trend of individualization. It satisfies the need for counseling and, in the sense of globalization it is a platform. This study shows the positive and critical points of this online tool. However, *mygoalin* will be optimized to be used by students as a helpful self-assessment for a self-reflection about their way of self-regulated learning. Exclusively an evaluation is missing concerning the analysis of the effects of *mygoalin* on the learning performance. Therefore, future research should focus on this issue.

References

1. Field, J.: Lifelong Learning and the New Educational Order. Trentham Books, United Kingdom (2000)
2. Mürner, B., Polexe, L.: Digitale Medien im Wandel der Bildungskultur–neues Lernen als Chance. In: Antretter, T., Dorfinger, J., Ebner, M., Kopp, M., Nagler, W., Pauschenwein, J., Raunig, M., Rechberger, M., Rehatschek, H., Schweighofer, P., Staber, R. (eds.) Videos in der (Hochschul-) Lehre, vol. 9(3), pp. S.21–S.38. Verein Forum neue Medien in der Lehre Austria (2014)

3. Óhidy, A.: Lebenslanges Lernen und die europäische Bildungspolitik: Adaptation des Lifelong Learning-Konzepts der Europäischen Union in Deutschland und Ungarn. VS Verlag für Sozialwissenschaften, Wiesbaden (2009)
4. Schober, B., Finsterwald, M., Wagner, P., Spiel, C.: Lebenslanges Lernen als Herausforderung der Wissensgesellschaft: Die Schule als Ort der Förderung von Bildungsmotivation und selbstreguliertem Lernen. Nationaler Bildungsbericht Österreich **2**, 121–139 (2009)
5. Boekaerts, M., Minnaert, A.: Self-regulation with respect to informal learning. Int. J. Educ. Res. **31**(6), 533–544 (1999)
6. Landmann, M., Perels, F., Otto, B., Schnick-Vollmer, K., Schmitz, B.: Selbstregulation und selbstreguliertes Lernen. In: Wild, E., Möller, J. (eds.) Pädagogische Psychologie, pp. 45–65. Springer, Heidelberg (2015)
7. Schiefele, U.: Selbstreguliertes Lernen im Kontext von Schule und Hochschule. Psychologie Schweizerische Zeitschrift Für Psychologie Und Ihre Anwendungen **17**(3), 1–2 (2003)
8. Schreiber, B.: Selbstreguliertes Lernen. Waxmann (1998)
9. Schunk, D.H., Zimmerman, B.: Handbook of Self-Regulation of Learning and Performance. Taylor & Francis, New York (2011)
10. Zimmerman, B. L.: Models of self-regulated learning and academic achievement. In: Schunk, D.H., Zimmerman, B.L. (eds.) Self-Regulated Learning and Academic Achievement. Theory, Research and Practice, pp. 1–25. Springer, New York (1989)
11. Wolters, C.A., Hussain, M.: Investigating grit and its relations with college students' self-regulated learning and academic achievement. Metacognition Learn. **10**(3), 293–311 (2015)
12. Zimmerman, B.J., Schunk, D.H.: Self-regulated learning and academic achievement: Theory, research, and practice. Springer Science & Business Media (2012)
13. Boekaerts, M.: Self-regulated learning: where we are today. Int. J. Educ. Res. **31**(6), 445–457 (1999)
14. Landmann, M.: Selbstregulation erfolgreich fördern: Praxisnahe Trainingsprogramme für effektives Lernen. W. Kohlhammer Verlag (2007)
15. Mayring, P.: Qualitative Inhaltsanalyse. Beltz Psychologie Verlags Union, Grundlagen und Techniken. Weinheim (2005)
16. Panadero, E.: A review of self-regulated learning: six models and four directions for research. Front. Psychol. **8**, 422 (2017)
17. Fill Giordano, R.: Zur Analyse und Förderung selbstregulierten Lernens - Entwicklung, Erprobung und Optimierung des online Self-Assessments *mygoalin*. Dissertation: Alpen Adria Universität Klagenfurt, Fakultät für Interdisziplinäre Forschung und Fortbildung (Klagenfurt, Graz, Wien), Austria (2017)
18. Kizilcec, R.F., Pérez-Sanagustín, M., Maldonado, J.J.: Self-regulated learning strategies predict learner behavior and goal attainment in Massive Open Online Courses. Comput. Educ. **104**, 18–33 (2017)
19. Mayring, P.: Qualitative content analysis: theoretical foundation, basic procedures and software solution (2014)
20. Kohlbacher, F.: The use of qualitative content analysis in case study research. In: Forum Qualitative Sozialforschung, vol. 7, No. 1 (2006)
21. Viswesvaran, C., Ones, D.S.: Meta-analyses of fakability estimates: implications for personality measurement. Educ. Psychol. Meas. **59**(2), 197–210 (1999)

Design of a Multi-agent Architecture for Implementing Educational Drama Techniques Using Robot Actors

Flor A. Bravo(✉), Alejandra M. Gonzalez, and Enrique Gonzalez

Engineering School, Pontificia Universidad Javeriana, Bogotá, Colombia
{bravof, agonzalez, egonzal}@javeriana.edu.co

Abstract. DramaBot, a platform for implementing drama techniques using robot actor that support the learning and teaching of non-technical school subjects, is presented. This platform provides non-programmer students and teachers with an easy-to-use programming environment that allows the intuitive creation of theatrical scripts for robot actors. Additionally, DramaBot enriches the script performance by adding an emotional interpretation of the defined users' actions and by also producing emergent emotional and life-like behaviors. The design of the architecture of DramaBot is based on multi-agent system (MAS) approach. The Agent-Oriented Programming based on Organizational Approach methodology (AOPOA) is used to design the multi-agent system of DramaBot architecture. After applying the AOPOA methodology, it was found that DramaBot has three main types of agents: Script Agent, Director Agent, and Actor Agent. The Actor Agent is mapped to be implemented a Belief-Desire-Intention (BDI) architecture that includes four subsystems: belief, cooperation, motivation, and action subsystems. In this paper, the agent design process and the Actor Agent architecture are presented.

Keywords: Educational robotics · Drama techniques with robots
Robotic theater · Multi-robot systems · Multi-agent systems · BDI agent

1 Introduction

Educational robotics has shown a great potential to enrich and enhance learning and teaching experiences [3, 5]. However, despite the potential of robotics in education, the scope and impact of robot-based activities have focused mostly on mathematics, physics, computer science, and foreign language subjects [3]. Few studies focus on expanding the use of robotics in other areas of knowledge, such as music, history, biology, human anatomy, and emotional intelligence [10, 12]. Dramatizations with robot actors can be a strategy to integrate educational robotics into the learning and teaching of non-technical school subjects [4]. The familiarity of teachers and students with drama techniques such storytelling, role play, and forum theatre may facilitate the use of drama with robots to support the developed the topics of non-technical subjects. However, it is essential to provide to non-programmer teachers and students of a hardware and software platform

T. Di Mascio et al. (Eds.): MIS4TEL 2018, AISC 804, pp. 172–180, 2019.
https://doi.org/10.1007/978-3-319-98872-6_21

that enable the intuitive creation and performance of theatrical plays. The educational use of dramatizations with robots is a recent research area [4]. Most of the applications on robotic theater focus on entertained field and social robots research [6]. Therefore, there are great research opportunities in the design of educational robot actors and programming environments to create and control theatrical scripts.

This paper presents the design of a multi-agent architecture called DramaBot for implementing theatrical plays with robots in educational settings. DramaBot is an improvement to the previous architectures developed and implemented by our research group SIRP-SIDRe [1, 15]. This new architecture focuses on the educational field and non-programmer users, which added new requirements in the architecture design. The paper is structured as follows: Sect. 2 shows the potential of dramatization with robot actors to enrich and enhance learning and teaching experiences. Section 3 analyzes the requirements that should meet a platform for implementing drama techniques with robots in educational settings. In Sect. 4, the design of DramBot is presented. The description of the identified agents and the general structure of the Actor Agent are introduced in Sect. 5. Finally, the conclusions and future work are given in Sect. 6.

2 Drama-Based Activities with Robots

Drama strategies such as role play, forum theater, hot seating, the mantle of the expert, and storytelling, are well-established pedagogical tools that support teaching and learning processes. Drama-based activities provide rich opportunities for students to collective knowledge building and foster skills such as teamwork, collaboration, creativity, communication, negotiation, critical thinking, decision-making, and, solve real-world problems. Drama-based activities can be used across the school curriculum to actively involve students in their own learning process actively. Drama activities also are ideal for cross-curricular learning [2].

The potential of drama in education can be increased through the integration of drama activities and educational robotics. The implementation of drama strategies using robots as actors of the play can provide students meaningful and relevant learning contexts that attract and keep students interested and motivated in the learning process [4, 11]. The use of robots in education has shown a great potential to enrich and enhance the learning experience. For example, the motion, behaviors, and appearance of robots can promote learning through analogies or metaphors. A robot can help students in understanding concepts that are abstract or unfamiliar by comparing them with behaviors and appearance of the robot [3, 5]. In the context of the educational drama, a robot actor can be a valuable tool for students to express their feelings, emotions, needs, and ideas without fear of being judged or criticized. A robot actor could also facilitate the expression of shy or introverted students, who do not enjoy performing in front of classmates [4].

The implementation of drama techniques with robot actors can be divided into four steps: The first step is planning the dramatization with robots, the topic of the dramatization with robots is selected. Then, a drama technique (e.g., role play, pantomime, forum theater, narration) to develop the selected topic is chosen. Subsequently, students develop the story of the theatrical play. It involves defining characters and their features, the main events in the story, and the places where these events take place. The constructing of the theatrical stage and customizing the robot actors is developed in the second step. The third step is to create the theatrical script that robot actors will perform. The last step is presenting the dramatization with robots to the class and performing a reflection activity. The process of theatrical script creation can be seen as a robot programming process. If a teacher or student does not have programming experience, the script creation could be a challenge. It is necessary to provide non-programmer teachers and students a tool that facilitates the creation of the theatrical script for robot actors. In the next section, the requirements of a platform that allows implementing drama with robots are analyzed.

3 Requirements of a Platform to Implement Educational Drama with Robots

In order to facilitate the use of drama with robots to support non-technical school subjects, it is essential to provide to non-programmer teachers and students of a hardware and software platform that allows the intuitive creation and performance of theatrical plays. Through a literature review and analysis of users' perceptions were identified the following requirements that a platform for implementing theatrical plays in educational settings should meet [4]: intuitive programming, multiple robot actors, believable characters, and interactive plays.

In the context of learning and teaching of non-technical school subjects, most of the teachers are non-programmers. In consequence, the platform for robotic drama should have a script-authoring environment (software) that allows non-programmer users to create and control dramatizations with robots intuitively. In most cases, the theatrical plays involve more than one character. Thus, an architecture for robotic drama should support multiple robot actors (hardware). The robot actors should be able to interpret believable characters in order to help the audience to understand what the character wants to convey and to identify its inner state. Additionally, believable robot actor behaviors can help capture the students' attention and help students to engage with the characters and their situations. According to the above, a platform for dramatization with robots should produce believable robot behaviors that are coherent with the character features and with the theatrical script description [4, 13, 14]. Moreover, this platform should generate life-like behaviors in the character because a character motionless give the feeling that it has no life. Finally, the platform should also be able to manage interactive scripts based on audience feedback. Interactive dramatizations could provide a valuable medium to reinforce learning, to adapt the play to the learning needs of students, and to monitor learners' progress and achievement [16]. Based on the identified requirements, a platform called DramaBot is proposed. The following section presents the design of DramaBot architecture using the multi-agent system approach.

4 Design of DramaBot Architecture

The design of DramaBot architecture is based on the MAS paradigm, which allows analyzing and modeling a problem as a society of rational agents that interact and cooperate to solve it. AOPOA methodology developed by SIDRe research group of the Pontificia Universidad Javeriana, Bogotá, Colombia, is used to design the MAS [9]. AOPOA provides a systematic procedure for designing a MAS based on a hierarchical decomposition of roles, which can be obtained through decomposition process of the objectives of the system. Each role is responsible for achieving a set of individual objectives by taking advantage of their skills and using the resources available in their environment.

The first step of the AOPOA methodology is to identify functional and non-functional requirements. The functional requirements are actions that the architecture must do. For example, DramaBot has the following functional requirements: (1) configure the theatrical play, (2) create theatrical scripts with one or more storylines intuitively, (3) interpret and enrich the theatrical script, (4) configure the robot actor platforms, (5) execute the theatrical script, and (6) manage scripts with multiple storylines. The non-functional requirements describe properties or qualities that a system must have to be acceptable to users. For example, DramaBot should be friendly, attractive and easy to use for non-programmer students and teachers. The platform should also be modular and flexible, allow the control of the script execution in real time, and have multi-language support.

The second step is to analyze the functionalities that are expected from the system and register them in use case diagrams. Use cases show how external entities such as people or things interact with the system. According to the use case diagram is shown Fig. 1, three types of users interact with DramaBot: non-programmer users, audience, and programmer-users.

The third step is to identify the general objectives of DramaBot based on previous analysis of functional and nonfunctional requirements and use cases. The objectives of the DramaBot architecture are also shown in Fig. 1.

The fourth step is to determine the abilities, external resources, and external entities that are required to achieve each goal. An ability refers to capacity, talent, or knowledge necessary to perform a task; for example, obtaining and interpreting environment information, determining actions, communicating with others, working cooperatively, and acting on the environment. In the case of DramaBot, the external resources that are necessary to meet the identified goals include the theatrical play configuration data, the theatrical script information, and the theatrical stage description. These resources can be obtained from external entities such as programmer users, non-programmer users, and the audience.

The fifth step is to identify the tasks and grouping these task into roles. Each task is characterized by the set of abilities and external resources required to meet the goal. Once the tasks are identified, they are grouped into a single role that represents the MAS and this role will be responsible for its implementation. Later, the role is decomposed into other roles to reduce the complexity of the tasks performed by this role (see Table 1). It is important to group similar tasks to eliminate dependencies or resources conflict.

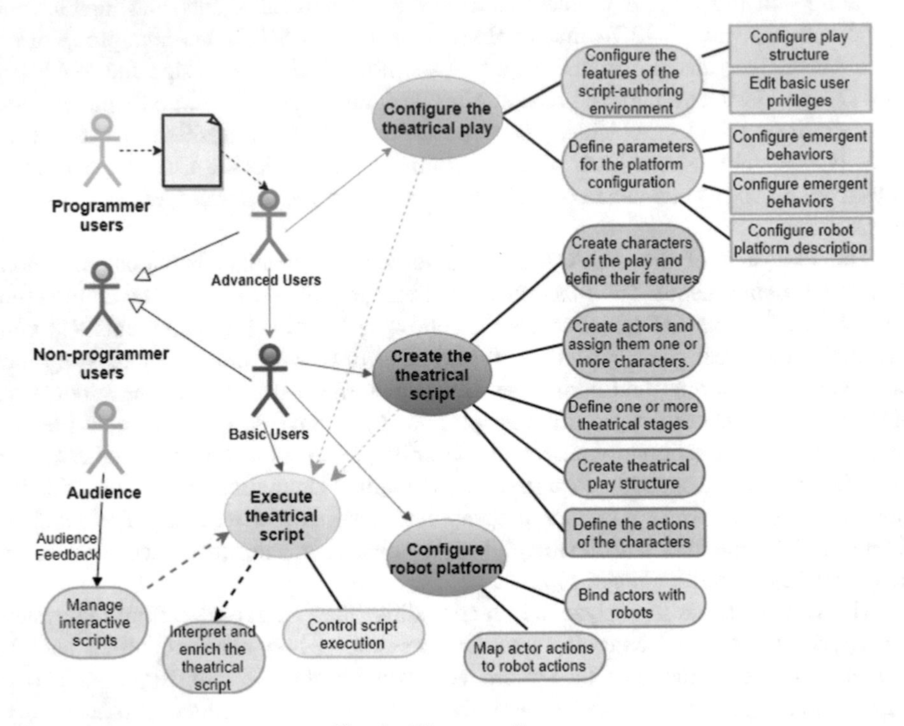

Fig. 1. Use case diagram

Table 1. List of roles of the MAS

Role	Description
SMA	Implement a robotic drama
Role 1-script information manager	Request and receive the necessary information that comes from the user to generate the script and configure the play
Role 2-script execution manager	Receive the necessary information that comes from the user to control the script performance
Role 3-play manager	Configure robot actors and coordinate the script execution
Role 4-audience manager	Get audience feedback to select the storyline of scripts with multiple storylines
Role 5-actor manager	Follow and enrich the script
Role 6-beliefs manager	Manage agent information about his environment, the other agents, and his own internal state
Role 7-cooperation manager	Manage communication with other agents
Role 8-motivation manager	Select what action the actor should perform
Role 9-action manager	Plan and execute the selected action

The role analysis allowed identifying three types of agent: Script Agent (roles 1, 2), Director Agent (roles 3, 4) and Actor Agent (roles 5–9). The description of the identified agents and the general structure of the actor agent are introduced in the following section.

5 Characterization of Agents

With the analysis performed, three agents are identified: Script Agent, Director Agent and Actor Agent (see Fig. 2). These agents are described below.

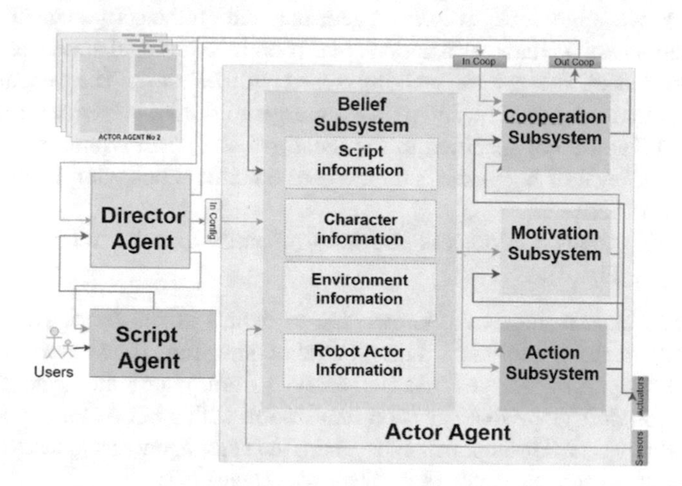

Fig. 2. DramaBot architecture

5.1 Script Agent

This agent is responsible for obtaining from the user the information related to the theatrical script, the play configuration, the robot platform features, and the actions mapping between the actors and the robot platforms. Through an intuitive script-authoring environment, non-programmer users provide information about the characters and their features, the cast of actors, and the theatrical stage. In this interface, users also create the theatrical script, where they specify which characters perform the action, what action they take and their attitudes, where the actions take place on the stage, and when actions happen. Another task of this agent is to obtain the commands to control the script execution (e.g., start, pause, stop the script performance) and the audience feedback that is used to manage the scripts with multiple storylines. The Script Agent sends the obtained information to the Director Agent.

5.2 Director Agent

Using the information provided by the Script Agent, the Director Agent informs each Actor Agent the assignment of the characters, the character's features, the theatrical script information, the theatrical stage description, the robot features and the mapping of actor actions to robot actions. The script execution is decentralized, that is, the Director Agent does not intervene in the way of executing the script. Each actor agent is responsible for synchronizing their actions with other actors. During the script execution, the Director Agent manages scripts with multiple storylines using the audience feedback and sends to Actor Agents the events related to the theatrical script execution.

5.3 Actor Agent

This agent is responsible for interpreting, enriching and executing the theatrical script. There are two ways in which an Actor Agent may enrich the script execution: the first one is to modulate the character's actions based on its emotional state; for example, if the character is sad, the actor's walking speed will be slow. The second one is to produce emergent emotional behaviors and emergent life-like behaviors (like breathing and blinking) that are not specified in the theatrical script; for example, if a character has a very high level of happiness, an emergent emotional behavior, such as a hooray sound, could be activated.

The role decomposition process allowed to identify that the Actor Agent has four subsystems (see Fig. 2):

- *Belief Subsystem* manages the knowledge or beliefs of the Actor Agent about the world, other actors, and itself. These beliefs come from the sensory readings, the theatrical script, or messages that are received from other Actor Agents.
- *Cooperation Subsystem* manages communication with other Actor Agents and with the Director Agent. Through this subsystem, the actor agent can synchronize actions and manage shared resources (e.g., theatrical stage).
- *Motivation Subsystem* deals with the deliberative process of the Actor Agent. The deliberation is based on BDI paradigm. The beliefs can activate one or more goals of the agent. The goals that the agent will carry out are selected according to an objectives prioritization and a resource conflict analysis [8]. The goals are categorized into five priority levels: At the survival level (high priority) are goals that are essential for the personal care and protection of the agent. At the obligation level are rules and norms that the agent must follow according to its context. At the opportunity level are goals that directly linked to the general objective of the system. At the facilitating level are goals that allow reaching the preconditions that should meet to activate opportunity objectives. Finally, at the life-like behaviors level (low priority) are goals that produce emergent life-like behaviors
- *Action Subsystem* is responsible for executing the selected goal by the motivation subsystem. Action Subsystem creates a step-by-step plan with actions that allow the actor to achieve the selected objective. This subsystem is also responsible for giving an emotional interpretation to the actions of the robot through the manipulation of the parameters of the action (e.g., walking speed).

6 Conclusions

This paper presented a multi-agent architecture for creation and performance of theatrical plays in educational settings. Most of the components of the proposed architecture are being implemented in BESA (Behavior-oriented Event-Driven Social-Base Agent) framework was selected for the architecture implementation [7]. Currently, we are implementing an intuitive interface that allows programming robot actors through the narration of the actions of the characters using commands in natural language and connecting words. For the design of this software, two experiments were carried out. In a first experiment, the children created a script using blocks with natural language commands that describe the actions (i.e., speak) and the inner state (i.e., happy) of the character. The greatest difficulty for children was to express concurrent actions. In a second experiment, blocks with connectors such as "at the same time" and "after that" helped the children to synchronize simultaneous and sequential actions. The next step is to validate whether the architecture meets the design requirements and the user needs through the measurement of the level of usability and the learning outcomes.

References

1. Avila, D.S., et al.: Plataforma de dramatización robótica modular (2016)
2. Baldwin, P.: With Drama in Mind: Real Learning in Imagined Worlds. Bloomsbury Publishing, London (2012)
3. Benitti, F.B.V.: Exploring the educational potential of robotics in schools: a systematic review. Comput. Educ. **58**(3), 978–988 (2012)
4. Bravo, F.A., et al.: Interactive drama with robots for teaching non-technical subjects. J. Hum. Robot Interact. **6**, 48–69 (2017)
5. Eguchi, A.: Educational robotics theories and practice: tips for how to do it right. In: Robotics: Concepts, Methodologies, Tools, and Applications, p. 193 (2013)
6. Fernandez, J.M.A., Bonarini, A.: Towards an autonomous theatrical robot (2013)
7. Garzón Ruiz, J.P., González Guerrero, E.: BESA/ME: plataforma para desarrollo de aplicaciones multiagente sobre dispositivos móviles con JME. Rev. Av. En Sist. E Informática. **6**, 3 (2009)
8. Gonzalez, A.M.: Diseño de sistemas embebidos complejos a partir de agentes BDI híbridos con migración de dominio. Pontificia Universidad Javeriana, Colombia (2014)
9. González, E., Torres, M.: AOPOA organizational approach for agent oriented programming. Presented at the 8th International Conference on Enterprise Information Systems-ICEIS (2006)
10. Hamner, E., Cross, J.: Arts & bots: techniques for distributing a STEAM robotics program through K-12 classrooms. In: ISEC 2013 - 3rd IEEE Integrated STEM Education Conference (2013)
11. Laamanen, M., et al.: Theater robotics for human technology education. In: ACM International Conference Proceeding Series, pp. 127–131 (2015)
12. Leite, I., et al.: Emotional storytelling in the classroom: individual versus group interaction between children and robots. In: HRI, pp. 75–82 (2015)
13. Leite, I., et al.: The influence of empathy in human–robot relations. Int. J. Hum. Comput. Stud. **71**(3), 250–260 (2013)

14. Mateas, M.: An Oz-centric review of interactive drama and believable agents. Artificial intelligence today (1999)
15. De La Peña, A.: RoboAct modelo de control autónomo y cooperativo para el Teatro Robótico. Pontificia Universidad Javeriana, Colombia (2014)
16. Shank, J.D.: Interactive Open Educational Resources: A Guide to Finding, Choosing, and Using What's Out There to Transform College Teaching. Wiley, Hoboken (2013)

Predictors of Performance in Programming: The Moderating Role of eXtreme Apprenticeship, Sex and Educational Background

Ugo Solitro[1(✉)], Margherita Brondino[2], Giada Vicentini[2], Daniela Raccanello[2], Roberto Burro[2], and Margherita Pasini[2]

[1] Department of Computer Science, Università degli Studi di Verona, Verona, Italy
ugo.solitro@univr.it
[2] Department of Human Sciences, Università degli Studi di Verona, Verona, Italy
{margherita.brondino,daniela.raccanello,roberto.burro, margherita.pasini}@univr.it

Abstract. Digital literacy and computer skills are considered a fundamental part of citizen education in Europe. University courses in general assume that the first year students possess adequate computational background and abilities. But unfortunately this is not always the case: freshmen experience troubles in analysing and solving problems with computation tools, in particular by means of programming activities. Therefore, it is an imperative task to find strategies that can mitigate initial difficulties and balance background deficiencies. In this work, we consider the effect of the eXtreme Apprenticeship teaching methodology and analyse the role of sex and background.

Keywords: Computational thinking · Algorithmic thinking
Problem solving · Computing education · eXtreme Apprenticeship
Programming learning · Moderation · Academic performance

1 Introduction

1.1 Informatics Education

Informatics Education should be a fundamental part of students instruction: a recent joint report by ACM and Informatics Europe [6] stressed that "citizens need to be educated in both digital literacy and informatics". Many European countries have already developed programs to introduce "digital skills" and "computational knowledge" starting from primary school or even earlier. Nevertheless, in many cases we are dealing with pilot projects, and it is under discussion which contents should be developed and also the didactic approach.

The general relevance of the scientific and educational importance of certain aspects of informatics has been emphasised in several works starting from the seminal papers about algorithmic thinking [5] and computational thinking [9,18].

T. Di Mascio et al. (Eds.): MIS4TEL 2018, AISC 804, pp. 181–189, 2019.
https://doi.org/10.1007/978-3-319-98872-6_22

Many projects focused on the development of the abilities included in the so-called computational thinking and in particular on problem analysis and solving, and on solution coding. Several implementation of these activities have been carried out in secondary school, and frequently these activities got a positive welcome and achieved successful outcomes (see, for instance, [13]).

1.2 Antecedents for a Good Performance in Programming

Performance in programming is connected with a series of antecedents which could help in the prediction of individuals' strength and difficulties in this area [12]. One of the most important antecedents is algorithmic thinking [3], that is the thought process towards formulating the steps that leads to the desired result [7,8]. Algorithmic thinking involves many basic skills, such as, for instance: reading comprehension, critical and systemic thinking, cognitive meta-components identification, planning and problem solving, creativity and intellectual curiosity, mathematical skills and conditional reasoning, procedural thinking and temporal reasoning, analytical and quantitative reasoning, as well as analogical, syllogistic and combinatorial reasoning. University curricula assume that students have a good confidence with digital software tools and hardware devices; moreover, it is not infrequent that through their academic life students have to make use of software tools that implicitly assume familiarity with the computational skills described above [15].

Exploring the conjoint effect of algorithmic thinking, teaching methodology and educational background in predicting a good performance in programming could be an interesting research topic, useful to define a more fruitful teaching and learning experience. The aim of this study is to test the relationship between some basic skills, connected with algorithmic thinking, and performance in programming in a group of university students. We also explore the role played by a specific teaching methodology - i.e. a lightened version of *eXtreme Apprenticeship* (**XA**) [16,17] - in enhancing this hypothesized relationship. At the same time, we also explore whether sex and educational background as well could have and effect in modifying the relationship between the basic skills and the performance in programming. We know that students' cultural background could impact the effectiveness of teaching programming. This cultural background, connected with the educational experience during the secondary school, could also define in some way the possibility to achieve good results in programming, even if the gap between humanities-oriented and sciences-oriented students. We conjecture that this difference can be reduced using XA methodology [14]. Finally, we explore the role of sex in the mentioned relationships, to verify whether male and female students showed the same pattern of relationship between algorithmic thinking and performance in programming.

1.3 The Case Study

In this work, we focus on the experience of first year students attending the course of *"Computer Programming with Laboratory"* which is a compulsory course in the first year for a bachelor curriculum in *Applied Mathematics* at the *University of Verona* (Italy). We applied the *eXtreme Apprenticeship* teaching methodology [17] in an adapted form, inspired to some experiences of a Bozen University team [1,2,4]. For this study the students of two different academic years has been considered. The students mainly come from *Liceo*, a kind of school that specifically prepares for university studies; also most of them choose to follow a science-oriented curriculum. Even if informatics is included as a subject in all curricula, it is treated in general as a marginal subject. Hence most of the students have just a pale idea of computational tools, both as a mental methodology and as a software device. In the first part of the course (roughly the first three months) we are required to deal about the difficulties of neophytes and the possible boredom of veterans. So we decided for an introduction to programming (algorithmic problem solving and coding) putting an emphasis on practical activities with a special attention to the programming style in conjunction with a rigorous analysis and characterisation of problems.

The practical activities are organized as follows:

- every week some exercises of increasing difficulty are proposed and partially developed in the computer laboratory with the support of some tutors;
- the support consists in suggestion and detection of errors, but in general no solution for the exercises is provided;
- a part of these exercises is submitted by students through a *Moodle* platform for evaluation and feedback;
- the laboratory staff (teacher, assistants and tutors) provide also further direct support to improve the solutions.

2 Method

2.1 Participants

Participants were 80 university students (50% female, mean age = 19.49 years, SD = 1.47). Considering their educational backgrounds, 50 students (62.5%) attended a scientific/technical high school (ST), 11 students (13.8%) attended an economic-business high school (EB), and 7 students (8.8%) attended a humanistic-artistic high school (HA). For 12 students this information was missing.

2.2 Procedure and Measures

Data were collected between October and December 2016 and 2017 during normal class. The research was conducted in the context of a larger project about the influence of some psychological characteristics (e.g. creativity, emotions and motivations) on programming performance [10,11].

All the students voluntary signed the informed consent form for participation. At the beginning of the course, the students had to take an entry test (TEST00) assessing basic skills connected with algorithmic thinking. After two months, students took a second test (TEST01) divided in two sections: the first one was a theoretical test on basic general notions about programming; the second part consisted of three exercises about problem comprehension and specification, algorithms understanding, coding and errors correction.

TEST00: Algorithmic Thinking. The entry test was used to assess basic algorithmic skills and consists in 4 exercises, proposed on-line through *Google Forms*:

1. characterization of an informally described problem;
2. comprehension of an informal algorithm;
3. completion of an informal algorithm;
4. detection of errors in an informal algorithm.

All exercises are described in natural language; also algorithms are described in a sort of natural language pseudo-code that requires no previous technical knowledge. The answers of students are provided through multiple choices or short text. Entry test scores were standardized to be in the interval from 0 to 1.

XA: Evaluation During XA. As described above, **XA** practice consisted in a set of practical activities, and at the end of each activity students were asked to deliver an exercise that summarizes the covered topics. In the considered period we assigned 3 of such exercises and the final value is the average of the assessments in the three exercises. On the basis of this score, students has been divided in three groups, using percentiles (from 1st to 33rd percentile = Low level of XA; from 33rd to 66th percentile = Medium level of XA; from 66th to 100th percentile = High level of XA). Only the two extreme groups (low level and high level of XA) has been considered for the moderation analysis. This measure gives us an idea of the actual effectiveness of the XA methodology.

TEST01: Performance in Programming Task. The test consisted of two parts: a general, theoretical section in which the knowledge of fundamental notions was verified; a practical, programming section, where students had to solve some exercises of increasing difficulty about programming competences and problem solving skills. Class performance was operationalized in terms of the score obtained in the two different parts of the partial exam: the theoretical score (**TH**), and the programming score (**PR**). Performance scores were standardized to be in the interval from 0 to 1.

2.3 Data Analysis

We calculated descriptive statistics, inter-correlations, and *t*-tests and *ANOVA* using *SPSS* version 21.0 for Windows; effect size was evaluated with Cohen's *d*. We ran moderation analyses using *AMOS*.

3 Results

3.1 The Effect of XA, Sex, and Educational Background

A first analysis concerns the effect of XA effectiveness on performance. Two one-way ANOVAs were run, one using TH as the dependent variable and one using the PR. The Independent variable was the evaluation during XA, with three levels (low, medium, high). The effect was significant both for TH ($F(2,77) = 24.6$, $p < .001$) and PR ($F(2,77) = 20.3$, $p < .001$). Post-hoc comparisons using Bonferroni's correction revealed that the low XA group performed worse than both the medium and the high XA groups in TH (low XA: M = .54, SD = .15, N = 26; medium XA: M = .76, SD = .17, N = 30; high XA: M = .83, SD = .12, N = 24). The effect size for both the significant difference was large (low vs medium: d = 1.37; low vs high: d = 2.13). Considering the performance in PR, all the three groups performed significantly differently (low XA: M = .34, SD = .20, N = 26; medium XA: M = .53, SD = .19, N = 30; high XA: M = .68, SD = .24, N = 24). All the effect sizes were large (low vs medium: d = .97; low vs high: d = 1.58; medium vs high: d = 0.711).

In a second step, we analysed the effect of sex on performance. No significant effect was found. Male and female students performed in the same way, both on TH (male students: M = .69, SD = .21; female students: M = .73, SD = .17) and on PR (M = .21, SD = .24).

The last analysis considered the potential effect of educational background on performance. No significant effect was found for TH (ST: M = .73, SD = .20; EB: M = .61, SD = .13; HA: M = .73, SD = .21). The effect on PR score was significant ($F(2,65) = 4.308$, $p = .018$), and the post-hoc analysis showed that the only two groups which performed differently on PR score were ST and EB (ST: M = .56, SD = .24; EB: M = .34, SD = .13; HA: M = .48, SD = .25). The difference was quite large: d = 1.12.

It is interesting to observe that the performance in the theoretical part was better than performance in programming ($t(79) = 9.072$, $p < .001$).

3.2 The Moderation Role of XA, Sex and Educational Background

First of all, we performed a path analysis, with TEST00 as the predictor and the two scores, TH and PR, as the dependent variables. TEST00 significantly predicts both TH and PR. It is important to note that TEST00 explains a larger amount of variance for PR than for TH (27% vs 8%). This means that this initial test is a better predictor of the programming performance than of the theoretical performance.

No one of the three moderators added to the base model had a significant effect. Nevertheless, regression coefficients changed in the different groups, as shown in Figs. 1, 2 and 3.

Looking at the XA as a moderator of the relationship between TEST00 and performance (Fig. 1), for the low XA group TEST00 was not a predictor of performance, while for the high XA group TEST00 predicted PR performance.

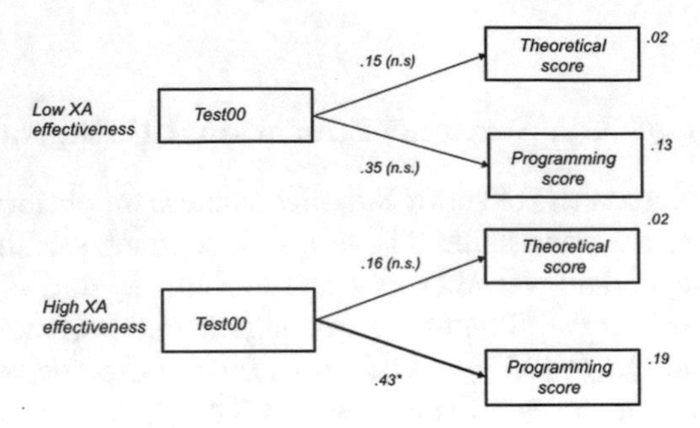

Fig. 1. XA as a moderator

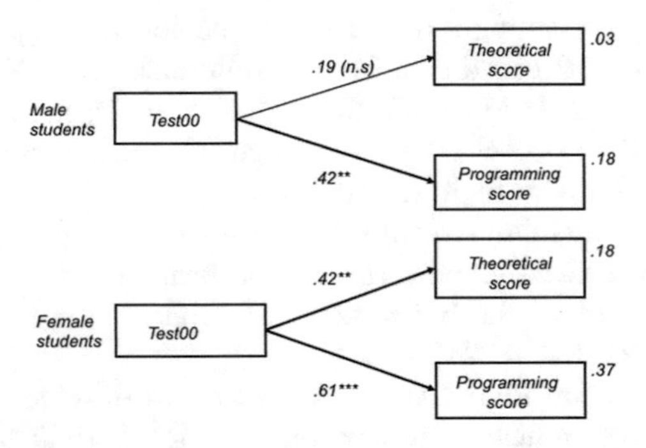

Fig. 2. Sex as a moderator

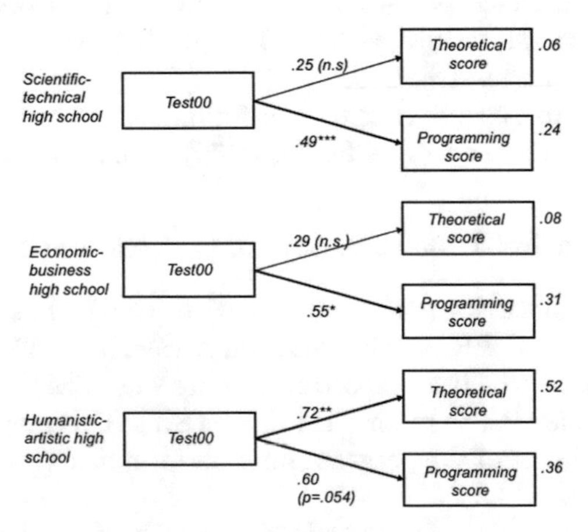

Fig. 3. Background as a moderator

Considering sex (Fig. 2), TEST00 was a stronger predictor for female students than for male ones. Actually, it was not a predictor of TH for male students.

The last moderation effect explored in this study concerns the educational background, that is which kind of high school students attended before enrolling at the university. As it is possible to see in Fig. 3, TEST00 was a stronger predictor for students with a humanistic/artistic educational background than for the other kind of schools.

4 Conclusion

These results show that a good level of effectiveness of XA, assessed with the evaluation during the XA experience, is connected with a good level of performance in programming, both for a theoretical test and for a practical test. On the contrary, nor sex neither educational background showed any effect on the performance. Data only revealed a difference in programming performance between students coming from scientific/technical high school and students coming from economic-business high school, with a better performance of the first group.

Considering the conjoint effect of algorithmic thinking, teaching methodology and educational background in predicting a good performance in programming, the first important result is that algorithmic thinking predicts performance, both in the theoretical test and in the programming part, with a larger amount of explained variance for the second test. Even if no one of the other three variables added one at a time to the model had a significant effect, it can be noted that algorithmic thinking predicts performance only for the group with high level of XA effectiveness, and only considering the programming part. Furthermore, the predicting value of algorithmic thinking was stronger for female students than for male students, both for the theoretical part and for the programming part. Finally, an interesting effect arises for educational background: scientific/technical students, as well as economic/business students showed the same pattern of relationship, with algorithmic thinking predicting programming score and not theoretical score, whereas humanistic/artistic students showed a different pattern, with algorithmic thinking predicting theoretical score more than programming score.

The limited number of participants did not allow to test all this variables at the same time, giving a real picture of the conjoint effect of all the considered potential predictors of a good performance in programming. For this reason, the research is still continuing, to gather new data. A promising research field connected with the same topic concerns a deeper understanding of the cognitive aspects of the programming abilities.

Finally, we are considering the possibility of an improvement of the XA methodology with a reinforced use of a partially automated on line support and an exploratory application for high school students.

References

1. Del Fatto, V., Dodero, G., Gennari, R.: How measuring student performances allows for measuring blended extreme apprenticeship for learning Bash programming (2015)
2. Del Fatto, V., Dodero, G., Lena, R.: Experiencing a new method in teaching Databases using Blended eXtreme Apprenticeship. Technical report, DMS (2015)
3. Denning, P.J.: The profession of IT Beyond computational thinking. Commun. ACM **52**(6), 28–30 (2009)
4. Dodero, G., Di Cerbo, F.: Extreme apprenticeship goes blended: an experience. In: Proceedings of the 12th IEEE International Conference on Advanced Learning Technologies, ICALT 2012, pp. 324–326 (2012)
5. Knuth, D.E.: Algorithmic thinking and mathematical thinking. Am. Math. Mon. **92**(3), 170–181 (1985). http://www.jstor.org/stable/2322871
6. Gander, W., Petit, A., Berry, G., Demo, B., Vahrenhold, J., McGettrick, A., Boyle, R., Mendelson, A., Stephenson, C., Ghezzi, C., et al.: Informatics education: Europe cannot afford to miss the boat. ACM (2013). http://europe.acm.org/iereport/ie.html
7. Hu, C.: Computational thinking – what it might mean and what we might do about it. In: Proceedings of the 16th Annual Joint Conference on Innovation and Technology in Computer Science Education, ITiCSE 2011 (2011)
8. Katai, Z.: The challenge of promoting algorithmic thinking of both sciences- and humanities-oriented learners. J. Comput. Assisted Learn. **31**(4), 287–299 (2015)
9. Papert, S.: An exploration in the space of mathematics educations. Int. J. Comput. Math. Learn. **1**(1), 95–123 (1996)
10. Pasini, M., Solitro, U., Brondino, M., Burro, R., Raccanello, D., Zorzi, M.: Psychology of programming: the role of creativity, empathy and systemizing. In: Advances in Intelligent Systems and Computing (2017)
11. Pasini, M., Solitro, U., Brondino, M., Raccanello, D.: The challenge of learning to program: motivation and achievement emotions in an eXtreme apprenticeship experience. In: 27th Annual Workshop of the Psychology of Programming Interest Group, PPIG 2016 (2016)
12. Plerou, A.: Algorithmic thinking and mathematical learning difficulties classification. Am. J. Appl. Psychol. **5**(5), 22 (2016). http://www.sciencepublishinggroup.com/journal/paperinfo?journalid=203&doi=10.11648/j.ajap.20160505.11
13. Solitro, U., Pasini, M., De Gradi, D., Brondino, M.: A preliminary investigation on computational abilities in secondary school. In: Lecture Notes in Computer Science (including subseries Lecture Notes in Artificial Intelligence and Lecture Notes in Bioinformatics), vol. 10696. LNCS, pp. 169–179 (2017)
14. Solitro, U., Zorzi, M., Pasini, M., Brondino, M.: A "light" application of blended extreme apprenticeship in teaching programming to students of mathematics. In: Methodologies and Intelligent Systems for Technology Enhanced Learning, 6th International Conference (MIS4TEL 2016), University of Sevilla, Sevilla (Spain), 1st–3rd June 2016. Advances in Intelligent System and Computing, vol. 478, pp. 73–80. Springer, Cham (2016)
15. Solitro, U., Zorzi, M., Pasini, M., Brondino, M.: Computational thinking: high school training and academic education. In: GOODTECHS Conference Proceedings (2016)
16. Solitro, U., Zorzi, M., Pasini, M., Brondino, M.: Early training in programming: from high school to college. In: Lecture Notes of the Institute for Computer Sciences, Social-Informatics and Telecommunications Engineering, LNICST (2017)

17. Vihavainen, A., Paksula, M., Luukkainen, M.: Extreme apprenticeship method in teaching programming for beginners. In: Proceedings of the 42nd ACM Technical Symposium on Computer Science Education, SIGCSE 2011, p. 93 (2011). http://portal.acm.org/citation.cfm?doid=1953163.1953196
18. Wing, J.M.: Computational thinking. Commun. ACM **49**(3), 33–35 (2006)

On the Design and Development of an Assessment System with Adaptive Capabilities

Angelo Bernardi[3], Carlo Innamorati[2], Cesare Padovani[2], Roberta Romanelli[1], Aristide Saggino[1], Marco Tommasi[1], and Pierpaolo Vittorini[2(✉)]

[1] University of Chieti-Pescara, Via dei Vestini, 66100 Chieti, Italy
{r.romanelli,aristide.saggino,marco.tommasi}@unich.it
[2] University of L'Aquila, P.le S. Tommasi 1, 67100 L'Aquila, Italy
{carlo.innamorati,cesare.padovani,pierpaolo.vittorini}@univaq.it
[3] Servizi Elaborazione Dati SpA, Loc. Campo di Pile, 67100 L'Aquila, Italy
a.bernardi@sedaq.it

Abstract. Individual assessment is an important tool in this society. Tests can be created according to the Classical Test Theory (CTT) or to the Item Response Theory (IRT), the latter giving the possibility to build Computerized Adaptive Testing (CAT) systems. In such a context, the paper introduces the available systems for CTT, IRT and CAT, highlights the main characteristics that are taken as initial requirements for the design of a novel system, called UTS (UnivAQ Test Suite), whose architecture and initial functionalities are presented in the paper.

Keywords: Assessment · CTT · IRT · CAT

1 Introduction

Individual assessment is an important tool in this society. It can be used for clinical diagnosis, for personnel selection, for vocational choice and in educational setting [21]. In the context of learning, assessment can essentially be divided into formative and summative assessment [18]. Formative assessment takes place at the end of a period of study and the results are used in order to determine examination outcome. Summative assessment instead takes place during a course as a means of checking on student learning, by the teacher or by the students themselves, in order to check on progresses. Classical assessment administration is based on the administration of tests, made up of several questions, in a paper-and-pencil format. On the other hand, online assessment is the method of using computers to deliver and analyse tests [15]. Within any assessment system, question types may vary. For example, questions may include short essay type questions, true or false type questions, or multiple-choice questions.

When limiting questions to have dichotomous or multiple-choice answers, they are usually called items and have been largely studied in psychometry.

T. Di Mascio et al. (Eds.): MIS4TEL 2018, AISC 804, pp. 190–199, 2019.
https://doi.org/10.1007/978-3-319-98872-6_23

In such a context, tests created according to the Classical Test Theory [12] (CTT) assign to an individual examinee an observed test score given as the unweighted sum of responses to test items, plus a casual error component. So, the examinee individual ability level is represented by this sum of correct answers. On the contrary, the Item Response Theory [4,27] (IRT) states that the examinee's probability of giving a correct answer depends on the interaction between two elements: a variable determining successful performance on a task, that is examinee's ability and item parameters such as difficulty level, discrimination, and guessing [2,14]. By taking advance of IRT, it is also possible an adaptive test administration, known as Computerized Adaptive Testing [25,34] (CAT). In brief, a CAT system starts by administering items and by estimating the initial examinee's ability. Then, further items are provided to the examinee, iteratively, until the examinee's ability is estimated within a given precision. Adaptation relies in how the CAT system selects the items needed for the estimation.

Despite both online assessment systems, IRT and CAT are not a recent concept, the progresses in psychometric research and in computer science make the conceptual design, technical development and adoption of advanced online assessment systems still a current research topic (see, e.g., [5,22]).

Therefore, the paper initially summarizes the state of the art concerning the process of assessment and the characteristics of the available systems, both under CTT, IRT and CAT (Sect. 2), then proposes an extensible architecture for performing both classical and adaptive testing (Sect. 3). The paper ends with a short discussion about the system and the future plans (Sect. 4).

2 Background

2.1 Online Assessment Systems

Online assessment systems usually focus on CTT and have their strength on the large variety of possible answers that can be used for the assessment. Among the many, in this short review we mention two well-known open-source solutions, i.e., TcExam [1] and MOODLE [11].

TcExam offers a high grade of adaptability to a variety of utilization scenarios and can be expanded to include new functionalities. Its interface conforms to different standards and has been also successfully used by blind people. TcExam supports up to four basic question types, i.e., MCSA (Multiple Choice Single Answer – the examinee can only specify one correct answer), MCMA (Multiple Choice Multiple Answers – the examinee may select all answers that apply, ORDER (the examinee must select the right order of the alternative answers) and TEXT (free-answer questions, essay questions, subjective questions, short-answer questions, i.e., answers that can consist in a word, a sentence, a paragraph or a composition). It also supports different remote authentication methods that make TcExam easy to be included in existing organizations.

MOODLE is a complete e-learning platform, more than a specific online assessment system. Nevertheless, it has a modern and responsive interface, and its assessment system is extensible due to a plugin architecture. On the other hand, it is a very large project, with many functionalities, which affect the simplicity in creating and delivering tests.

As a further effort in this context, we also mention the MWBTS software [20]: developed within the University of L'Aquila, it has been used since 2006 by circa 30 professors and 4000 students/year. It can handle open, MCSA, MCMA, file uploads and code snippets as questions' types. However, its interface is not responsive and suffers of all changes that took place over the years. However, this system actually represents the starting point of our project: starting from the lessons learned during its development and use, we decided to join our efforts and develop a fresh system, with an improved interface and architecture, with adaptive capabilities.

2.2 IRT and CAT

IRT has been thoroughly used in psychology. In contrast with classical psychological test – made up of a fixed number of items – IRT permits to build tailored tests starting from a collected set of items, known as item bank. Within this item bank, items are grouped according to their item parameter estimates to build as many tests as levels of ability. Difficulty level (b_i) is the ability level at which the probability of correctly answering the item is 0.50 (plus half the probability of correctly guessing the answer). Discrimination (a_i) is the item's capacity to discriminate between examinees with different ability levels, and the guessing (c_i) parameter is the probability of guessing a correct answer for examinees with very low ability levels. One could model the item bank choosing how many item parameters have to be used. Three IRT models exist: the one parameter model or Rasch model (1PL, [33]); the two parameters model (2PL, [37]) and the three-parameters or Birnbaum model (3PL, [4]).

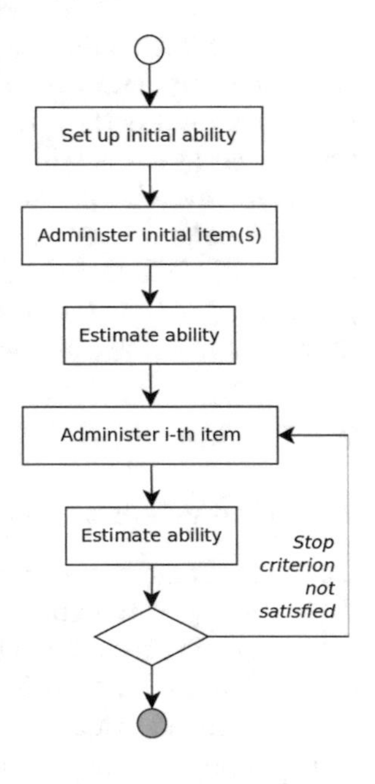

Fig. 1. Flowchart followed by a CAT system

After fitting the best IRT model, items within the item bank are organized according to the IRT item parameter estimates. In this way, it is possible to select specific items for an examinee – given the most current information about the examinee's proficiency and the items available in the pool – and therefore each examinee can have a tailored test. Such a process can be automatized within CAT systems. CAT improves measurement with respect to the classical test administration because (i) adopts the appropriate item response model, (ii) takes advance of an accurate item parameter estimate, (iii) uses a very large item bank and (iv) can use specific and efficient procedure for adaptive testing [39]. However, a drawback is that before administering items in a CAT format, it is necessary to create a large item bank. Initially, these items are administered in a conventional paper-and-pencil format and then calibrated, i.e., estimating the

a_i, b_i and c_i parameters. Once one obtained the IRT parameters' estimates, items are located within the item bank according to their item parameter estimates and how an item contributes to the examinee's ability estimation.

A classical CAT algorithm consists of some basic components: (i) an item selection criteria for the first item; (ii) the scoring process of the examinee's responses to obtain a location estimate; (iii) an item selection criteria for other items; and (iv) the stopping criteria so to end test administration. In details, see Fig. 1, a CAT system starts assuming that the examinee is of average ability. Other starting assumptions could be: (i) use ancillary information to have an initial ability estimate; (ii) randomly choose an initial ability estimate. Then, once an initial ability estimate is obtained, a first item is selected. If one has not information about individual ability level, an item with average b_i is chosen for a theta/ability level equal to 0. After the examinee's answer, the item may be scored using several estimation method: the maximum likelihood (MLE), the expected a posteriori (EAP) and the maximum a posteriori (MAP) [31,37,38]. Once a new ability estimate is obtained, a new item is administered, so that it is the most informative within the item bank for the current ability estimate, or gives the greatest weighted information, or determines the greatest reduction in the variance of the posterior distribution. These phases are repeated until a stopping criterion is reached, usually based on Standard Error Estimation.

2.3 Systems for Adaptive Testing

Several commercial and open-source software are available for adaptive administration of tests. A brief review on them is reported below.

With regard to the commercial solutions, FastTest is a configurable assessment system through which it is possible to create and edit items, to assembly test forms and to create a package for adaptive administration [43]. Prometric is an environment that presents test delivery to testing organizations [6]. McCann Associates works on the development and distribution of assessment, certification, business intelligence, and personal development solutions. In adaptive learning and assessment, their platform offers different tests for diagnostic testing; in adaptive placement testing, their system supports administrators to place students in courses suited to their skill sets [29].

By focusing instead on open-source software, the leading platform is Concerto [36]. It permits to create online assessments with textual and graphical feedback, uses different R statistical packages [41] as back-end for all calculations and provides a modern user interface to the end users. IRT-CAT [23] is a system with limited functionalities, not under active development. In the specific context of online learning, CAT has been discussed in [35] to adapt the proposed learning material, in [17] to support formative evaluations in the SIETTE project (also available as a MOODLE plugin), in [30] it is seen as a potential component in MOOCs, in [32] to develop a flexible online platform, in [44] how to help solve several common problems encountered in educational settings, in [19] for the development and implementation of an adaptive testing system.

It is also worth noting the MIRT [9], mirtCAT [8] and catR [28] packages that provide IRT and CAT functions in R. The Open Source Computerized Adaptive Testing System (OSCATS) [45] is instead a C library that implements the IRT and item selection algorithms used in CAT. It provide ready-to-use code for running the CAT item selection and ability/classification estimation.

2.4 Summary

Table 1 summarizes the major characteristics of the systems mentioned above, in terms of the following characteristics:

Question types. If the questions include only dichotomous or polytomous answers, or other types of answer are available (e.g., open, file upload, code snippets);
Flexibility. It regards the testing strategy, how items are selected, and if different termination criteria can be chosen;
Integrability. If the system can interact with other systems, e.g., for import/export of data, as a plug-in in existing system;
Development. If the system is still under active development or not;
Interface. If the system offers at least a modern or even a responsive interface.

Table 1. Characteristics of the reviewed systems, adapted and extended from [32]. Y = Yes, N = No, L = Limited. Note that only the characteristics explicitly discussed in the cited papers, or verified by the authors (e.g., in the system web site) are reported in the table.

	Question types			Flexibility			Integrability with other platforms	Under active development	Modern/ responsive interface
	Dichotomous answers	Polytomous answers	Other	Testing strategy	Item selection	Termination criterion			
Online assessment systems									
TcExam	Y	Y	L	N	N	N	Y	Y	Y
MOODLE	Y	Y	Y	L	L	L	Y	Y	Y
MWBTS	Y	Y	Y	N	N	N	N	Y	N
CAT systems									
Concerto	Y	Y	Y	Y	Y	L	N	Y	Y
IRT-CAT	Y				L	N	N	N	N
OSCATS	Y	Y	Y	Y	Y	N	Y	N	N
FastTest	Y	Y	Y	Y	Y	Y	N	N	N
Prometric	Y	Y		Y	Y	N			
McCann	Y	Y		Y	Y				
Learning systems with CAT support									
Salcedo et al	Y			N	N	N	N		N
SIETTE	Y			L	L	L	L	Y	N
Huang et al	Y			N	N	N	L		N
Oppl et al	Y	Y	L	Y	Y	Y	L	Y	Y

Such a characteristics are also the set of requirements of the system under development, called UTS (UnivAQ Test Suite) whose architecture is discussed in the next section.

3 System Architecture

The overall UTS system architecture is depicted in Fig. 2. Its main blocks are (i) the actual web application, (ii) the artificial intelligence (AI) engine, (iii) the plugin engine, and (iv) the adaptivity engine. Each of them are briefly discussed below, with respect to the aforementioned requirements:

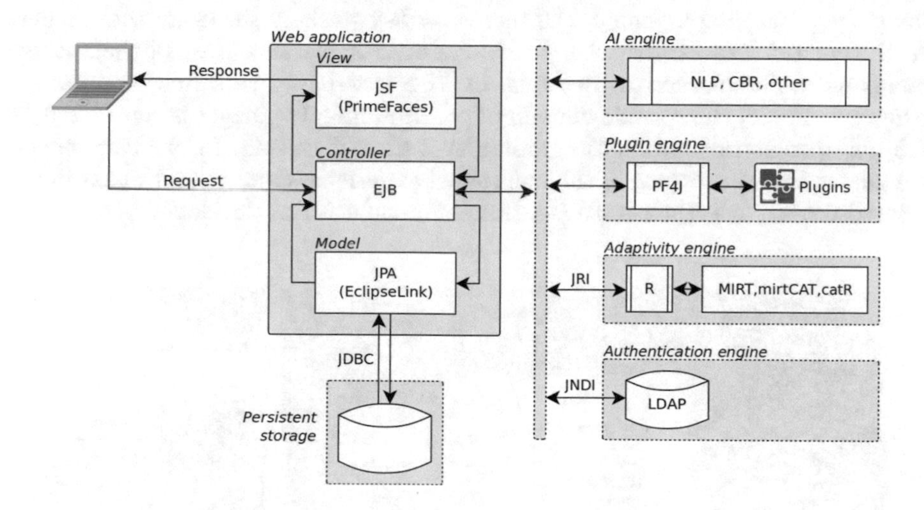

Fig. 2. UTS system architecture

Web Application. The web application follows the well-know MVC design pattern [24]. In our project:

- the views are developed by using PrimeFaces [40], one of the leading implementation of the Java Server Faces (JSF) standard, which will enable us to develop modern and responsive interfaces;
- the model relies on the EclipseLink implementation of the Java Persistence API (JPA) [16], which permits a seamless mapping between the Java classes and their persistence, independently from the underlying DBMS;
- the controller layer is implemented through Enterprise Java Beans (EJB) [3]. It interacts – besides with the view and model layers – also with LDAP directory servers for user authentication (to facilitate the integration of our system in existing companies or universities), and the AI, plugin and adaptation engines which are discussed below in detail.

AI Engine. The AI engine is included in the architecture mainly for the automated grading of essay questions. The need for such a component is in connection with open questions or code-snippets questions. For example, within a course regarding the analysis of biomedical data through R held in the University of L'Aquila, students are required to issue R commands and provide the correct interpretation of the results. The course is attended by more than 200 students and includes 17 code-snippets questions for summative assessment and 11 different exercises for formative assessment. Therefore, an automated method able to grade the answers could help the lecturer in screening the submitted answers and reduce the time for defining the final grades. It is worth noting that the task of automatic grading essay questions has been already discussed in the scientific literature. The majority of the approaches use natural language processing techniques to search for keywords in the essay with respect to the correct answer given by the evaluator. For instance, in [10] the authors compared different scoring methods, in [7] a case-based reasoning approach is proposed, in [26] the specific domain of programming languages is considered, in [13] an open-source automatic grading system is discussed. In our component, we aim at taking advance of the mentioned experiences and the advances in the related fields. Nevertheless, so far, this component is not developed yet;

Fig. 3. Screenshots regarding: login, users' management, IRT items' calibration and management

Plugin Engine. The plugin engine allows the system to be extended in terms of its functionalities, e.g., further types of questions. At the time of writing, the plugin engine has been developed by taking advantage of the PF4J framework, but no plugins are available yet;

Adaptivity Engine. The adaptivity engine takes care of supporting all CAT functionalities within the system. Because of the need for fast calculations, we decided – similarly as for the Concerto software and given the experience of the authors in similar projects [42] – to implement it as a wrapper to R and the MIRT, mirtCAT and catR packages mentioned above. For example, the adaptivity engine could help in transforming several classical paper-and-pencil tests in a CAT format, by reducing the testing time, providing an immediate scoring with low probability of scoring errors, standardized instructions and procedures, as well as an increased score accuracy and increased test security. So far, test calibration and ability estimation are implemented within the system, whereas item selection and termination criteria are still under development.

4 Discussion and Conclusions

The paper presented the initial steps concerning the development of an online assessment system with CAT functionalities, called UTS, to be primarily used in course assessment activities within the University of L'Aquila and for psychometric research in the University of Chieti.

With respect to the online assessment system currently used in our University, our proposal adopts a more modern architecture, introduces AI and CAT into the assessment process, and improves on usability. With regard to psychometric research, the project will permit to administer a battery of psychological tests in an adaptive format, with reduced time administration sessions but also with informative ability estimates. Moreover, each adaptive test that could be created from an item bank would permit us to have a total score or several subtest scores in order to provide detailed examinee profiles useful in making decisions or specific diagnosis.

The project is at its beginning. So far, the analysis of the requirements and the design of the architecture are complete, as well as the implementation of the administrative interface, the calibration and management of IRT items.

Figure 3 shows the interfaces for logging into the system, manage the users, calibrate and manage the IRT items, respectively. The interface regarding the users' management is shown within a mobile browser on purpose, so to demonstrate the interface responsiveness.

References

1. Asuni, N.: TCExam - Open Source Computer-Based Assessment Software (2017). https://tcexam.org/
2. Baker, F., Seock-Ho, K.: Item Response Theory. Dekker Media, New York (2012)
3. Bergsten, H.: JavaServer Faces: Building Web-Based User Interfaces. O'Reilly Media, Inc., Sebastopol (2004)

4. Birnbaum, A.: Some latent trait models and their use in inferring an examinee's ability. In: Statistical Theories of Mental Test Scores, pp. 395–479 (1968)
5. van Boxel, M., Eggen, T.: The Implementation of Nationwide High Stakes Computerized (adaptive) Testing in the Netherlands (2017)
6. Brannick, M.: Prometric: Trusted Test Development and Delivery Provider (2017). https://www.prometric.com/en-us/Pages/home.aspx
7. Briscoe-Smith, A., Evangelopoulos, N.: Case-based grading: a conceptual introduction. In: Proceedings of ISECON 2002, vol. 19 (2002)
8. Chalmers, P.: mirtCAT: Computerized Adaptive Testing with Multidimensional Item Response Theory, July 2017. https://cran.r-project.org/web/packages/mirtCAT/index.html
9. Chalmers, R.P., et al.: mirt: a multidimensional item response theory package for the R environment. J. Stat. Softw. **48**(6), 1–29 (2012)
10. Chang, S.H., Lin, P.C., Lin, Z.C.: Measures of partial knowledge and unexpected responses in multiple-choice tests. J. Educ. Technol. Soc. **10**(4), 95–109 (2007)
11. Cole, J., Foster, H.: Using Moodle: Teaching with the Popular Open Source Course Management System. O'Reilly Media, Inc., Sebastopol (2007)
12. DeVellis, R.F.: Classical test theory. Med. Care **44**(11), S50–S59 (2006)
13. Edwards, S.H., Perez-Quinones, M.A.: Web-CAT: automatically grading programming assignments. In: ACM SIGCSE Bulletin, vol. 40, pp. 328–328 (2008)
14. Embretson, S.E., Reise, S.P.: Item Response Theory. Psychology Press, New York (2013)
15. Gikandi, J.W., Morrow, D., Davis, N.E.: Online formative assessment in higher education: a review of the literature. Comput. Educ. **57**(4), 2333–2351 (2011)
16. Goncalves, A.: Java persistence API. In: Beginning Java EE 7, pp. 103–124. Springer (2013)
17. Guzmán, E., Conejo, R.: Self-assessment in a feasible, adaptive web-based testing system. IEEE Trans. Educ. **48**(4), 688–695 (2005)
18. Harlen, W., James, M.: Assessment and learning: differences and relationships between formative and summative assessment. Assess. Educ. Principles Policy Pract. **4**(3), 365–379 (1997)
19. Huang, Y.M., Lin, Y.T., Cheng, S.C.: An adaptive testing system for supporting versatile educational assessment. Comput. Educ. **52**(1), 53–67 (2009)
20. Innamorati, C.: Multimedia Web-Based Testing System (2018). http://test.med.univaq.it/
21. Knight, P., Yorke, M.: Assessment, Learning and Employability. McGraw-Hill Education, Maidenhead (2003)
22. Kröhne, U.: Multidimensional Adaptive Testing Environment (MATE) - Software for the Implementation of Computerized Adaptive Tests (2011)
23. Lee, S.Y., Mott, B.W., Lester, J.C.: Real-time narrative-centered tutorial planning for story-based learning. In: Intelligent Tutoring Systems, pp. 476–481. Springer (2012)
24. Leff, A., Rayfield, J.T.: Web-application development using the model/view/controller design pattern. In: 2001 Proceeding of the Fifth IEEE International Enterprise Distributed Object Computing Conference, EDOC 2001, pp. 118–127 (2001)
25. Van der Linden, W.J., Glas, C.A.: Computerized Adaptive Testing: Theory and Practice. Springer, Heidelberg (2000)
26. Lingling, M., Xiaojie, Q., Zhihong, Z., Gang, Z., Ying, X.: An assessment tool for assembly language programming. In: 2008 International Conference on Computer Science and Software Engineering, vol. 5, pp. 882–884 (2008)

27. Lord, F.M., Novick, M.R.: Statistical Theories of Mental Test Scores. IAP, Charlotte (2008)
28. Magis, D., Raîche, G.: catR: an R package for computerized adaptive testing. Appl. Psychol. Meas. **35**(7), 576–577 (2011)
29. McCann Associates: McCann Associates - Changing the Way the World Learns (2017). http://www.mccanntesting.com
30. Meyer, J.P., Zhu, S.: Fair and equitable measurement of student learning in MOOCs: an introduction to item response theory, scale linking, and score equating. Res. Pract. Assess. **8**, 26–39 (2013)
31. Mislevy, R.J.: Recent developments in the factor analysis of categorical variables. ETS Res. Rep. Ser. **1985**(1) (1985)
32. Oppl, S., Reisinger, F., Eckmaier, A., Helm, C.: A flexible online platform for computerized adaptive testing. Int. J. Educ. Technol. High. Educ. **14**(1), 2 (2017). https://doi.org/10.1186/s41239-017-0039-0
33. Rasch, G.: Probabilistic Models for Some Intelligence and Attainment Tests. Danmarks Paedagogiske Institut, Copenhagen (1960)
34. Reckase, M.D.: An interactive computer program for tailored testing based on the one-parameter logistic model. Behav. Res. Methods **6**(2), 208–212 (1974)
35. Salcedo, P., Pinninghoff, M.A., Contreras, R.: Computerized adaptive tests and item response theory on a distance education platform. In: Mira, J., Álvarez, J.R. (eds.) Artificial Intelligence and Knowledge Engineering Applications: A Bioinspired Approach, pp. 613–621. Springer, Heidelberg (2005)
36. Scalise, K., Allen, D.D.: Use of open-source software for adaptive measurement: concerto as an R-based computer adaptive development and delivery platform. Br. J. Math. Stat. Psychol. **68**(3), 478–496 (2015)
37. Swaminathan, H., Gifford, J.A.: Bayesian estimation in the two-parameter logistic model. Psychometrika **50**(3), 349–364 (1985)
38. Tsutakawa, R.K.: Estimation of two-parameter logistic item response curves. J. Educ. Stat. **9**(4), 263–276 (1984)
39. Urry, V.W.: Tailored testing: a successful application of latent trait theory. J. Educ. Meas. **14**(2), 181–196 (1977)
40. Varaksin, O.: PrimeFaces Cookbook. Packt Publishing Ltd., Birmingham (2013)
41. Venables, W.N., Smith, D.M.: An Introduction to R. Network Theory Ltd., Bristol (2009)
42. Vittorini, P., Michetti, M., di Orio, F.: A SOA statistical engine for biomedical data. Comput. Methods Programs Biomed. **92**(1), 144–153 (2008)
43. Weiss, D.J.: FastTest — Computerized Adaptive Testing, Educational Assessment (2017). http://www.assess.com/fasttest/
44. Weiss, D.J., Kingsbury, G.: Application of computerized adaptive testing to educational problems. J. Educ. Meas. **21**(4), 361–375 (1984)
45. Wietsma, T.: OSCATS: Open Source Computerized Adaptive Testing System, March 2016. https://github.com/tristanwietsma/oscats

Designing a Personalizable ASD-Oriented AAC Tool: An Action Research Experience

Tania Di Mascio(✉), Laura Tarantino, Lidia Cirelli, Sara Peretti, and Monica Mazza

Università degli Studi dell'Aquila, L'Aquila, Italy
{tania.dimascio, laura.tarantino}@univaq.it

Abstract. ICT tools are considered very promising in the treatment of people with Autism Spectrum Disorder (ASD), who experience - among others - severe difficulties in communication and interaction. A number of Augmentative and Alternative Communication (AAC) tools have been proposed to compensate for severe speech-language impairments in the expression or comprehension of spoken/written language for people with a variety of conditions that hamper their communication ability. In the case of ASD people, domain experts underline the necessity of high personalization of assistive tools, given the extreme individuality of the condition. In this paper we report on an on-going design process of an ASD-oriented AAC tool able to support operators in personalization tasks. The design is being carried out as an action research experience aimed at developing a prototype and at defining an underlying AAC model.

Keywords: Action research · Technology-Enhanced Treatment
Autism

1 Introduction

Autism Spectrum Disorders (ASD) are characterized by restricted, repetitive and stereotyped behavior and core deficits in social communication and interaction [1]. Among others, children with ASD suffer from impairments in early-developing social abilities that are considered precursor to language development and, as a consequence, they have significantly delayed language [15]. Studies report that between 25% and 50% of ASD individuals never acquire functional language, though early diagnosis and interventions may make this estimate decrease [5, 9, 15, 20].

Since '90s research has explored a variety of ICT-based approaches to ASD treatment as *cognitive rehabilitation tools*, as special *education tools* and as *assistive technologies*, to counteract the impact of autistic sensory and cognitive impairments on daily life [4]. The positive results encouraged, e.g., the exploitation of ICT for *communication* and *speech therapy* (see, e.g., [20]). A number of applications and tools have been proposed to support high-tech aided Augmentative and Alternative Communication (AAC), a term which embraces tools and strategies that an individual with speech/language impairments – not necessarily due to ASD – can use to supplement or replace speech or writing (AAC systems are defined 'unaided' or 'aided' depending on

T. Di Mascio et al. (Eds.): MIS4TEL 2018, AISC 804, pp. 200–209, 2019.
https://doi.org/10.1007/978-3-319-98872-6_24

whether they do not or do require an external tool, and, in turn, aided AAC systems are classified as 'low-tech' or 'high-tech', depending on whether they do not or do utilize electronic devices) [5, 13]. In general, an aided AAC message can include gestures, photographs, pictures, line drawings, letters and words, alone or in combination. A widely diffused AAC form which, in low tech versions, may not require complex or expensive materials is the Picture Exchange Communication System (PECS), a six phases teaching protocol to teach ASD children or any individual with a communication impairment a way to communicate within a social context [8].

The efficacy of AAC methods based on visual cues (see, e.g., [10]) along with the availability of low cost interactive devices (such as tablet and smartphone) which proved to be effective technological aids for ASD individuals [11, 12, 14], fosters the investigation on ICT-based AAC tools specifically designed for ASD treatment. It has to be observed that ASD is a heterogeneous disorder and that the severity of symptoms – including communication and language difficulties – widely varies across affected individuals. Such extreme variability, along with the goal to achieve a solution with ecological validity, suggests to aim at highly personalizable tools, in which content mirrors specific needs, knowledge, and experiences of the ASD person using it. The need of customization raises the necessity to aid not only the ASD person but also professional caregivers (e.g., medical doctors, psychologists, rehabilitation technicians) responsible for his/her treatment: continuous assessment, monitoring, and dynamic reconfiguration of the tool have to be supported.

This objective is being pursued by an on-going project carried out by TetaLab (Technology-Enhanced Treatment for Autism Laboratory), a multidisciplinary laboratory based on the cooperation among the Department of Information Engineering, Computer Science & Mathematics, the Department of Applied Clinical Sciences & Biotechnology, and the Center for Autism of the University of L'Aquila. Furthermore, TetaLab regularly cooperates with the Abruzzo Regional Reference Center for Autism. In particular, the project is aimed at developing the prototype of a personalizable ASD-oriented high-tech aided AAC tool, including a front-end for the ASD person and a back-end for the professional caregiver (generically denoted as *operator* in the remainder on the paper) responsible for the front-end customization. The project has three main milestones: (M1) a back-end with customization functionality, (M2) personalized front-end with logging capability, (M3) back-end with monitoring functionality. In this paper we focus on the first project milestone related to customization oriented operator's needs and tasks; in particular we will emphasize the methodological design approach adopted with the specific objective of pursuing scientific knowledge and generalizable results while developing the prototype.

2 The Methodological Framework

As underlined in [19], one of the methodological challenges in developing ICT tools for ASD people is to coordinate and harmonize the diverse and sometimes divergent perspectives of researchers, designers, domain experts, and other stakeholders involved in multidisciplinary projects. Incremental, practice-driven approaches are suggested to foster close collaboration between researchers and practitioners and to increase the chances that project results have a real impact on the organization [19]. *Action research* (AR) seems to provide an appropriate answer to such demands for its juxtaposition of action (practice) and research (theory), its iterative nature, and its commitment to the production of new knowledge through the seeking of solutions or improvements to "real-life" practical problem situations and interventions in ecologically valid contexts [2, 18, 19]. Differently from other software development methodologies, AR is motivated by scientific prospects and committed to the production of scholarly knowledge along with the solution of a specific problem at hand [2].

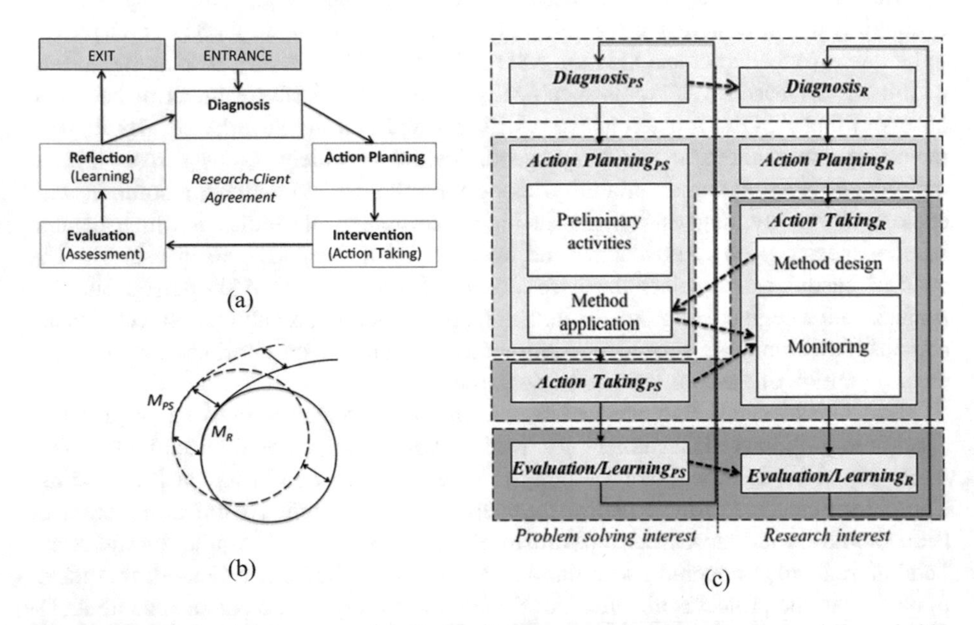

Fig. 1. Models for AR processes: (a) the Cyclical Process Model by Susman and Evered [23], (b) the tandem model [18], (c) our view for a structured tandem-based CAR process [7]

AR is performed collaboratively by researchers and an organizational "client", under the principle that social processes can be studied best by introducing changes into the processes and observing the effects of the changes, within the framework of a cyclical process repeated until a satisfactory outcome is achieved [2, 18]. According to the view in [23], after the establishment of a researcher-client agreement, five phases are iterated (Fig. 1(a)): *diagnosis* corresponds to the identification of primary problems causing the organization's desire for change and develops a theoretical framework to guide the process; *action planning* specifies actions that, guided by the theoretical framework, should relieve the organizational problem; *action taking* implements the planned action; *evaluation* determines whether the theoretical effects of the action were realized and produced the desired results; *learning* formalizes the knowledge gained throughout the process w.r.t. the problem situation and the scientific community. To enforce the AR mission of production of scholarly knowledge, [18] proposed a model including two cycles running in tandem (Fig. 1(b)): one cycle addresses the client's problem solving interest while the other one addresses the researcher's scholarly interest. Two methods are hence used: M_R is Action Research itself used to investigate on a real-world problem situation A, while M_{PS} is the method adopted for the problem solving of a real-world example of A.

Though aimed at overcoming the lack of direct guidance on "how-to-do" AR, the work in [18] leaves anyhow to researchers the burden of structuring the AR process. Based on our direct AR experiences, in [7] we proposed the structured model in Fig. 1(c) which refines the tandem approach in the tricky case in which the outcome of the research cycle is the problem-solving method of the real-world problem. We singled out a regular structure that rules time scheduling (what happens before/while what) and exchange of information to steer the relationships between the two cycles and the actors involved; following the structure of ACTION-PLANNING and ACTION-TAKING on the two sides, only once the newly designed problem-solving method (or intermediate versions of it) is available on the research side (on the right), the team can start to use it in the problem-solving side (on the left). The project discussed in this paper is being carried out following such AR scheme.

3 The Organization Situation and the RC Agreement

In our case, operators from the Abruzzo Regional Reference Center for Autism (CRAA) asked TetaLab consultancy for introducing ad hoc AAC visual tools to support two specific non-verbal low-functioning ASD individuals in the age range 18–22[1]. Initially CRAA requested simple ad-hoc low-functional high-tech tools, tailored in term of pictorial and audio content, to give these two ASD persons the ability to communicate specific needs by familiar images and voices belonging to their real-life experience. Computer scientists from TetaLab suggested that a much more effective approach was to develop a general advanced high-tech interactive tool customizable towards any possible ASD individual; a solution of this kind could also allow operators to log and analyze the usage of the tool by the ASD non verbal person, to be guided in an incremental personalization. CRRA operators agreed with such a view.

A number of requirements were elicited with CRAA people (see 1st column of Table 1) and selected commercial AAC tools conceived for non verbal people were assessed against these requirements. In particular, we considered tools for the iPad platform because both literature results [11, 12, 14] and CRAA operators' experience confirmed usability and acceptability of this platform by ASD people (ethical issues advised against investigation of uncertain solutions). From Table 1, which summarizes results of this analysis, one can see that none of the existing tools met all the requirements. More importantly, two additional considerations led to the decision of studying a novel solution. Firstly, these tools – not specifically conceived for the ASD and its levels of severity – are too difficult to use by individuals with level 3 or low-functioning autism

[1] The ASD diagnosis was provided by experienced clinicians according to the new criteria of the DSM-5 [1]. ASD diagnosis was confirmed using the Autism Diagnostic Observation Schedule, Second Edition [14]. Verbal mental age was assessed with the Test for Reception of Grammar, Version 2 [3, 22].

(as the two persons of our study), characterized – among others – by sever impairments in cognitive functioning and verbal communication and in some cases also by level 2 or medium-functioning autism, characterized – among others – by ecolocalic language and reduced or absent social reciprocity. Secondly, all tools are aimed at supporting the development of a structured language with structured sentences, while in severe ASD cases the primary objective is to support basic communication of needs and feelings independently from a formal language. It was hence decided to proceed according to an AR approach aimed at both implementing a novel high-tech ASD–oriented solution and defining an associated visual AAC model.

Table 1. The table assesses selected commercial tools against initial requirements requested by the CRAA. Tools are listed in the following order: Niki Talk (T1), Immaginario (T2), IoParlo (T3), Proloquo2Go (T4), AACTalkingTab (T5), and Blu(e) (T6). *Legenda:* NV = Non Verbal, ID = Intellectual Deficit, CD = Communication Disorder

Requirements/Tools[a]	T1	T2	T3	T4	T5	T6
Target deficit	NV & ASD	ID & ASD	CD	NV	CD	CD
PECS-based interaction	Yes	Yes	Yes	Yes	Yes	Yes
Speech synthesis	Yes	No	Yes	Yes	Yes	No
Voice recording	Yes	Yes	No	No	No	Yes
Content personalization	Yes	Yes	Yes	Yes	Yes	Yes
Image personalization	Yes	No	No	Yes	Yes	Yes
Layout personalization	No	No	No	Yes	Yes	No
Off-line mode	Yes	Yes	Yes	Yes	Yes	/
Multi-user management	No	Yes	No	Yes	No	/
Account sharing	Yes	No	No	No	No	Yes
Cloud computing	Yes	No	No	Yes	Yes	Yes
Agenda	No	Yes	No	No	No	No
Free	Yes	No	Yes	No	No	No

[a]Official sites of the reviewed tools (last accessed 2018/02/21): T1 - http://www.nikitalk.com/ T2 - http://www.fingertalks.it/immaginario/ T3 - https://itunes.apple.com/it/app/ioparlo/id406247136 T4 - http://www.assistiveware.com/product/proloquo2go T5 - http://www.aactalkingtabs.com/ T6 - http://www.tabletautismo.it/

We based our project on the approach in Fig. 1(c), where "method design" is to be intended as "visual AAC model definition", and "method application" as "model instantiation". Overall, we conducted three iterations (see Sect. 4) to achieve milestone M1. The reflection phase discussed so far is actually the entrance of the cyclical AR process and reflection on the organization situation and the evaluation in Table 1 are – de facto – part of DIAGNOSIS_{PS} and DIAGNOSIS_{R} of the 1st iteration, respectively.

Following principles and criteria of canonical Action Research, the Researcher-Client Agreement (RCA) establishes focus and objectives of the project, defines roles and responsibilities of the participants as well as measures to evaluate the results. The *working team* comprises two panels: the *CRAA panel*, i.e., the client of the AR, including one medical doctor and ASD operators, and a *panel of experts* from the TetaLab team including two psychologists, two action researchers with background in Computer Science and HCI, and a SW developer. The team agreed that the *focus of the AR project* is "needs and tasks of operators handling communication with ASD non-verbal people" and that, based on literature and operators' experience, the novel tool has to be compatible with the iPad platform. As to *client commitment, roles and responsibilities,* there is a word-of-mouth agreement related to the participation of team members into all stages of the AR process: action researchers guides the overall process while the client – according to User Centered Design – informs design choices and assesses subsequent versions of the prototype under development. As to *objectives* and *methods* of the AR intervention, Table 2 provides a summary based on the general schema proposed by [18] for tandem AR. As to AR *evaluation measures*, in order to provide a qualitative assessment of AR results, prototypes have to undergo expert-based evaluations against the initial requirements and their refinements throughout the iterative process, with particular emphasis on content personalization capabilities (milestone M1) and on monitoring capabilities (milestone M3).

Table 2. Elements of our AR, according to [18]: A is a real world problem situation, P is a real-world example of A that allows the researcher to investigate A, F is a theoretical premise declared by the researcher prior to any intervention in A, M_R is the research method, and M_{PS} is the method (M) which is employed to guide the problem solving (PS) intervention.

A	Issues/challenges in supporting operators for communication with ASD non-verbal people
P	Setting up a personalized AAC tool of specific ASD non verbal persons
F	Aided high-tech personalizable AAC tools may improve the efficiency of operators' work
M_R	Action research
M_{PS}	The new multisensorial AAC model

4 The Cyclical Process

So far, our cyclical process required three iterations related to the achievement of milestone M1, respectively focused on (1) assessment of main features of the ASD person's front-end, (2) trial and (3) assessment of operator's customization back-end.

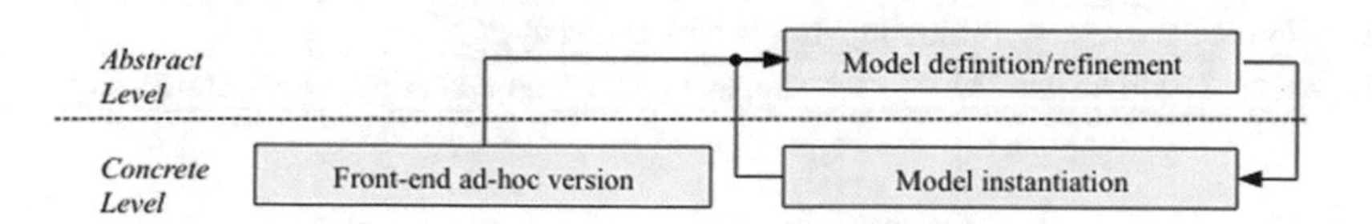

Fig. 2. The iterative alternation of design focuses

Process Overview. Before going into reporting the iterations we want to discuss results from the ACTION-PLANNING$_R$ stage of the 1st iteration, aimed at designing the overall research project and coordinating AR actors. It was decided to model the whole process as the iterative alternation of concrete and abstract levels (Fig. 2): the first step is the creation of a low-fidelity prototype of a front-end conceived for one specific ASD non verbal person, to allow an expert-based evaluation of the guiding ideas (1st iteration); results from this evaluation would be the basis for a generalization of the specific example, leading to the definition of an AAC model and the implementation of a back-end prototype based on it (to be done in ACTION-TAKING$_R$ of the 2st iteration); once the model and the back-end are available, they can be used by CRAA operators on the problem-solving side, respectively in ACTION-PLANNING$_{PS}$ and ACTION-TAKING$_{PS}$ of the 2nd iteration, to conceptualize the personalized AAC environment (model instantiation) and setting it up by the back-end; then results from the EVALUATION/LEARNING stages would guide the next iteration aimed at refining the model and the related prototype; the process is iterated until results are satisfactory.

Activities and Main Outcomes. We can now give a closer look to main results of the three iterations from the two sides of the CAR project.

First Iteration. As planned, the 1st iteration was aimed at designing, building and evaluating a simple low-fidelity front-end conceived for one specific ASD person (we randomly started from one of the two persons of the study, a 22-years old non-verbal female). The mockup was realized in PowerPoint with the following characteristics: (1) all communication elements are represented by a multisensorial triplet including an image, a text, and a recorded voice reading the text; and (2) all communication intentions are articulated as a sequence of simple actions on the screen (tapping on visual elements): a first tap always selects a category on the home page (Fig. 3(a)) and is followed by one or two taps corresponding to the actual message, depending on whether this message is an atomic one (e.g., “to sleep”) or a composite one (e.g., “to eat a biscuit”), as in Figs. 3(c) and (d). Activities in the EVALUATION/LEARNING stages assessed the mockup and refined requirements as in Table 3.

Table 3. List of requirements as produced by the 1st iteration of the AR

Multisensoriality	Image-based communication, speech synthesis, voice recording
Personalization	Content, image, layout, categories
Extras	Agenda, entertainment
Operating Modes	Operator mode, user mode, offline mode
Monitoring	Click logging, statistical analysis
Technical Requirements	Multi-user management, account sharing, cloud computing, free

Second Iteration. ACTION-TAKING$_R$ focused on (1) the generalization of the assessed concrete example leading to the AAC model sketched in Fig. 4, able to shape composite messages made up of any number of levels, and (2) the implementation of a first prototype for the backend as a responsive web application. Figure 3(f) shows a screen in which the operator can set up a category content by specifying triplets, possibly clicking on the tree icon to access an additional level associated with such node in case of composite message. As planned, the ACTION-PLANNING$_{PS}$ and ACTION-TAKING$_{PS}$ stages then focused on the instantiation of the model by CRAA operators and the setup of the personalized environment by the backend, respectively (Fig. 3(e) shows a personalized category screen). In the EVALUATION/LEARNING the prototype underwent an expert-based usability evaluation, which highlighted efficiency drawbacks in the overall procedure needed to customize from scratch a new user front-end, while no problem was revealed as to efficacy and satisfaction.

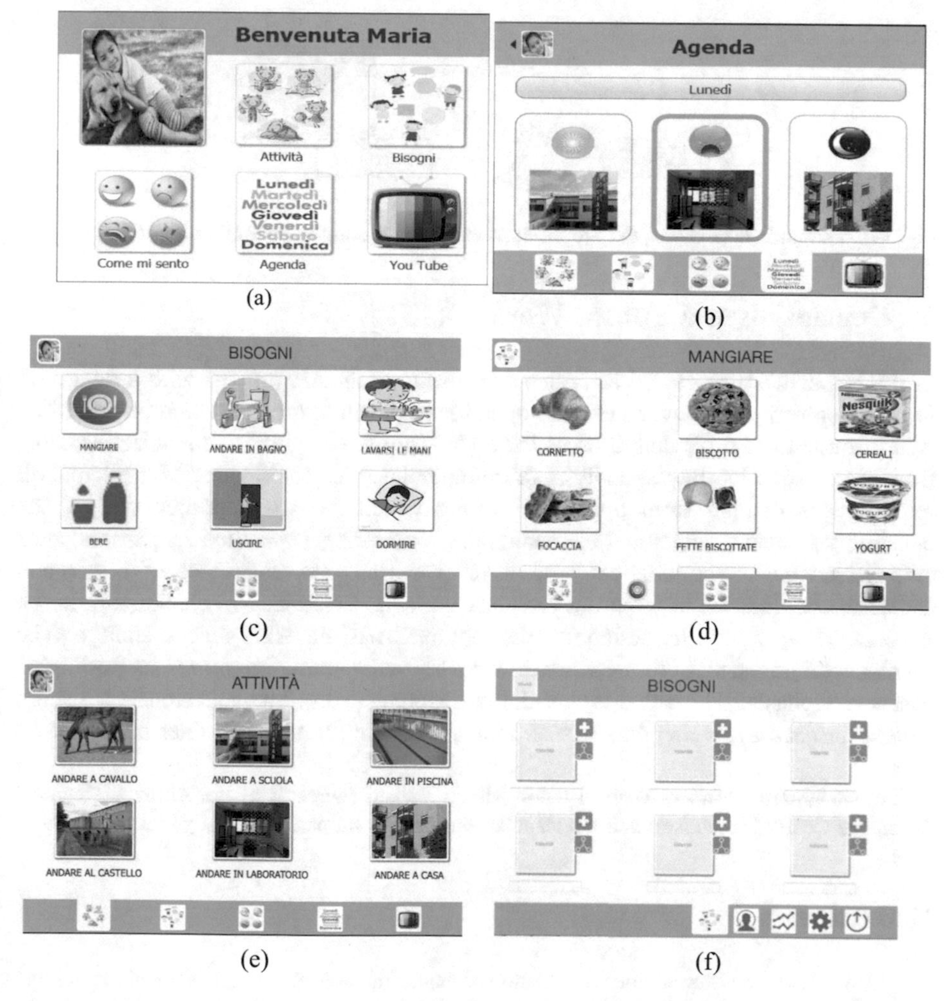

Fig. 3. From (a) to (e) Screenshots from the front-end: (a) the home page listing communication categories; (b) the agenda; (c) the BISOGNI ('Needs') category, which includes both atomic requests (e.g., DORMIRE – 'to sleep') and requests articulated by composite messages (e.g., MANGIARE BISCOTTO – 'to eat a biscuit') by accessing the second screen in (d); (e) the Activity category with personalized ecological elements; (f) a screenshot from the backend.

Third Iteration. Based on such results, the activities in ACTION-TAKING$_{R}$ focused on refining the AAC model and the associated backend prototype; at the model level we introduced the concept of "preset elements" both for Categories and Category elements, which, at backend level, led to the possibility of setting, saving, later re-using, and possibly customizing AAC front-end templates. Activities in ACTION-PLANNING$_{PS}$ and ACTION-TAKING$_{PS}$ then focused on a new instantiation of the model by CRAA operators, and the setting of example personalized AAC environments by the new release of the back-end prototype. In the EVALUATION/LEARNING the prototype underwent a second expert-based evaluation, which approved all design choices related to the functionalities so far introduced. Milestone M1 is then achieved.

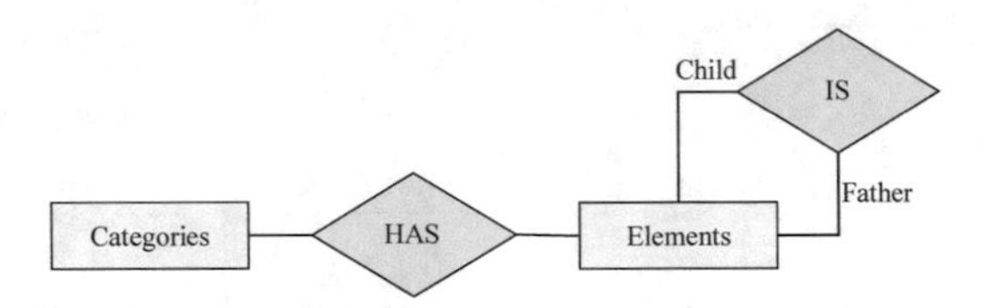

Fig. 4. A simplified sketch of the proposed model, highlighting main entities and relationships.

5 Conclusion and Future Work

In this paper we overviewed activities and results of an AR process with a dual aim: (1) development of a back-end for an operator responsible for the customization of an AAC front-end, and (2) definition of the underlying AAC model. This objective came from a lack revealed by the analysis of commercial AAC tools: none of them met all requirements coming from operators, among which an ASD oriented design, the possibility of using ecological elements, and monitoring capability. In particular we emphasized the methodological process to share our AR experience of actors and activities coordination in a multidisciplinary context, recognized as challenging in the ASD case [19]. It is worth noting that the achieved AAC model is general enough to be specialized/integrated, in future work, in PECS-based learning programs, supporting its customary phases. The achieved results correspond to the first milestone of a more general project, for which current activities are running to achieve other milestones.

Acknowledgment. Authors wish to thank Marco Valenti (Director of the Abruzzo Regional Reference Center for Autism) and his team for the fruitful cooperation throughout the project.

References

1. APA, American Psychiatric Association: Diagnostic and Statistical Manual of Mental Disorders: DSM-V. American Psychiatric Publishing, Arlington (2013)
2. Baskerville, R.L.: Investigating information systems with action research. J. Commun. Assoc. Inf. Syst. **3**(Article 4) (1999)

3. Bishop, D.: The Test for Reception of Grammar (TROG-2)—Version 2. The Psychological Corporation, London (2003)
4. Boucenna, S., Narzisi, A., Tilmont, E., Muratori, F., Pioggia, G., Cohen, D., Chetouani, M.: Interactive technologies for autistic children: a review. Cogn. Comput. **6**, 722–740 (2014)
5. Brignell, A., Song, H., Zhu, J., Suo, C., Lu, D., Morgan, A.T.: Communication intervention for autism spectrum disorders in minimally verbal children. Cochrane Database Syst. Rev. **8**(1) (2016)
6. Davison, R.M., Martinsons, M.G., Kock, N.: Principles of canonical action research. J. Inf. Syst. **14**, 65–86 (2004)
7. Di Mascio, T., Tarantino, L.: New design techniques for new users: an action research-based approach. In: Caporarello, L., Cesaroni, F., Giesecke, R., Missikoff, M. (eds.) ITAIS 2015. LNISO, vol. 18, pp. 83–96. Springer, Heidelberg (2015)
8. Frost, L.A., Bondy, A.S.: The Picture Exchange Communication System Training Manual, 2nd edn. Pyramid Educational Products, Newark (2002)
9. Gillberg, C., Colement, M.: The Biology of the Autistic Syndromes, 3rd edn. Mac Keith Press, London (2000)
10. Grandin, T.: Thinking in Pictures, Expanded Edition: My Life with Autism. Blumsburt, USA (1996)
11. Hourcade, J.P., Natasha, E.B., Hansen, T.: Multitouch tablet applications and activities to enhance the social skills of children with autism spectrum disorders. J. Pers. Ubiquit. Comput. **16**(2), 157–168 (2012)
12. Hourcade, J.P., Williams, S.R., Miller, E.A., Huebner, K.E. Liang J.L.: Evaluation of tablet apps to encourage social interaction in children with autism spectrum disorders. In: CHI Conference on Human Factors in Computing Systems, CHI 2013, pp. 3197–3206. ACM, New York (2013)
13. International Society for Augmentative and Alternative Communication. https://www.isaac-online.org/. Accessed 21 Feb 2018
14. Lord, C., Rutter, M., Di Lavore, P.C., Risi, S., Gotham, K., Bishop, S.L.: Autism Diagnostic Observation Schedule (ADOS-2): Manual, 2nd edn. Western Psychological Services, Los Angeles (2012)
15. Mazza, M., Mariano, M., Peretti, S., Masedu, F., Pino, M.C., Valenti, M.: The role of theory of mind on social information processing in children with autism spectrum disorders: a mediation analysis. J. Autism Dev. Disorder **47**(5), 1369–1379 (2017)
16. McKay, J., Marshall, P.: The dual imperatives of action research. J. Inf. Technol. People **14**(1), 46–59 (2001)
17. Kagohara, D., et al.: Using iPods and iPads in teaching programs for individuals with developmental disabilities: a systematic review. Res. Dev. Disabil. **34**, 147–156 (2013)
18. Klinger, L., Dawson, G., Renner, P.: Autistic disorder. In: Mash, E., Barkley, R. (eds.) Child Psychopathology, 2nd edn. pp. 409–454. Guilford Press, New York (2002)
19. Porayska Pomsta, K., et al.: Developing technology for autism: an interdisciplinary approach. Pers. Ubiquit. Comput. **16**(2), 117–127 (2012)
20. Ramdoss, S., Lang, R., Mulloy, A.: Use of computer-based interventions to teach communication skills to children with autism spectrum disorders: a systematic review. J. Behav. Educ. **20**(1), 55–76 (2011)
21. Rutter, M.: Diagnosis and definition. In: Rutter, M., Schopler, E. (eds.) Autism - A Reappraisal of Concepts and Treatment, pp. 1–25. Plemun Press, New York (1978)
22. Suraniti, S., Ferri, R., Neri, V.: Test For Reception of Grammar-TROG-2, 2nd edn. Curatori edizione italiana, Giunti O.S., Firenze (2009)
23. Susman, G.I., Evered, R.D.: An assessment of the scientific merits of action research. J. Adm. Sci. Q. **23**, 582–603 (1978)

Workshop on Social and Personal Computing for Web-Supported Learning Communities (SPeL)

The Future of Learning Multisensory Experiences: Visual, Audio, Smell and Taste Senses

Aleksandra Klašnja-Milićević[1(✉)], Zoran Marošan[2], Mirjana Ivanović[1], Ninoslava Savić[2], and Boban Vesin[3]

[1] University of Novi Sad, Faculty of Sciences, Department of Mathematics and Informatics, Novi Sad, Serbia
{akm,mira}@dmi.uns.ac.rs

[2] Department of Informatics, Novi Sad School of Business, Novi Sad, Serbia
{zomar,ninas}@uns.ac.rs

[3] Department of Computer and Information Science, Trondheim, Norway
vesin@ntnu.no

Abstract. This paper describes an investigative study about the sense of smell, taste, hearing and vision to assist process of learning. This will serve to start the design and construction of a multisensory human-computer interface for different educational applications. The most important part is understanding the ways in which learners' senses process learning and memorize information and in which relation these activities are.

Though sensory systems and interfaces have developed significantly over the last few decades, there are still unfulfilled challenges in understanding multisensory experiences in Human Computer Interaction. The researchers generally rely on vision and listening, some of them focus on touch, but the taste and smell senses remain uninvestigated. Understanding the ways in which human's senses influence the process, learning effects and memorizing information may be important to e-learning system and its higher functionalities. In order to analyze these questions, we carried out several usability studies with students to see if visual or audio experiences, different smells or tastes can assist and support memorizing information. Obtained results have shown improvement of learning abilities when using different smells, tastes and virtual reality facilities.

Keywords: Learning · Human-computer Interface · Virtual reality
Sense of smell · Sense of taste

1 Introduction

Virtual Reality (VR), as an immersive environment, need to be similar to the real world in order to create a lifelike and believable experience grounded in reality [1]. A person using virtual reality equipment can look around the artificial world, move around in it, and interact with virtual features or items. The effect is commonly created by VR headsets consisting of a head-mounted display that provides an intense user experience. The inherent head-mounted display can easily produce the instinctive feeling of being

T. Di Mascio et al. (Eds.): MIS4TEL 2018, AISC 804, pp. 213–221, 2019.
https://doi.org/10.1007/978-3-319-98872-6_25

in the simulated world. For a complete immersion in a virtual world, all our senses should be involved. Nevertheless, most VR environments today do not actually address all of them. The sense of vision and hearing are the most commonly used, increasingly harnessing touch, while taste and smell are insufficiently explored [1–4]. Our research aims to contribute integration the sense of smell and taste in an educational human-computer interface (HCI).

In the field of education, the immersive experience that VR provides can be used by instructors and teachers to improve learners' intention of engaging in learning activities. VR can be applied in the learning of various subjects, such as: physics, medicine, chemistry, linguistics, demographics, etc. [5].

In this paper, we analyse learning processes within virtual learning environment and multisensory learning experiences with smell and taste senses. It is essential to determine what audio, visual, taste and olfactory experiences we can design for, and how we can meaningfully stimulate such experiences when interacting with technology. Importantly, we need to determine the contribution of different taste and smell senses, as well as their combination, to design more effective and attractive digital multisensory experiences in educational environments. We carried out several usability studies with students to see if visual or audio experiences, different smells or tastes can assist and support memorizing information. A usability study was carried out to assess the experiences and the effects of VR headset in a learning process, using an application for VR learning about the Planet system, named VR Solar system. In order to evaluate the effects of different smells on learning process in normal healthy learners, we selected traditional herbal medicines for memory and for brain functions, in the form of essential oils: Citrus, Rosemary and Mint (*Citrus sinensis* L., *Rosmarinus officinalis* L. and *Mentha x piperita* L.). Likewise, there are flavouring substances that affect better memory and brain functions. Our choice was chocolate and coffee (*Socolata* L., *Coffeum L.*) to assess the current state of knowledge in the literature regarding the effects of these substances on learning process and on learners' brain functions. In our experiments, we made various combinations of senses: smell, taste, vision and hearing.

The rest of the paper is organized as follows. Section 2 describes some previous research related to our proposal. Section 3 explains the design and implementation of the research process and usability studies. Experimental results to prove the validity of the multisensory experiences in learning activities are described in Sect. 4. Finally, Sect. 5 provides the concluding remarks.

2 Related Work

In last several decades, TEL (Technology Enhanced Learning) is contemporary trend at different levels of education and in wide range of educational environments and tools. Important aspects of TEL that enrich educational activities are multifold and include: improvements in instructional design, usage of virtual and augmented reality (AR) in better HCI, different human senses like smell and taste.

Accordingly, recently different papers that take care of these TEL aspects have been published. For example, paper [6] brings comprehensive overview of 69 studies, overall effect and the impact of selected instructional design principles in using virtual reality technology instruction in K-12 or higher education environments.

In the paper [7] authors concentrated on use of augmented reality in engineering education. They used Physics Playground tool for explaining physical experiments in forms of animations. Students have challenge to interact and practice with virtual objects and learn in interesting and in entertaining way. But in their study authors did not combine AR facilities with additional kind of information like olfactory or taste.

In [8] authors assess the current state of knowledge in the literature regarding the general effects of green tea on neuropsychology i.e. cognition and brain functions in humans. Different studies considered in [8] suggested that sage spp., rosemary and lemon balm could improve cognitive performance and could be effective in improving mood or cognition for patients with different levels of cognition disorders and diseases.

Some performed experiments shown that using more than one human sense in multimodal human-computer interfaces could be an effective way to better acquire knowledge [4]. For example, olfactory information can be used as additional component to learners' auditory, visual or tactile channels.

In the paper [9] authors tried to prove that mint odor assists memorization and information recall during and after reading an educational Web page.

In her thesis Kaye, a more than a decade and a half ago, experimented with use of computer-controlled olfactory display [10]. She concluded that olfactory display must rely on differences between smells, but not differences in intensity of the same smell. A theoretical framework for scent in human-computer interactions, but also concepts of olfactory icons have been proposed in her thesis. Main conclusion of the thesis was connected to the rhythm of presentations, i.e. scent (rose, lemon and mint) is better suited for display slowly changing, continuous information than discrete events.

In our study we went a step ahead and performed experiments with students including in HCI elements all three aspects: virtual reality but also several smells and tastes.

3 Study Design

The study involved 135 students, 46 of them are attending the Faculty of Sciences at our University of Novi Sad, while 89 are enrolled at the Novi Sad School of Business.

The goal of the study was to investigate how the sense of smell, taste, hearing and vision can improve the memorization. The effect of the sense of smell was tested by exposing the students to vapors of three different essential oils: rosemary, citrus and mint. In order to examine the effect of taste students had the chance to consume coffee and/or chocolate during learning process. The influence of hearing and vision was tested using VR headset.

Regarding the smell and taste enquiry students were divided in nine different groups. Three of them were only exposed to each of the essential oils we mentioned before. The next three, in addition to the essential oil's exposure tasted a piece of chocolate. The last three groups combined the effect of each of the essential oil with the consumption of chocolate and drinking coffee.

Each of the 135 students agreed to participate in the experiment which was conducted in two steps. First, each student had to complete four consecutive short-memory tests. For each test, they had 30 s to look and try to remember the items that were presented to them. After that, the items were removed, and the students had one minute to write in a form as much items as s/he remembered. The first of the four tests presented to the students consisted of 16 images, the second of 10 words and the third of 10 numbers. For the fourth test, based on 10 words, the form in which the students were supposed to write the memorized words was different. For each of the 10 words, two letters were already written to help the students to remember a greater number of words.

The effect of smell and taste was measured by comparing the number of memorized items while students were exposed to motivating factors with the number of memorized items while students were not exposed to them. According to that, the result achieved under the influence of certain motivating factor was classified as "better", "equal" or "worse". The results are presented in percentages, showing the distribution of all students from the specific motivating factor group among those three possible outcomes.

Virtual learning environment was analyzed with 60 students. The students were similar in terms of age (ranging from 20–22) and educational background, which was confirmed with a pretest. None of the students did know the answer to any question. All participants of the experiment were divided into four groups, in a random manner. The first two groups were the control groups where the students learned in traditional manner. In the second two groups the learning process was realized with VR headset. The application was of educational type designed to teach about the planetary system, named *VR Solar System*. The educational materials were very picturesque, visually attractive, especially in a quaint style. According to the results of the experiments concerning smell and taste testing, we selected combination of citrus oil and flavor of chocolate and coffee, to verify if multisensory experiences can assist learning process.

At the classroom, an experimenter explained the purpose and procedure of the test. First, the experimenter administered a pretest to the all participants with 10 questions about the planetary system. Then, students of the experimental group were divided into two groups and separated into two classrooms. Students of the first group put VR headsets and accessed to the learning material. Students of the second group were exposed to vapor of citrus oil and consumed a piece of chocolate and a cup of coffee before putting VR headsets.

On the other hand, students of the control group were also divided into two groups and separated into two classrooms. Students of the first group watched video material with the same content as the experimental group. Each student of the second group was exposed to vapor of citrus oil and consumed a piece of chocolate and a cup of coffee before watching video material. Each learning phase lasted exactly 12 min in duration. After the task was completed, a post-test with the same questions of the pretest was administered to each participant of four groups.

4 Experimental Results

The first three diagrams depict the effect of rosemary oil, combination of rosemary oil with chocolate and at last the combination of rosemary oil with chocolate and coffee. Each diagram shows results for the picture test, textual test, numerical test and the textual test with suggestions in a form of two letters for each word. The last test is marked as “Text (S)” in all the diagrams that will follow. The last data column in the diagram shows the average value of all four types of tests.

Figure 1 shows that the exposure to the rosemary essential oil helped 51.67% students to get better results in memorizing items considering all four types of tests. The best improvement (66.67%) was noted in the case of pictures test. The consumption of chocolate as an addition to the inhalation of rosemary essential oil didn’t show any significant changes to the previous results (Fig. 2). Results related to students that were exposed to the influence of rosemary oil and consumed in the same time chocolate and coffee, showing however a low decrease (45.83%) for the “better” category (Fig. 3).

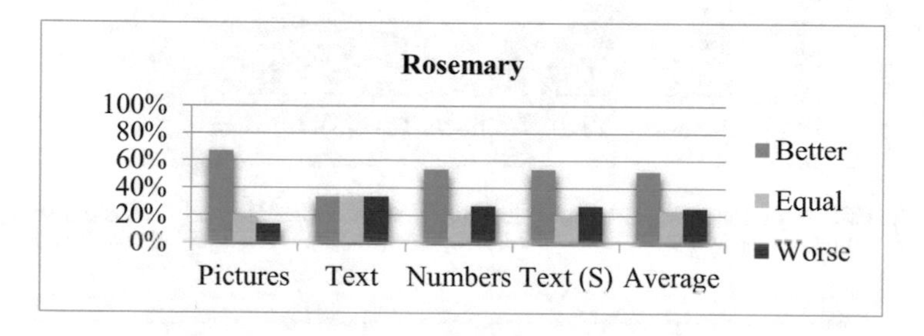

Fig. 1. Effects of rosemary oil vapor

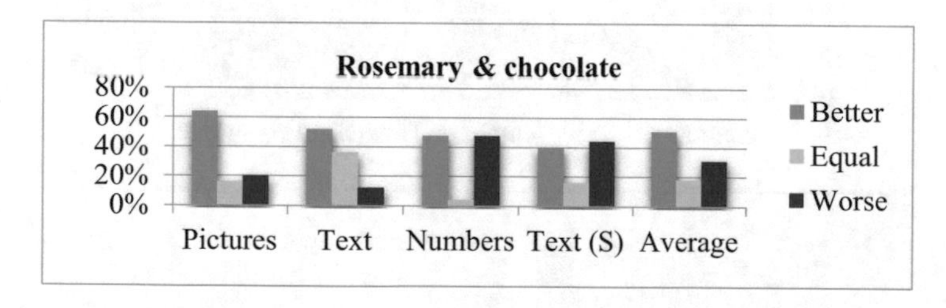

Fig. 2. Effects of rosemary oil vapor combined with chocolate consumption

The second set of diagrams show that the separate use of mint oil vapor (Fig. 4), as well as the use of mint oil vapor combined with chocolate (Fig. 5) didn’t help students to achieve significantly better results taking into consideration all four tests. The inhalation of mint oil helped 39.29% of students to reach better results, while the combination of mint oil inhalation and chocolate consumption led 47.73% of them to the same results. However, the group of students exposed to the inhalation of mint oil and the consumption of chocolate and coffee (Fig. 6) had a considerably higher percentage (73.08%) of those who succeeded to memorize more items comparing to the situation in which they were not exposed to any of previously mentioned motivating factors. It is interesting to point out that in the case of picture test, in all three groups related to the use of mint and its combination with chocolate and coffee the percentage of students with better results was very high (71.43%, 72.72% and 76.92%).

The best results in the experiment concerning smell and taste testing were achieved with citrus oil and its combinations with chocolate and coffee. When students were exposed only to the citrus essential oil vapor, 53.57% of them accomplished better results, considering all four types of tests (Fig. 7). That percentage was higher (60.94%) when citrus oil inhalation was combined with the chocolate consumption (Fig. 8), reaching the highest value (65.00%) when students were exposed to citrus oil

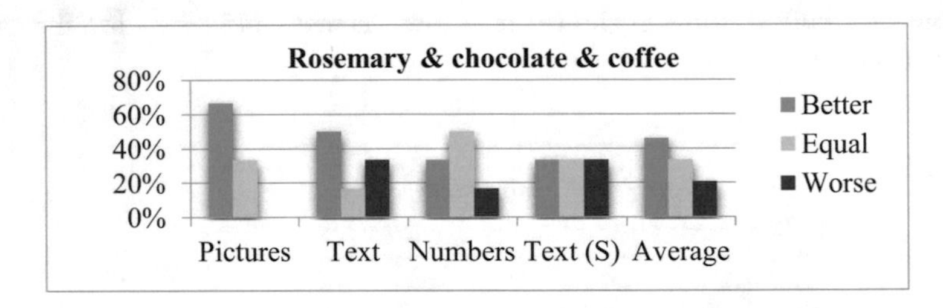

Fig. 3. Effects of rosemary oil vapor combined with chocolate and coffee consumption

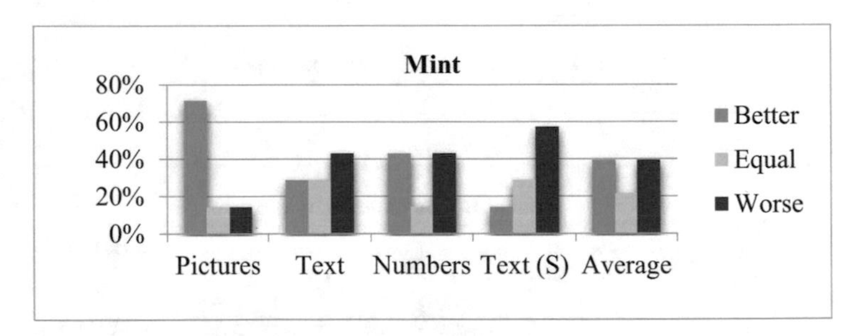

Fig. 4. Effects of mint oil vapor

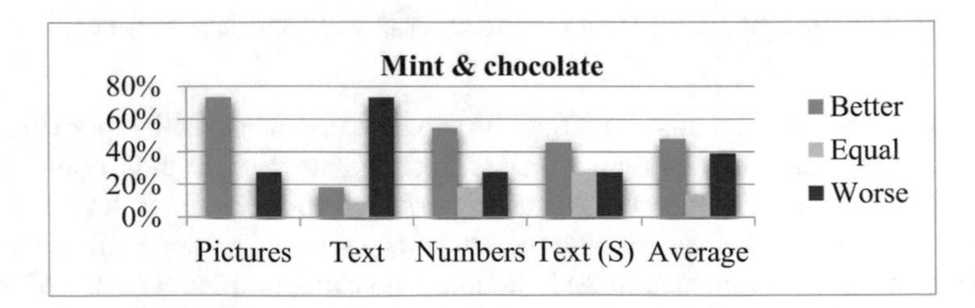

Fig. 5. Effects of mint oil vapor combined with chocolate consumption

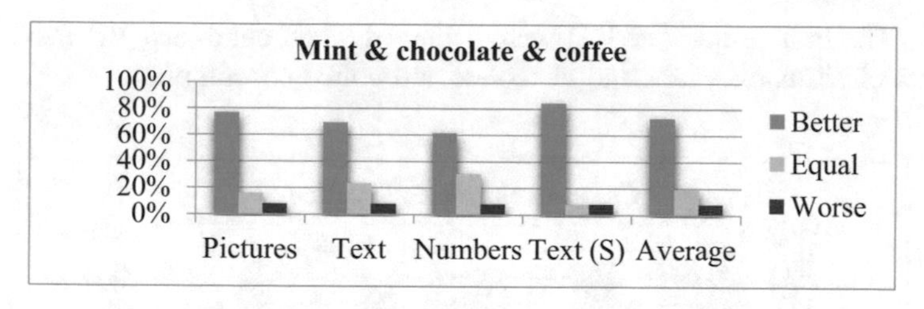

Fig. 6. Effects of mint oil vapor combined with chocolate and coffee consumption

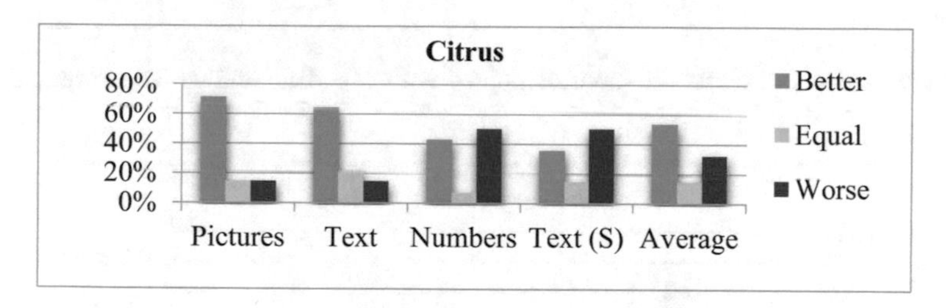

Fig. 7. Effects of citrus oil vapor

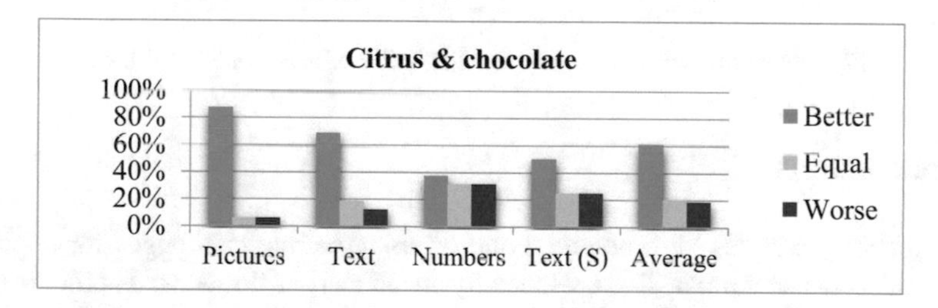

Fig. 8. Effects of citrus oil vapor combined with chocolate consumption

inhalation combined with the consumption of chocolate and coffee (Fig. 9). The significant positive effect of citrus oil and its combination with chocolate and coffee was this time registered not only in the case of picture tests (71.43%, 87.50% and 80.00%), but also regarding the textual test (64.29%, 68.75% and 66.67%).

The results of the latest experiment provide visual and auditory access, as well as the combination with the citrus essential oil vapor, as the smell of which proved to be the most successful, and the combination of flavors of chocolate and coffee. These all effects were measured by comparing the number of correct answers while students using VR headset with the number of correct answers while students learn in traditional manner. The results showed that the involvement only VR facilities improve learning abilities. Namely, when students learned with VR headset, 71.67% of them accomplished better results, then learners from the group, which learned in traditional way

(Fig. 10). The best results (79.1%) were achieved when combined VR facilities with the effects of citrus oil vapor and chocolate and coffee consumption.

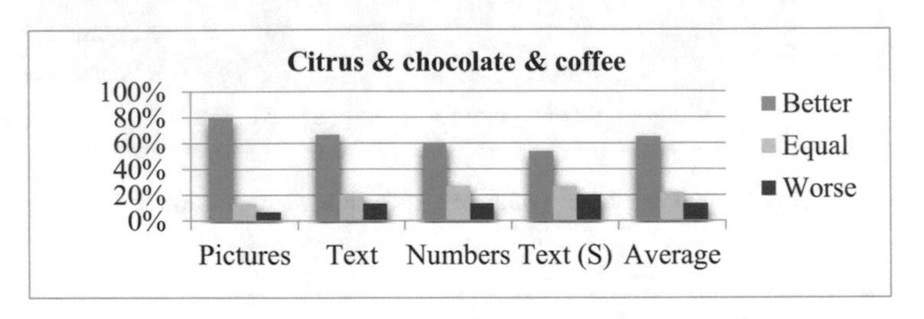

Fig. 9. Effects of citrus oil vapor combined with chocolate and coffee consumption

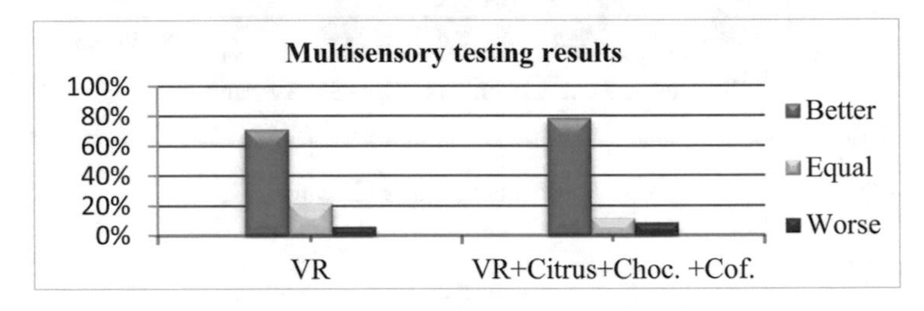

Fig. 10. Multisensory testing results: VR facilities, smell and tastes

5 Conclusions

Virtual reality possesses high potential and offers attractive challenges for application in different areas of human lives, especially in education domains. Facilitating ubiquitous learning, VR applications have become portable and widely available on mobile devices and therefore very accessible to learners at all levels of education. Digital assets, such as textual, audio and video files, but also olfactory or tactile information can be integrated into learners' perceptions of the real world.

We have carried out several experiments to analyze different combinations of fragrances and flavors, as well as the possibility of combining them with visual and verbal effects. We have chosen an educational application, in order to emphasize the possibility of influencing multi-sensory experiences on e-learning process. Students feel comfortable about such approach and have gained a sense that tools are nice and useful for learning contents. Our experimental results show that visual and audio experiences, different smells and tastes, as well as their combination have the potential to support memorizing information, improve learning abilities and encourage design of more effective and attractive digital multisensory experiences in educational environments.

Future work aims at repeating the test by carrying out additional studies, to see if learning process changes over different periods of time. Likewise, it would be useful for studying usability and technical issues on the integration of olfactory information into multimodal interfaces.

References

1. Mansor, N., Jamaluddin, H., Shukor, Z.: Concept and application of virtual reality haptic technology: a review. J. Theor. Appl. Inf. Technol. **95**(14), (2017)
2. Obrist, M., Velasco, C., Thanh, C., Cheok, D., Spence, C., Ponnampalam, G.: Sensing the future of HCI: touch, taste, and smell user interfaces. Interactions **23**(5), 40–49 (2016)
3. Gallace, A., Spence, C.: In Touch with the Future: The Sense of Touch from Cognitive Neuroscience to Virtual Reality. OUP Oxford, Oxford (2014)
4. Thalmann, D.: Sensors and actuators for HCI and VR: a few case studies. In: Frontiers in Electronic Technologies, pp. 65–83. Springer (2017)
5. Shen, C.W., Ho, J.T., Kuo, T.C., Luong, T.H.: Behavioral intention of using virtual reality in learning. In: International Conference on World Wide Web Companion, pp. 129–137 (2017)
6. Zahira, M., Ernest, G., Cifuentes, L., Wendy, K., Trina, D.: Effectiveness of virtual reality-based instruction on students' learning outcomes in K-12 and higher education: a meta-analysis. Comput. Educ. **70**, 29–40 (2014)
7. Martín-Gutiérrez, J., Fabiani, P., Benesova, W., Meneses, D., Mora, W.: Augmented reality to promote collaborative and autonomous learning in higher education. Comput. Hum. Behav. **51**, 752–761 (2015)
8. Mancini, E., Beglinger, C., Drewe, J., Borgwardt, S.: Green tea effects on cognition, mood and human brain function: a systematic review. Phytomedicine **34**, 26–37 (2017)
9. Garcia-Ruiz, M., El-Seoud, S.A., Edwards, A., Aquino-Santos, R.: Integrating the sense of smell in an educational human computer interface. In: Interactive Computer Aided Learning Conference (2008)
10. Kaye, N.: Olfactory Display. MSc thesis, Massachusetts Institute of Technology (2001)

Developing Learning Scenarios for Educational Web Radio: A Learning Design Approach

Evangelia Triantafyllou[1(✉)], Effrosyni Liokou[1], and Anastasia Economou[2]

[1] Department of Architecture Design and Media Technology, Aalborg University Copenhagen, Copenhagen, Denmark
evt@create.aau.dk

[2] Department of Educational Technology, Cyprus Pedagogical Institute, Nicosia, Cyprus

Abstract. Radio has been used in education since the beginning of last century. With the rise of the internet, web radio came into life and provided new possibilities for web radio content (e.g. video apart from voice and music), asynchronous broadcasting, and cooperation between students from different schools. In this paper, we present a survey study that aimed to evaluate a Visual Learning Design (VLD) approach for developing educational scenarios in web radio. The study results indicated that the VLD approach helped teachers to think about the educational aspects of the web radio production and helped in communicating their ideas. Moreover, the teachers highlighted the possibilities and the impact such web radio implementations could have for student learning and collaboration. The only aspect that received some criticism was the provision of technology, and this was partially due to the lack of previous experience in web radio production. These results will guide the application of the VLD approach in a bigger scale.

Keywords: Web radio · Visual Learning Design · Educational scenarios

1 Introduction

The use of radio as an educational tool has already begun at the beginning of the previous century. In the United States, educational radio was used in both public schools and higher education institutions since the 1920s [1]. With the rise of the digital era, radio is no longer linear broadcasting, but also associated with metadata, synchronized slideshows and even short video clips [2]. Web radio has been used for enhancing and motivating learning in different subjects and at different educational levels. Levine and Franzel proposed a framework for using radio in the teaching of writing [3]. They argued that employing writing for radio in English high school classes contributes to student understanding of the relevance of school writing assignments. Moreover, they found that new technologies, such as smartphones and social media, make the introduction of radio at schools easier, since no school radio station, dedicated recording equipment, or specialized teacher skills are needed to use radio.

T. Di Mascio et al. (Eds.): MIS4TEL 2018, AISC 804, pp. 222–229, 2019.
https://doi.org/10.1007/978-3-319-98872-6_26

Apart from enhancing learning, web radio provides new possibilities for cooperation between students in the same class and between schools [4]. McGroathy identified benefits of planning and implementing cooperative learning activities in acquiring English as a second language by employing a radio show [5]. Similarly, Lemos Tello found that there is a positive correlation between the participation of students in an online radio show with the aim to foster speaking confidence, and the use of a cooperative learning strategy [6]. Piñero-Otero and Ramos investigated the potential of web radio for the sense of belonging creation and cohesion in higher education communities as perceived by students and professors at Aveiro University, Portugal [7]. They concluded that both students and professors believe that web radio can foster development of a sense of belonging, unity, and communication in the university community (a new channel of communication internal or external) by allowing participation in and dissemination of content production. Finally, Güney et al. investigated how the interactive services of an online children radio network (Radijojo) affect participation strategies and may nurture participation [8]. For this study, they adopted two analysis frameworks and collected qualitative data by analyzing web radio content analysis and conducting semi-structured interviews. Their results revealed that despite the lack of detailed interactive and participative tools, the model based on the development of collective content provided a potential for the inclusion of children in the communication.

In this paper, we present research carried out in the NEStOR (Networked European School Web Radio) project[1], which aims at developing the necessary tools and skills in order to successfully incorporate web-radio activities into educational settings. The project provides an online platform that will attract schools from all around Europe to produce radio productions with educational value. The process of implementing such productions may enable various literacies (such as media and information), and various skills (such as critical thinking, collaboration, creativity, etc.). The project has also developed a framework for supporting the educational aspects of the web radio (e.g. guides, learning scenarios and good practices), and uses the experience of the European School Radio (ESR) that operates successfully with more than 400 schools in Greece and Cyprus in the last 5 years. Our observations on activities and interactions on the web radio portal during these years confirm the literature presented above that claims that web radio may provide context for collaborative and social interaction and learning.

2 Learning Design

Web radio provides more opportunities regarding content transmission and collaboration for content production compared to traditional radio. However, educational activities taking place in web radio should be supported and justified by a pedagogical foundation, in order for them to have teaching and learning potential. In this project, we have chosen to follow a Learning Design (LD) approach for developing learning scenarios for web radio. LD is an approach to describe a learning and teaching process, which can be applied at different levels of educational practice, e.g. lesson, course or

[1] http://europeanschoolradio.eu/.

even curriculum level [9]. Moreover, it may be motivated by different pedagogical theories and employ different (technological) tools and resources. Literature on this subject and this paper refers to LD both as a product and as a process [10].

In our project, we adopted a Visualized Learning Design (VLD) methodology, which facilitates the design process and the sharing of LDs [11]. The VLD procedure incorporates three levels of design: the macro level, the meso level, and the micro level. The macro level (Course Map View) is the level where teachers/designers discuss their initial ideas and get into a general discussion of their LD. The meso level (Learning Outcomes View) is the second stage of the VLD methodology where teachers/designers group and refer to their LD activities and explicitly set the learning outcomes and expected outputs. Lastly, the third stage of the VLD methodology, the micro level, is the more detailed level, which includes specific tools, resources, methodologies and roles for each activity [11]. In this project, we followed a simplified version of the VLD approach, where the macro and meso level are combined to one level. We expect that the VLD approach will help the project partners to guide and support the participating school teachers in the designing of learning scenarios that will lead to successful web radio broadcasts, and at the same time enhance communication between teachers and stimulate innovative pedagogical solutions while designing.

In the first year of the project, we applied the VLD approach for developing learning scenarios at the three pilot schools participating in the project[2]. The schools had to use three different templates to describe their learning scenario for a web radio broadcast, namely a macro level design template, a micro level design template and a learning scenario template. The participating schools consist of one lower secondary school in Cyprus, one upper secondary school in Greece, and one upper secondary school in Lithuania. The school in Cyprus designed a learning scenario on literature and adolescence, which was implemented in four teaching hours, the school in Greece a learning scenario on climate change, which was implemented in six teaching hours, and the school in Lithuania designed a learning scenario on Lithuanian tales of magic, which was implemented in two teaching hours. All the learning scenarios developed by the three partner schools were implemented as web radio productions during the period of May – June 2017. All the web radio productions were broadcasted on the ESR portal, since the NEStOR portal was under development during that period.

In this article, we present a survey study that aimed to evaluate the VLD approach for developing educational scenarios in web radio, and addressed the following research questions:

- What are the teachers' perceptions on the LD development?
- What are the teachers' perceptions on the class implementation of the LD with focus on the process?
- What are the teachers' perceptions on the final web radio product?

[2] https://www.e-epimorfosi.ac.cy/nestor/o4a1.pdf.

3 Methods

As mentioned in the previous paragraph, teachers from the three school had to develop and implement a learning scenario for developing a web radio broadcast. All three schools had to follow the following procedure based on the VLD approach: (1) Investigate through their curriculum and their students' interests an area that could serve as a learning scenario for the project, (2) Share the macro level design of their scenario and discuss it with the project team, (3) Develop a detailed learning scenario, (4) Develop each activity of their learning scenario (optional), (5) Implement the learning scenario, (6) Evaluate the learning scenario and the web radio broadcast using the provided tools.

For the evaluation part, and in order to answer the aforementioned research questions, we designed an online survey study among the teachers of the three partner schools, who participated in the pilots of the first year of the project. The study employed an online questionnaire consisting of 5-point Likert scale with 25 items and 3 open-ended questions. The items of the Likert scale are covering three themes: the LD development, the class implementation, and the web radio output. The items on the LD development refer to the LD elements, process and support. The items on the class implementation refer to the LD aspects of implementation, its impact on students, as well as to the critical aspects of the implementation. The items on the web radio output refer to the content of the output, the audio quality as well as the students' learning outcomes (knowledge and skills). The questionnaire was distributed to the teachers via Survey Monkey and all eight participants from the three partner schools responded.

4 Results

The first part of the questionnaire consisted of twelve questions on the LD development (Fig. 1). Teachers' responses overall showed a positive stance towards the project and the VLD approach. Teachers' thought that the aims of the LD were related to the national curriculum, and were also related to web radio production. Moreover, teachers believed that the LD supported the development of students' transversal, and media and information literacy skills. It was overall indicated that the supporting material on LD and web radio production was satisfactory and seemed to be adjusted to teachers' and students' needs, and that the LDs included ways to assess the process of web radio production and the final outcome. Finally, almost all teachers (seven agreed, one neutral) agreed that they were provided with pedagogical support to develop LD.

In the open-ended questions, one teacher indicated also that: *"The most important was to develop the theme of the scenario, means to make it educational and close to the national curriculum. Therefore, it had to be interesting and attractive both for national and international audience."* This implies the teachers devoted time and effort to reach outcomes that would be both educational for the web radio producers and entertaining for the audience.

The second part of the questionnaire consisted of nine questions on the class implementation of the LD (Fig. 2). Most of the participating teachers indicated that the implementation run smoothly, and that the students reacted positively on the

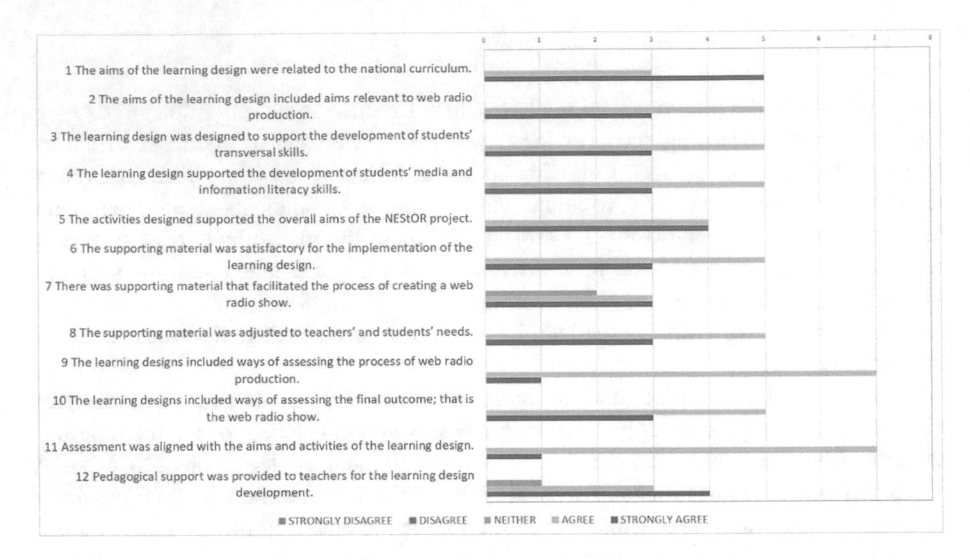

Fig. 1. Teachers' perceptions on learning design

implementation and seemed to have enjoyed this process. Specifically, teachers identified students' reactions as the most important aspect of the implementation for them. In the open-ended questions, one of the teachers suggested that through this implementation "*…students had the opportunity to express their thoughts and feelings.*" while another teacher discussed "*…the joy of making and the heartily participation of the students*". Another important statement made by a participating teacher was the following: *"The students were willing and eager to participate in the writing of dialogues and the recording of the radio show. After the radio show students agreed that the use of the web radio in the teaching process was innovative."* Moreover, all teachers agreed that students' transversal skills, and media and information literacy skills seem to have developed. Six teachers indicated that students did not seem to have faced major difficulties during the development of the web radio scenario, while the same amount of teachers stated that their students' digital skills were enough for the support of the web radio production. For the statement asking whether there was technology available to facilitate the production of the web radio show, there was six who agreed and one who strongly disagreed. Additionally, five teachers agreed that there was technological support for the production of the web radio show.

The last part of the questionnaire consisted of four questions on the final product of the LD, which was the web radio show (Fig. 3). All teachers agreed that the content of the web radio show was satisfactory in regards to the chosen curriculum unit, and that its audio quality was also satisfactory. Finally, teachers agreed that the web radio show demonstrated students' skills and reached the aims set in the LD. In the open-ended questions, one teacher mentioned that *"the most important was the conversion of the students['] written speech into a radio speech and also the production and recording of the broadcast."*

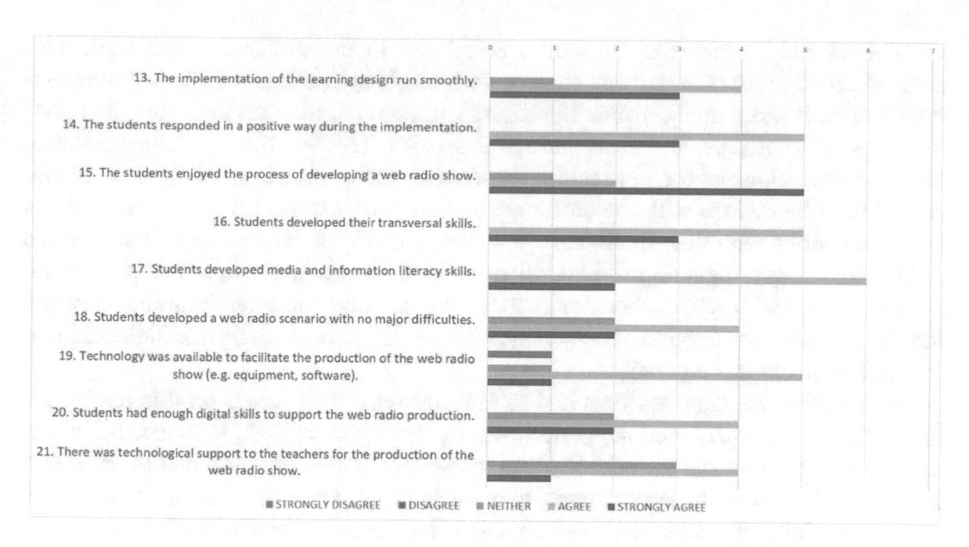

Fig. 2. Teachers' perceptions on the implementation of NEStOR

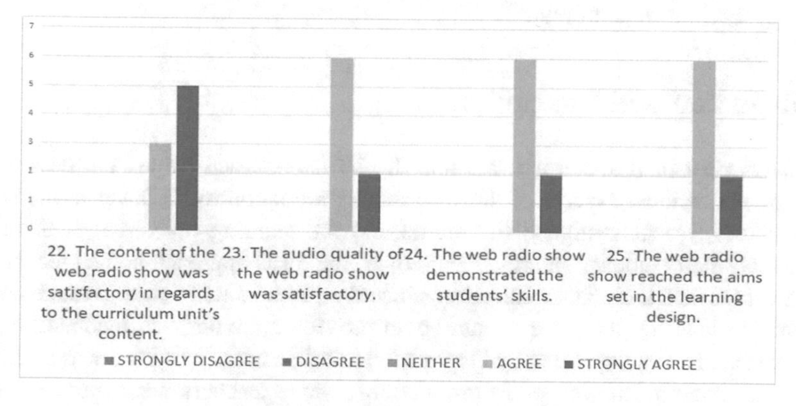

Fig. 3. Teachers' perceptions on the final product

Regarding the open-ended questions, some of the answers were discussed in the previous paragraphs to enhance the results of the close-ended questions. In the following, we will refer to the main themes mentioned by teachers when answering these questions. In the first question, where the teachers had to indicate the most important aspect of this implementation for them, the answers varied including the scenario, students' reactions, the production and the cooperation. One teacher particularly said that the most important aspect was: *"That the students cooperated actively, worked critically and responsibly with creativity and imagination, and, finally, acquired knowledge with quality and duration, in a really attractive way. This kind of implementations help the school go beyond its conventional framework, with substantial learning outcomes."*

The second open-ended question was related to the difficulties and challenges teachers faced during the implementation. Two teachers mentioned the time limitations involved, as it was a time consuming process and required teachers to devote time prior to the lesson to prepare, as well as during the lessons. This was often a challenge due to the time limitations of the curriculum. Moreover, there were also teachers who connected time limitations with the challenge to find participants at the beginning. Apart from difficulties with time limitations, there were a few comments on challenges and difficulties related to guiding and coordinating students. As this was a new process for both students and teachers, there were challenges involved into working this way and providing the necessary guidance and support. Finally, another challenge identified was the lack of necessary equipment.

In the third question, teachers had to indicate what they would do differently in a future implementation. Four suggested better planning in regards to when the implementation takes place in the school. Another time-related suggestion had to do with a shorter implementation. Furthermore, two teachers proposed alternative ways for the implementation. One suggested implementing the same scenario to another school for evaluating its impact in another class, while the other suggested that students in the same class could develop in groups several web radio shows covering different aspects of the selected curriculum unit.

5 Discussion and Conclusion

The results presented above indicate that the VLD approach helped teachers to think about the educational aspects of the web radio production to be developed. As mentioned before, teachers aligned the theme and the learning objectives with the curriculum and their students' interests. Moreover, the VLD approach helped teachers and the project team in their communication, since the teachers had a way to make concrete their initial ideas and the project team could provide more targeted feedback on these initial plans. The statement that had the most neutral answers was the one regarding the supporting material on web radio productions. Since teachers were provided with a guide covering all the technical aspects of this process, we assume that teachers felt they needed more support due to the lack of previous experience in this field.

As far as the class implementation of the LD is concerned, the teachers reported very positive experiences and perceptions, which are also reflected in their answers in the open-ended questions. Moreover, the teachers highlighted the possibilities and the impact such implementations could have for student learning and collaboration. Such possibilities include engaging students in learning scenarios aligned with their interests, providing them the space and time to develop their skills (digital, transversal, media and information literacy skills), collaborate in groups, and reach *'substantial learning outcomes'*. The aspect that seems to be weak in this part is again the technological support, and the provision of technology (software, equipment). Since the schools were responsible for the latter, the only thing that could be improved in the project is to provide more clear guidelines on the necessary technology.

The teachers' answers regarding the final web radio product showed also positive results. The teachers reported that it was overall an interesting experience for both teachers and students, and reached their goals and expectations. The aspect that received the most positive response was the alignment of the web radio output to the curriculum unit. We believe that this is due to importance of this alignment for schools, which was also evident during the LD development phase.

It has to be noted here, that the aforementioned results come from a small study, therefore they are mainly indications. Taking into consideration the experiences and the evaluation results of the first year, we aim at employing the VLD approach in a bigger scale implementation during the second year of the project. In the coming evaluation, we will also investigate student perceptions on the approach (mainly on the implementation of the LD and the final output) since this is an aspect lacking in the current evaluation due to time limitations.

Acknowledgements. The research presented in this paper was conducted under the "NEStOR: Networked European School Web Radio" project. This project is funded by the Erasmus+ programme of the European Commission. The European Commission support for the production of this publication does not constitute an endorsement of the contents which reflects the views only of the authors, and the Commission cannot be held responsible for any use which may be made of the information contained therein.

References

1. Lamb, T. R.: The emergence of educational radio: Schools of air. TechTrends **56**(2), 9–10 (2012)
2. Kozamernik, F., Mullane, M.: An introduction to internet radio. EBU Tech. Rev. (2005)
3. Levine, S., Franzel, J.: Teaching writing with radio. Engl. J. **104**, 21–29 (2015)
4. Coccoli, M.: The use of web-radio in mobile-learning (2014)
5. McGroarty, M.: The benefits of cooperative learning arrangements in second language instruction. NABE J. **13**, 127–143 (1989)
6. Lemos Tello, N.C.: On air: Participation in an online radio show to foster speaking confidence. A cooperative learning-based strategies study. Profile Issues Teach. Prof. Dev. **14**, 91–112 (2012)
7. Piñero-Otero, T., Ramos, F.: Radio 2.0 in higher education communities. An approximation of Aveiro university members perceptions (2012)
8. Güney, S., Rizvanoglu, K., Öztürk, Ö.: Web radio by children? An explorative study on an international children's radio network, pp. 61–79 (2013)
9. Dalziel, J.: Implementing learning design: a decade of lessons learned, pp. 210–220 (2013)
10. Avraamidou, A., Economou, A.: Visualized learning design: the challenges of transferring an innovation in the cyprus educational system. Teach. Engl. Technol. **12**, 3–17 (2012)
11. Conole, G., Brasher, A., Cross, S., Weller, M., Clark, P., Culver, J.: Visualising learning design to foster and support good practice and creativity. Educ. Media Int. **45**, 177–194 (2008)

Pragmatic Game Authoring Support for an Adaptive Competence-Based Educational System

Florentina Condrea[1(✉)], Luca Cuoco[2], Tony-Alexandru Dincu[1], Elvira Popescu[1], Andrea Sterbini[2], and Marco Temperini[2(✉)]

[1] Computers and Information Technology Department, University of Craiova, Craiova, Romania
[2] Computer, Control, and Management Engineering Department, Sapienza University, Rome, Italy
marte@diag.uniroma1.it

Abstract. Educators can be encouraged to use games for their teaching, and this would likely lead them to positive results. However, while a general model for effective educational games requires the collaboration of several experts, entailing a high amount of work and interaction among them, teachers are often alone in their task of developing learning experiences. Hence, a significant research issue in TEL is the simplification of the task of developing an educational game: by "simplification" we mean that the requirements of the visual- and mechanics-related aspects of a game are left to a minimum for the teacher, while she could concentrate on the learning-related aspects of the game. In this paper we present a framework aimed to simplify the teacher's task, through the use of game editors. In particular, we designed and implemented two editor tools, aimed to allow a teacher to produce game learning resources, while mainly focusing on the subject matter of interest, rather than on the technical development aspects. The editors allow defining games with one of two possible fixed mechanics; they produce games as HTML5 web applications, that can function standalone, or can be integrated in an adaptive e-learning platform (DEV).

Keywords: Game-based learning · Editor-based game development
Personalized learning · Adaptivity · Student model

1 Introduction

The game-based approach in Technology Enhanced Learning (TEL) has been surging ahead for many years, and is one of the most significant current approaches [1]. The occurrence of "meaningful play" [2] is profitable in TEL, when the actions performed during the play match with the educational aims of the game-based learning activity. During play, events such as the recognition that a decision is needed, and the "practical" execution of an action, as opposed to its detailed verbal explanation, are features increasing the learning performance, especially in applicative fields [3, 4]. Personalization and adaptivity are two important topics in TEL research, and they have

T. Di Mascio et al. (Eds.): MIS4TEL 2018, AISC 804, pp. 230–238, 2019.
https://doi.org/10.1007/978-3-319-98872-6_27

counterparts in game-based learning: as it happens in traditional adaptive learning systems [5], in an adaptive game-based learning system the personalization of the learning offer is based on a *Student Model* (SM, a collection of personal traits of the individual student). *Competence*, *learning style*, *player style* are instances of such personal characteristics, and the learner can be allowed to experience different games, or adapted instances of the same game, according to her current SM. Such a timely personalization of the learning-gaming activity [6] can be effective, making the experience more acceptable, more rapidly usable, and more likely to bring the learner in a "flow state" [7], and ultimately more fruitful.

Instructors can be encouraged to use games for their teaching, and their students' learning, and it is likely that this would lead them to positive results. However, a general model for effective educational games requires the collaboration of game designers, educational experts and subject-matter experts, entailing a large amount of work, and interaction among experts. But teachers are often alone in their task of developing learning experiences; moreover, several distinct sub-topics in the course, though small they might be, would need dedicated games; and, additionally, the need for such games might be noticed too late to fit in the workflow, with consequences on delayed, if not canceled development. Hence, we might expect that only few teachers would embark, without strong support, into the task of defining multiple educational games, for possibly small multiple sub-topics in their teaching plan, and for possibly small numbers of students. So, a significant research issue is the simplification of the development task of a game, meant as a way to allow the teacher: (i) to deal with requirements reduced to a minimum, in regard to the visual and mechanics-related aspects of a game, and (ii) to concentrate on the pedagogical aspects of the educational game, directly related to the subject matter.

In this paper we present an approach for the above simplification issue. In particular, we designed and implemented two editor applications, aimed to allow a teacher to produce games where the mechanics are basically fixed, the visual aspects are managed by simple images uploaded to the system, and the teacher-author can focus on knowledge issues of the subject matter only. The editors are easily usable by a teacher with limited technical skills and they generate games as HTML5 web applications, that can function standalone, or can be integrated in an adaptive e-learning platform (DEV). Before describing the editors (in Sect. 4), we briefly present some related work (in Sect. 2) and the DEV system (in Sect. 3). DEV [8] is an adaptive educational platform which supports the definition of courses as collections of learning experiences (LE), where: (i) each LE is supposed to provide assessment means; (ii) a LE can consist of a game, imported in DEV according to a specification protocol; and (iii) the learner can build adaptively her path of LEs, as freely as this is allowed by the current state of her SM.

2 Related Work

Game–based learning (GBL) is one of the hot topics for research in TEL [9, 10]. It is often accompanied by the conceptual frameworks of *Gamification* [11], where game-inspired elements are integrated in a non-game environment, and *Open Social Learner*

Modeling [5], where the learner is supported by the possibility to access her model and compare it with others, or with the average model.

A comprehensive analysis of GBL trends is beyond this paper's scope, so here we focus on *adaptivity* of games and of learning paths, *game-based assessment* supporting student model update, and use of *game editors* easing teacher's work. "Adaptive games" [12] are game applications in which interface and mechanics can depend on player's *gamer* type: knowledge can be dealt with, in different ways, by either privileging high score accomplishments, or adapting topics visualization, or managing time limits, according to the gamer type (*Explorers*, *Achievers*, *Winners*). Paper [10] addresses personalization for students with little or no access to teachers; a learner is proposed game learning activities according to the student model (based on previous choices of the learner, and general acceptance of the game by the others). The system in [6] models the student by learning style and player (gaming) style. Paper [13] discusses assessment in a game as the design of *information trails*, tracking learner's behaviour and measuring accomplishments.

Regarding the task of game development, it usually encompasses the implementation of game mechanics (deemed to guide the player towards learning aims), game visual scenes, game logic and capability to interact with the player. Such a task becomes quite challenging when it falls on the teacher alone and can be discouraging, even if just small learning activities were to be produced. Game scenes, for instance, are usually managed by means of complex software systems (e.g., OGRE [14], Unity3D [15]), and some research work has been proposed in order to ease this kind of activity [16]. On the other hand, while commercial games have complex artificial intelligence taking care of the interaction with player, good results can be achieved with low inferential models, based on simple question/answer mechanics [17].

An alternative to building a whole educational game is in "modding" commercial games, which entails dealing with dedicated, and not easy-to-use, applications. *Civilization* is a well-known commercial game used to support learning, especially in History [18, 19]. Similarly, in [20], "World of Warcraft" (WoW) scenarios are modified, by using the modding software *World Editor* [21] and *JASS* as a scripting language [22], in order to obtain a learning activity in Physics.

3 DEV System

DEV [8] is an e-learning system where the teacher can integrate basically any kind of resource in her course, under the form of Learning Experiences (LE). The SM, representing the individual learner, is competence based [23]; it is the set of *skills* currently possessed by the learner. A skill $<k,c>$ denotes that the *competency* k (an item of knowledge) is possessed with *certainty* c ($c \in (0,1]$). A *Glossary* collects the definitions (textual) of the competencies used in DEV at any moment.

In DEV a Learning Resource (LR) is any application that can be accessed by a learner while performing a learning experience (a textual document, a video or audio clip, in principle any resource available via a web address, and of course a game application). We name Formal Learning Resources (FLR) those that include a way to assess the learning outcome of their use, and provide such outcome in the form of a set of *statistics* (STAT). All the other learning resources are called *informal* (ILR). An ILR can be made to function as a FLR, in DEV, by augmenting it with a questionnaire, that would provide the assessment part missing in the ILR.

A LE embeds a FLR (or a ILR + questionnaire) in a pedagogically complete frame, comprised of: (i) Textual definitions; (ii) Set of skills required to reasonably undertake the experience: LE.RS = $\{<r_1,rc_1>, \ldots, <r_n,rc_n>\}$; (iii) Set of skills that can be recognised as possessed, LE.AS = $\{<a_1,ac_1>, \ldots, <a_m,ac_m>\}$, after successful completion of the experience; (iv) Mapping *m*: FLR.STATS → P(LE.AS) associating some statistics of the FLR to some elements of LE.AS (this is used in the SM update phase).

A course C in DEV has a Target Knowledge (C.TK, a set of skills) denoting its pedagogical aims, and is comprised of a set of LEs, that are rendered by a graph (a LE is connected to another if the former's AS contains skills present in the latter's RS). DEV student interface (Fig. 1a) allows the learner to navigate the graph of LEs of a course, see their characteristics (AS, RS), select an LE of her choosing, and undertake it (*play*, if it is a game resource). Currently, the system allows a learner to select an LE only among those whose RS are covered by the learner's SM. After the LE completion, an updating process is performed on the learner's SM, according to the STAT coming from the LE undertaking/play. This adds in the SM those skills of LE.AS that are not yet present in the model, or modifies the certainty values of those already in it. The SM updating process takes care of the fact that the same LE could be repeated by the learner, and that some skills could likely be gained from different LEs chosen by the learner: in these cases the certainty computation for a skill is tuned according to the number of LE's repetitions, and the number of times the skill was already gained, so that the certainty will grow at a progressively lower rate. The student interface is completed by visualizations of the SM and course TK (Fig. 1b), so as to allow the learner to create her own learning path, built by the sequence of chosen LEs.

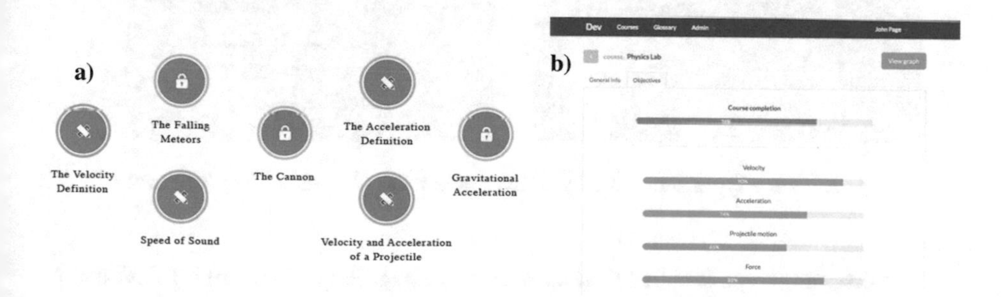

Fig. 1. (a) LE graph: the learner already took the leftmost LE; of the others, some are locked (not yet selectable by the learner), and some are available for selection; (b) SM visualization.

4 Game Authoring Support

As mentioned in the Introduction, our goal is to provide the teacher with a simple and agile tool for quickly creating educational games to feed the DEV system. More specifically, the authoring tool should provide predefined game mechanics, so that the teacher can focus on the educational, subject-related issues; in particular, the main tasks of the instructor would be to define sets of questions/answers, as well as basic interactions with objects and non-player characters (NPC), with a specific learning purpose. In addition, the developed games should be fully compatible with DEV formal resource protocol, being able to provide updates to the student model.

Thus, two editors were proposed: (i) a more comprehensive one, called *Canvas editor*, which is aimed at creating games from scratch (as described in Subsect. 4.1) and (ii) a more basic one, called *Forms editor*, which is meant for creating games starting from an existing prototype (as described in Subsect. 4.2). From a technical point of view, the editors' implementation is based on ASP.NET Web API & Microsoft SQL Server (for the back-end), and ReactJS[1] & FabricJS[2] Javascript libraries (for the front-end).

4.1 Canvas Editor

The *Canvas editor* provides the possibility to develop a game based on a predefined logic, by customizing and combining existing objects. Two types of games are currently supported by the editor, but more could be subsequently included.

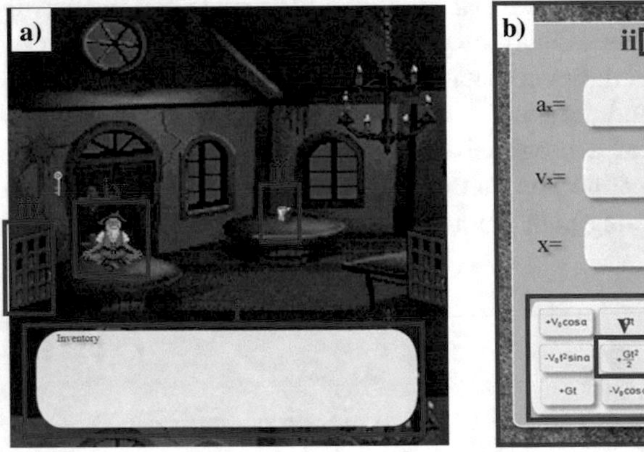

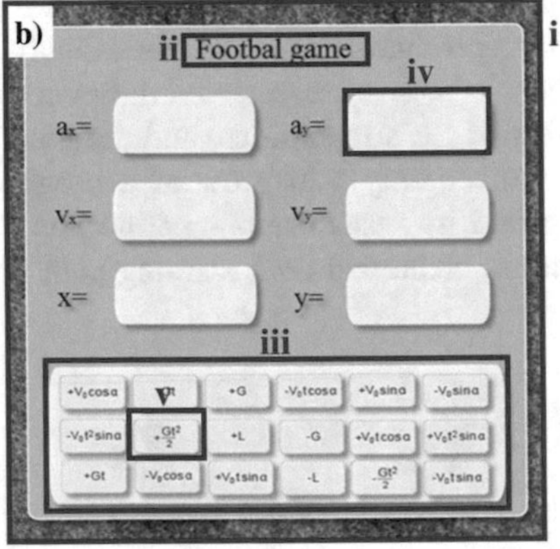

Fig. 2. Game types supported by Canvas editor: (a) Collect & Use; (b) Puzzle

The first game type is called *Collect & Use* (see Fig. 2a) and its aim is to ask and/or answer several questions coming from a NPC, and collect one or more objects from multiple environments; the learner is rewarded for her answers and selection of objects, which leads to a corresponding update of the student model in DEV.

The second game type is called *Puzzle* (see Fig. 2b) and its aim is to pair appropriate items by using drag and drop actions (e.g., match questions and solutions, or build a solution by sequencing available elements); again, the learner receives a number

[1] https://reactjs.org.
[2] http://fabricjs.com.

of points for the correctly matched items, which is subsequently reflected in her SM. For each of these games, the *Canvas editor* allows the teacher to define the layout (background) as well as multiple objects with their locations and properties. The main components of the editor are: (i) *toolbox* - which contains all the objects that can be used to create a game, grouped by game type; (ii) *canvas* – the game area, visible to the student, on which the teacher can drag and drop various objects from the toolbox; (iii) *properties* – a section which allows the teacher to configure various attributes of the game and objects; (iv) *toolbar* – which provides canvas editing options. A screenshot of the editor is presented in Fig. 3.

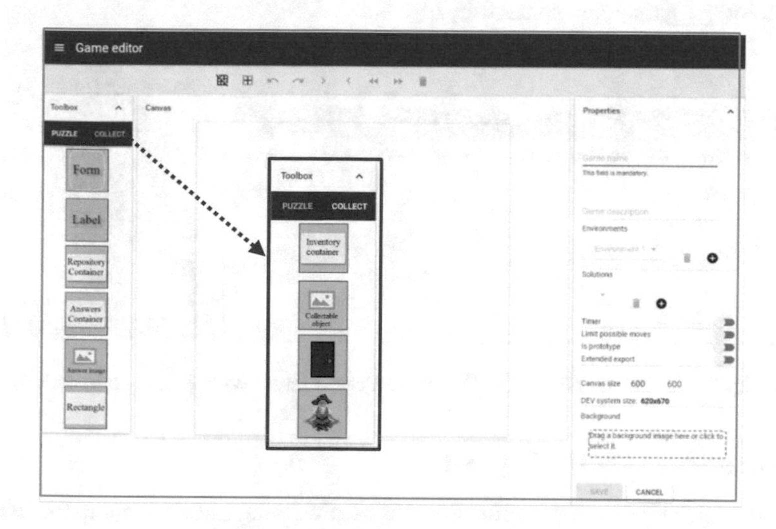

Fig. 3. Canvas editor main page

The range of objects which can be added to a *Collect & Use* game includes: (i) *inventory container* - repository for all the items collected by the player; (ii) *collectable objects* - items spread in the game environment that the student may collect in order to gain points; (iii) *door* - mechanism for navigating through games' different environments; (iv) *guru* - a NPC which can ask the student various questions, and at the same time offer hints and suggestions. Some of these objects may be seen in Fig. 2a.

Similarly, the range of objects which can be added to a *Puzzle* game includes: (i) *form* - for grouping the main game elements of the environment; (ii) *label* - a static object that can be used to insert a game description or as a placeholder for questions; (iii) *repository container* - used as a holder for all the available elements allowing to build the answers in the game; (iv) *answer container* - used to hold the elements composing an answer to a specific question; (v) *answer-element image* - used to compose a possible answer to a question; (vi) *rectangle* - a static object used to create layout structures. Some of these objects may be seen in Fig. 2b.

The game may be further configured through various properties, such as: multiple environments, limited time or number of moves allowed for the student.

4.2 Forms Editor

The *Forms editor* provides an even simpler and quicker way of creating a game, starting from an existing *prototype*; this is a template game, which has predefined canvas and objects structure. Four steps need to be completed by the teacher in order to develop a *Puzzle* game (which is currently supported by the *Forms editor*): (i) Select a game prototype; (ii) Define game properties; (iii) Define exercise details (i.e., questions & repository of answers); (iv) Assign answers (i.e., map the answers from the repository to the corresponding questions). Once the game is created, it can be further customized using the *Canvas editor*. A screenshot of the *Forms editor*, together with a basic prototype game are included in Fig. 4.

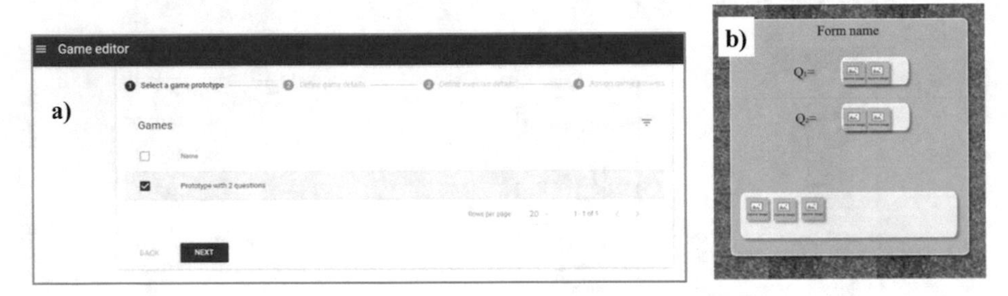

Fig. 4. (a) Forms editor; (b) Prototype Puzzle game with 2 built-in questions

4.3 Integration with DEV System

Both editors provide an *Export* functionality, which generates two components: (i) the actual game (*<Game name>.zip*), a web resource which includes game interface and logic; (ii) the game statistics (*<Game name>.devscript*), which will be used to update the student model after the game is played. Thus, once the games are developed, they can be imported in DEV system and used as formal resources in the *Course builder*, in order to create the interface supporting the learner to build her adaptive learning path.

5 Conclusions

A general model for the development of effective educational games requires the collaboration of game designers, educational experts and subject-matter experts, entailing a high amount of work and sustained interaction. However, in many cases, such a comprehensive team may not be available, especially when basic games are required, which address small chunks of the syllabus. Hence, in this paper we have shown an approach to supporting teachers in the development of such simple games by means of two easy to use editors. The games produced by these editors can be seamlessly integrated in the DEV system for personalized learning, and can be effectively used as simple assessment tools, supporting the management of the student model in a distance learning course.

As future work, we aim to extend the editors to support more types of games, with more complex mechanics. Furthermore, we plan to evaluate the editors in real world settings, with teachers and students from various disciplines and levels of study.

References

1. Wu, W.H., Chiou, W.B., Kao, H.Y., Hu, C.H.A., Huang, S.H.: Re-exploring game-assisted learning research: the perspective of learning theoretical bases. Comput. Educ. **59**(4), 1153–1161 (2012)
2. Salen, K., Zimmermann, E.: Rules of play: game design fundamentals. MIT Press, Cambridge MA (2004)
3. Coller, B.D., Scott, M.J.: Effectiveness of using a video game to teach a course in mechanical engineering. Comput. Educ. **53**, 900–912 (2009)
4. Pasin, F., Giroux, H.: The impact of a simulation game on operations management education. Comput. Educ. **57**, 1240–1254 (2011)
5. Liang, M., Guerra, J., Marai, G.E., Brusilovsky, P.: Collaborative e-learning through open social student modeling and Progressive Zoom navigation. In: Proceedings of the CollaborateCom, pp. 252–261 (2012)
6. Lindberg, R.T.S., Laine, T.H.: Detecting play and learning styles for adaptive educational games. Proc. CSEDU **1**, 181–189 (2016)
7. Mirvis, P.H.: Flow: the psychology of optimal experience. Acad. Manag. Rev. **16**(3), 636–640 (1991)
8. Cuoco, L., Sterbini, A., Temperini, M.: An adaptive, competence based, approach to serious games sequencing in technology enhanced learning. Proc. CSEDU **1**, 589–596 (2017)
9. Van Eck, R.: Digital game-based learning: it's not just the digital natives who are restless. EDUCAUSE Rev **41**(2), 16–30 (2006). EDUCAUSE
10. Metawaa, M., Berkling, K.: Personalizing game selection for mobile learning. Proc. CSEDU **2**, 306–313 (2016)
11. Deterding, S., Khaled, R., Nacke, L., Dixon, D.: Gamification: toward a definition. In: Proceedings of the CHI 2011 Gamification Workshop, pp. 12–15. ACM (2011)
12. Magerko, B., Heeter, C., Fitzgerald, J., Medler, B.: Intelligent adaptation of digital game-based learning. In: Proceedings of the FuturePlay 2008, pp. 200–203. ACM (2008)
13. Loh, C.S.: Designing online games assessment as information trails. In: Gibson, D., Aldrich, C., Prensky, M. (eds.) Games and Simulations in Online Learning: Research and Development Frameworks, pp. 323–348. InfoSci (2007)
14. OGRE. Accessed 12 Feb 2018. https://www.ogre3d.org/
15. Unity 3D. Accessed 12 Feb 2018. https://unity3d.com/
16. Cai, L., Chen, Z.: Design and implementation of OGRE-based game scene editor software. In: Proceedings of the CiSE, pp. 1–4. IEEE (2010)
17. Khan, Z., Maddeaux, M., Kapralos, B.: Fydlyty: a low-fidelity serious game for medical-based cultural competence education. In: Proceedings of the INTETAIN, pp. 195–199. IEEE (2015)
18. Wender, S., Watson, I.: Using reinforcement learning for city site selection in the turn-based strategy game Civilization IV. In: Proceedings of the Symposium on CIG 2008, pp. 372–377. IEEE (2008)
19. Pagnotti, J., Russell, W.B.I.I.I.: Using Civilization IV to engage students in world history content. Soc. Stud. **103**(1), 39–48 (2012)

20. Sun Lin, H-Z, Chiou, G-F: Modding commercial game for physics learning: a preliminary study. In: Proceedings of DGITEL, pp. 225–227. IEEE (2010)
21. WoW, World Editor. Accessed 12 Feb 2018. https://wow.gamepedia.com/World_Editor
22. JASS. Accessed 12 Feb 2018. https://en.wikipedia.org/wiki/JASS
23. Ilahi-Amri, M., Cheniti-Belcadhi, L., Braham, R.: A framework for competence based e-assessment. Interact. Des. Archit. J. (IxD&A) **32**, 189–204 (2017)

Generator of Tests for Learning Check in Case of Courses that Use Learning Blocks

Doru Anastasiu Popescu[1(✉)], Daniel Nijloveanu[2], and Nicolae Bold[1]

[1] Department of Mathematics and Computer Science, University of Pitești, Pitești, Romania
dopopan@gmail.com, bold_nicolae@yahoo.com

[2] Faculty of Management, Economic Engineering in Agriculture and Rural Development, University of Agronomic Sciences and Veterinary Medicine Bucharest, Slatina Branch, Romania
nijloveanu_daniel@yahoo.com

Abstract. Block learning is a useful and efficient way of learning currently used in many learning systems in both pre-university and university environments. Based on the key concept of learning by focusing, this style of learning is based on the progressive way of gaining knowledge and acquiring different types of skills. In this paper we will present a method of assessing learners by indirectly encouraging them to use block-based solving problem strategies by solving tests obtained from a genetic algorithm. After the generation of sequences of blocks that solve the problem is made by using a genetic algorithm, the resulted tests are used to assess learners in order to familiarize them with sequential learning.

Keywords: Block · Learning · Genetic · Generation · Test

1 Introduction

Changes in the technological, economic and social environments appear to lead to modifications in the individual learning strategies [16]. Today students obviously have different ways of gaining knowledge and solving problems. Teachers started to foresee this change, thus they started to combine the traditional methods of teaching with diagrams, presentations, graphics, images, audio or video to structure the presentation of information [13]. Furthermore, learning has been driven to other differently-structured means of learning in the form of electronic learning, [11] and [12], leading to the formation of educational online platforms [10]. In practice, even if the teaching follows a traditional path, students search lateral modalities of learning [15] and seem to understand better if the information is structured in more creative and interactive interfaces (sources) than the plain text-based learning [14].

Learning by blocks is one of the interactive methods that can be successfully used to complete a programming course. Blocks are used currently as basic methods for courses related to programming, robotics and domains which use algorithms and scenarios/information that can be structured in steps, especially for creating the basic knowledge in a certain domain. As a confirmation, the usage of this method of learning

T. Di Mascio et al. (Eds.): MIS4TEL 2018, AISC 804, pp. 239–244, 2019.
https://doi.org/10.1007/978-3-319-98872-6_28

has led to the development of learning software tools, such as Blockly [5] or Alice [6] for showing the importance of the algorithmic thinking or Lego™ Mindstorms [4] for programming robots that can solve practical problems experienced in real life.

2 Blocks Used for Assessment Tests

Basically, learning by blocks refers to a type of learning where the teacher forms learning paths for a specific course which can be conceptualized using learning blocks. Learning blocks can find their utility in the process of learning either by being used by teachers to structure their courses or by students in order to solve exercises and problems. Structuring information in blocks and learning by creating relations between them develop specific skills and the logic and rational components of thinking. Furthermore, the block-based learning develops lateral thinking skills such as spatial and temporal organization of activities. The solution of developing learning paths based on learning blocks can be extended to assessing students. In short words, students can be assessed by asking them to form paths that solve a problem.

In a short definition, a block is a unit of learning which contains a unit of information that transform an input data to an output based on several requirements. Basically, it has three components: input, output and requirements. A general form of a block used for assessment is shown in Fig. 1.

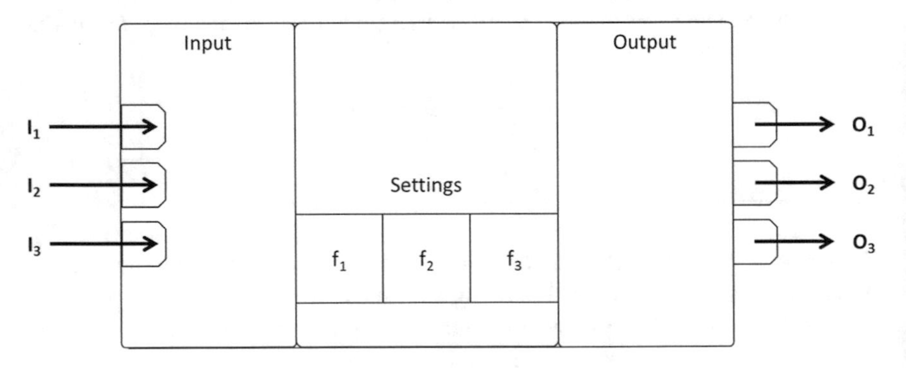

Fig. 1. General form of a learning block

The general idea of the assessment is the usage of a generator of paths which gives a solution to a given problem. Based on the generated solution, the student must create a path formed of a set of blocks instances that are known.

Based on such blocks, the students can be assessed related to the concepts learned during a specific course. Basically, the problem that must be answered can be mathematically reduced and enunciated as follows: given B_1, B_2, …, B_n blocks by their input, output and parameter components, it is required to determine permutations of these blocks B_{p1}, B_{p2}, …, B_{pn} so that a number of k requirements to be respected, denoted by f_1, f_2, …, f_k.

$$
\begin{gathered}
f_1 \overset{require}{\longrightarrow} B_{p_1}, \ldots, B_{po_1} \\
f_2 \overset{require}{\longrightarrow} B_{po_1+1}, \ldots, B_{po_2} \cdots \\
\cdots f_k \overset{require}{\longrightarrow} B_{po_{k-1}+1}, \ldots, B_{p_n}
\end{gathered} \tag{1}
$$

Shortly, for the ith requirement there must be generated a sequence of blocks which would meet that requirement. Also, there is a global requirement f_0 that must be met by the entire generated sequence, which reunites all the k requirements.

3 Model of Generating Tests

In this paper, we describe a model that solves the problem which consist in the generation of the correct solution, based on which the student must solve the problem. In other words, the given problem is solved using a genetic algorithm (several solutions are generated) and then the student is asked to solve the same problem having as checking base the generated solutions from above (after solving the problem, the student can confront the obtained solution with the solutions generated by the model).

We have created a model for generating block used for assessment with respect to several requirements, based on a genetic method. The operations and structures used are:

- The genes are represented by blocks which form permutations or chromosomes (sequences of blocks).
- A chromosome is a test. Basically, the model has as result a variety of tests formed of blocks that can be used for assessment. From this variety, a test that respects all the k requirements is chosen.
- The fitness function consists in the number of block connections that respect the given requirements.
- The mutation consists in swapping a gene with a random generated one from a random chromosome.
- The crossover operation consists in mixing parts of two chromosomes. Given two chromosomes and a random position within them, the second part of the first chromosome is swapped with the second part of the second chromosome and the first part of the second chromosome is swapped with the first part of the first chromosome with respect to the generated position.

The scheme of the model is presented in Fig. 2. The genetic algorithm takes place in the third step.

As a short example, using the concepts and notations presented in Sect. 2, we will take an example of describing a problem according to which a robot must follow a set path (for example, a rectangular path). If the robot must follow such a path, then the number of requirements is equal to 5 (four requirements for the rectangle sides and one for the global problem). If the rectangular path is denoted by ABCD, then f_1 will be the requirement for the side AB, f_2 for BC, f_3 for CD, f_4 for DA. The requirement f_0 stands

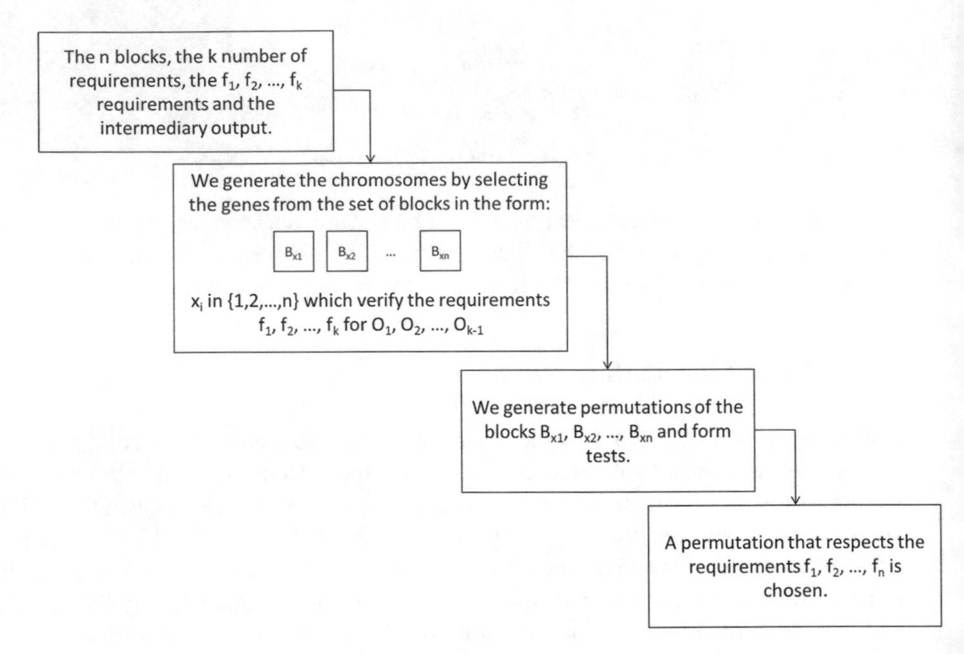

Fig. 2. The general scheme of the model

for the global movement path, including the rotations in turning points and the sound alerts made during the path.

4 Practical Aspects of the Described Model

Using this modality of learning, the gain of knowledge and the development of skills is combined with the reasoning and the logic of activity, developing in the same time the spatial and temporal organizational skills of the learners. Another important result of the usage of this method is represented by the development of problem solving skill, successfully applicable in real life. Also, besides the fact that the course information can be structured using blocks, the traditional way of block learning, i.e. the organization of course activity, can be obviously made using block-based method.

In order to support this, we will show an example of solving a practical problem using Lego Mindstorms: "*Randomly generate the number of rotations that a robot must make on the sides of a rectangle, in terms of the side lengths measured in motor revolutions, the lengths from the set {4, 5, 6} and the widths from the set {2, 3, 4}, then move the robot from one corner then returning at the same position. All the operations must be made using the same program.*" As solution, we will use two numerical values *a* and *b* for length, respectively for width. Their creation will be made using two blocks *Variable* and for the initialization there will be used *Random* blocks, with the given requirements of number limit. The following blocks are the motor ones, adding the

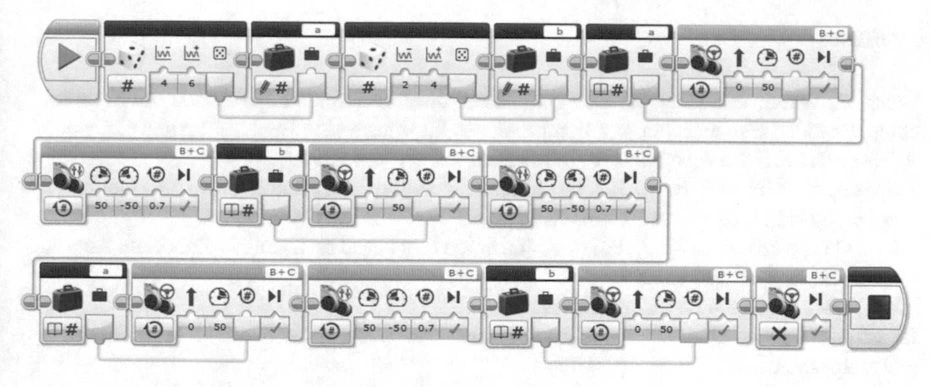

Fig. 3. Sequence of blocks in the Lego Mindstorms tool that solves the given problem

block *Variable* for creating the number of rotations, in order for the robot to move. Figure 3 presents the sequence of blocks that solves the given problem.

By using the generator described in Sect. 3, a great variety of the generated test is built. Besides that, tests can vary in complexity, based on the key parameters of the model: n (number of blocks), k (number of requirements), f (the actual requirements) and O (number of intermediary output). This generator can be successfully used for courses from various domains of activity. Many software tools have been developed in order to develop skills in areas of first-step programming (Scratch, Blockly, Alice) or robot programming (Lego Mindstorms).

5 Conclusions

From the experience gained at the courses and seminaries, students responded positively to this type of learning and from this fact results the necessity of such a generator. In addition, the usage of genetic algorithms is appropriate for situations as the one presented in our paper where the solutions set is extremely large and their generation would use an exponential runtime, in solving problems related to education [1], transport [2] or design [3]. As parallel structures, we have also used genetic algorithm in the educational area, more specific in assessment, for generating tests with certain requirements, either isolated or connected [9], using arborescence or genetic structures [7], with requirements given by solving time, degree of difficulty or closeness of answers [8]. As a future continuation of the present work, in the next period we propose to create an online tool for the generation of tests for courses that can be adapted to block teaching and learning.

References

1. Wang, F., Wang, W., Yang, H., Pan, Q.: A novel discrete differential evolution algorithm for computer-aided test-sheet composition problems. In: Information Engineering and Computer Science, ICIECS 2009, Wuhan, 19–20 December, pp. 1–4 (2009)
2. Rahmani, S., Mousavi, S.M., Kamali, M.J.: Modeling of road-traffic noise with the use of genetic algorithm. Appl. Soft Comput. **11**(1), 1008–1013 (2011)
3. Gama, C.L., Norford, L.K.: A design optimization tool based on a genetic algorithm. Autom. Constr. **11**(2), 173–184 (2002)
4. https://www.lego.com/en-us/mindstorms
5. Blockly-Games - series of educational games that teach programming. https://blockly-games.appspot.com/
6. Alice - innovative 3D programming environment. http://www.alice.org/index.php
7. Popescu, D.A., Bold, N., Domsa, O.: Generating assessment tests with restrictions using genetic algorithms. In: ICCA 2016, pp. 696–700 (2016)
8. Popescu, D.A., Domsa, O., Popescu, I.A.: Determining the degree of similarity of answers given at a test using the editing distance. In: ITHET 2016, pp. 1–4 (2016)
9. Popescu, D.A., Bold, N., Domsa, O.: A generator of sequences of hierarchical tests which contain specified keywords. In: SACI 2016, pp. 255–260 (2016)
10. Baron, C., Şerb, A., Iacob, N.M., Defta, C.L.: IT infrastructure model used for implementing an E-learning platform based on distributed databases. Qual. Access to Success J. **15**(140), 195–201 (2014)
11. Gaytan, J., McEwen-Beryl, C.: Effective online instructional and assessment strategies. Am. J. Distance Educ. **21**(3), 117–132 (2007)
12. Popescu, E.: Adaptation provisioning with respect to learning styles in a web-based educational system: an experimental study. J. Comput. Assist. Learn. **26**(4), 243–257 (2010)
13. Mazoue, J.G.: The MOOC Model: Challenging Traditional Education, Educause Review Online January/February (2013). Accessed 2 April 2013
14. O'Neil, H.F., Perez, R.S.: Web-Based Learning: Theory, Research, and Practice. Routledge, 5 September 2013
15. Joo, Y.J., Bong, M., Choi, H.J.: Self-efficacy for self-regulated learning, academic self-efficacy, and internet self-efficacy in web-based instruction. Educ. Technol. Res. Develop. **48**(2), 5–17 (2000)
16. Seman, L.O., Hausmann, R., Bezerra, E.A.: On the students' perceptions of the knowledge formation when submitted to a Project-Based Learning environment using web applications. Comput. Educ. **117**, 16–30 (2018). ISSN 0360-1315

Workshop on TEL in Nursing Education (NURSING)

Impact of an Electronic Nursing Documentation System on the Nursing Process Accuracy

Fabio D'Agostino[1(✉)], Valentina Zeffiro[1], Antonello Cocchieri[2], Mariangela Vanalli[1], Davide Ausili[3], Ercole Vellone[1], Maurizio Zega[2], and Rosaria Alvaro[1]

[1] University of Rome Tor Vergata, Rome, Italy
fabio.d.agostino@uniroma2.it
[2] University Hospital Agostino Gemelli, Rome, Italy
[3] University of Milan Bicocca, Milan, Italy

Abstract. While the nursing process provides a framework for documenting nursing practice and delivering patient-focused care, nurses often have difficulty applying the nursing process in clinical practice. Fortunately, electronic nursing documentation using clinical decision support systems (END-CDSS) can improve the accuracy of recorded nursing process. To date, however, no study has evaluated nursing documentation accuracy over time following END-CDSS implementation. Accordingly, the aim of this study was to evaluate if the Professional Assessment Instrument (PAI) END-CDSS improved the accuracy of nursing documentation in an Italian hospital cardiology inpatient unit. A quasi-experimental longitudinal design was conducted. A random sample of 120 nursing documentations was collected and evaluated using the D-Catch instrument. A significant improvement ($p < .001$) in nursing documentation accuracy scores was shown after PAI implementation. These results suggest that an END-CDSS can support nurses in the nursing process by improving their clinical reasoning skills.

Keywords: Nursing documentation accuracy · Nursing process
Electronic nursing documentation · Clinical decision support system
Longitudinal design

1 Introduction

The nursing process is the recognised framework for nursing practice to deliver patient-focused care; it also constitutes a framework for documentation [1, 2]. The use of the nursing process has shown an improvement in both patient outcomes and documentation accuracy [3, 4]. Nursing documentation accuracy is important for patient safety and care [5].

Nevertheless, several studies have shown deficiencies in the recording of the nursing process, including low accuracy in nursing documentation [6, 7]. These findings have shown that nurses have difficulties adequately recording the nursing process due to several factors (e.g., lack of diagnostic education, restrictive hospital policies) [8].

T. Di Mascio et al. (Eds.): MIS4TEL 2018, AISC 804, pp. 247–252, 2019.
https://doi.org/10.1007/978-3-319-98872-6_29

Electronic nursing documentation should include the nursing process phases, such as patient needs assessment, nursing diagnoses, planning and delivering interventions, and patient outcomes evaluation; furthermore, these nursing care elements should be expressed using a standardised nursing taxonomy [9]. Electronic nursing documentation that includes technologies like clinical decision support systems (END-CDSSs) can help nurses in the use of the nursing process by improving the accuracy of nursing documentation in clinical settings [8]. To our knowledge, no study has yet been conducted to evaluate nursing documentation accuracy over time after the implementation of an END-CDSS.

1.1 Professional Assessment Instrument (PAI)

The Professional Assessment Instrument (PAI) is an END-CDSS that can be used in hospital settings to record nursing care according to the phases of the nursing process (e.g., nursing assessment, nursing diagnosis, interventions) [10]. The PAI implements standardised nursing taxonomies to record nursing diagnoses (NANDA-International) and nursing interventions (Italian Nomenclature of Nursing Care Performance) as well as an END-CDSS to support nurses in the identification of nursing diagnoses and interventions following the nursing process phases [11, 12]. Beginning with the clinical findings recorded during nursing assessment, the END-CDSS suggests possible nursing diagnoses and interventions that must then be confirmed by the nurse.

In February 2013, the PAI was implemented in the university hospital 'Agostino Gemelli' in Rome, Italy.

1.2 Aim of the Study

The aim of this study was to evaluate if the use of the PAI END-CDSS improved the accuracy of nursing documentation in a cardiology inpatient unit of the 'Agostino Gemelli' hospital.

2 Methods

2.1 Design, Setting, Sample, and Data Collection

A quasi-experimental longitudinal design was conducted. The cardiology inpatient unit had 31 beds. Nursing documentation accuracy was audited at four different time points: one month before PAI implementation (Time 0), four months after PAI implementation (Time 1), eight months after PAI implementation (Time 2), and one year after PAI implementation (Time 3).

From a total number of 352 nursing documentations a random sample of 120 nursing documentations was extracted (30 nursing documentations at each time point) in the cardiology inpatient unit. Computer-generated random numbers were used for a simple randomization of nursing documentations. The selection criteria for the collected documentations were that they concerned patients discharged from the inpatient unit who were hospitalised for a minimum of three days.

Data were collected between January 2013 and February 2014 using the D-Catch instrument [13].

2.2 Instrument

D-Catch is an instrument to measure nursing documentation accuracy. It is composed of six items that measure (1) nursing documentation structure, (2) patient nursing assessment on admission, (3) nursing diagnoses, (4) nursing interventions, (5) nursing-sensitive outcomes, and (6) nursing documentation legibility. The D-Catch uses qualitative and quantitative criteria to evaluate nursing documentation accuracy. Each criteria uses a four-point Likert scale, with 4 = very good and 1 = poor for qualitative criteria and 4 = partially complete and 1 = none for quantitative criteria. Items 1 and 6 were evaluated using only qualitative criteria, with possible scores from 1 to 4 for each item. Items 2 through 5 were evaluated using both qualitative and quantitative criteria, with possible scores from 2 to 8 for each item. The D-Catch has shown adequate psychometric properties. Explorative and confirmative factorial analysis resulted in one factor, 'chronologically descriptive accuracy' (Items 1, 2, 4, 5, and 6), and one item, 'diagnostic accuracy' (Item 3) [13, 14].

Two scores were computed from the D-Catch instrument: a score for the chronologically descriptive accuracy factor (obtained from the sum of Items 1, 2, 4, 5, and 6) and a score for Item 3 (diagnostic accuracy). A higher score indicated better nursing documentation accuracy. After obtaining Institutional Review Board and study site approval, eight nurses were instructed in using D-Catch. Acceptable inter-rater reliability was assessed prior to data collection. For each nursing documentation, the length of hospital stay was also collected.

2.3 Data Analysis

SPSS version 21 was used to analyse the data. Descriptive statistics (mean, median, and range) were used to describe length of hospital stay and documentation accuracy scores. Since the data did not conform to the assumptions of parametric tests, such as analysis of variance, a Kruskal-Wallis H test for independent groups was run to assess if the use of PAI improved nursing documentation accuracy. Distributions of chronologically descriptive and diagnostic accuracy scores were not similar for all groups; therefore, mean rank was regarded as describing accuracy scores among time groups. Pairwise comparisons were performed using Dunn [15] procedure, with a Bonferroni correction used for multiple comparisons, to test differences in nursing documentation accuracy scores over time. Probabilities equal to and less than 0.05 were regarded as significant.

3 Results

A sample of 120 nursing documentations of patients admitted to a cardiology inpatient unit were investigated. The median length of stay was 7 days, ranging from 3 to 62 days.

Both chronologically descriptive ($\chi^2 = 43.61$, $p < .001$) and diagnostic accuracy ($\chi^2 = 18.01$, $p < .001$) scores were statistically significant different among the time groups. Regarding chronologically descriptive accuracy, post-hoc analysis revealed statistically significant differences ($p < .001$) in scores between Time 0 (*mean rank* = 27.07) and the other three time groups (*mean rank* = 60.50 for Time 1, *mean rank* = 72.47 for Time 2, and *mean rank* = 81.97 for Time 3), but not between any other group combinations (Fig. 1). In terms of diagnostic accuracy, post-hoc analysis revealed statistically significant differences ($p < .01$) in scores between Time 0 (*mean rank* = 34.85) and Time 1 (*mean rank* = 67.34) as well as between Time 0 and Time 3 (*mean rank* = 63.80), but not between any other group combinations (Fig. 2).

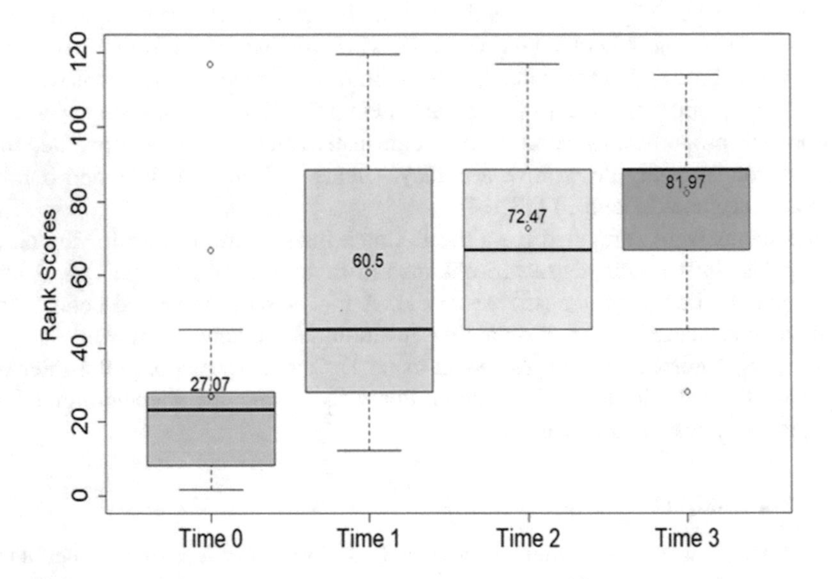

Fig. 1. Mean rank scores of chronologically descriptive accuracy by time

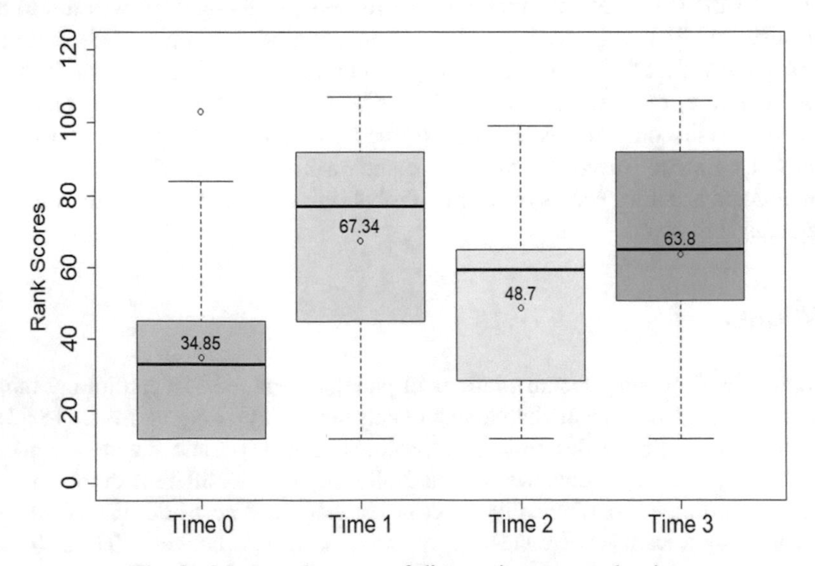

Fig. 2. Mean rank scores of diagnostic accuracy by time

4 Discussion

The primary finding of the present study was that the implementation of PAI improved nursing documentation accuracy. Therefore, the PAI was able to improve the learning and use of the nursing process in a clinical setting, showing its potentiality and suitability as field training technology-based tool.

There was a significant difference between Time 0 and all other time points for chronologically descriptive accuracy scores, indicating that the PAI improved the accuracy of nursing documentation and that this improvement was maintained over time. Indeed, even if a significant difference had not been found among other chronologically descriptive accuracy scores by time, such as between Time 1 and Time 2 or between Time 2 and Time 3, the findings still showed a linear trend of improvement over time.

A similar result was found for diagnostic accuracy, with significant accuracy score differences between Time 0 and Time 1 as well as between Time 0 and Time 3. Although there was a decrease in documentation accuracy at Time 2, this did not affect the general improvement in diagnostic accuracy shown after one year of PAI utilisation. Further, several factors related to nursing staffing and hospital organisation may have affected the diagnostic accuracy decrease found at Time 2, although analysing these variables was outside the scope of the present study.

The results also found lower scores for diagnostic accuracy than for chronologically descriptive accuracy. This means that additional interventions, such as education to improve nurses' diagnostic reasoning skills and strategies to improve the working conditions of hospital nurses, are needed to further improve diagnostic accuracy [16].

5 Conclusions

This study showed that the PAI was a useful field training technological tool to improve the use of the nursing process and, consequently, documentation accuracy in the long term. From a clinical point of view, the use of an END-CDSS, such as the PAI, can improve nurses' clinical reasoning skills by connecting the reasoning process to the nursing process.

However, further research is needed to investigate what factors in addition to the PAI, such as staffing or hospital organisation variables, can affect nursing process documentation accuracy. Other studies could investigate if improvements on the nursing process accuracy are associated with a better clinical communication among nurses, better patient outcomes and less missed nursing care.

Acknowledgements. This work was funded by the Italian Centre of Excellence for Nursing Scholarship, Rome, Italy.

References

1. American Nurses Association. http://www.nursingworld.org/EspeciallyForYou/What-is-Nursing/Tools-You-Need/Thenursingprocess.html
2. Peres, H.H., de Almeida Lopes Monteiro da Cruz, D., Lima, A.F., Gaidzinski, R.R., Ortiz, D.C., Mendes e Trindade, M., Tsukamoto, R., Batista de Oliveira, N.: Conceptualization of an electronic system for documentation of nursing diagnosis, outcomes, and intervention. Stud. Health Technol. Inform. **160**, 279–283 (2010)
3. Perez Rivas, F.J., Martin-Iglesias, S., Pacheco del Cerro, J.L., Minguet Arenas, C., Garcia Lopez, M., Beamud Lagos, M.: Effectiveness of nursing process use in primary care. Int. J. Nurs. Know. **27**, 43–48 (2016)
4. Darmer, M.R., Ankersen, L., Nielsen, B.G., Landberger, G., Lippert, E., Egerod, I.: Nursing documentation audit–the effect of a VIPS implementation programme in Denmark. J. Clin. Nurs. **15**, 525–534 (2006)
5. Wang, N., Hailey, D., Yu, P.: Quality of nursing documentation and approaches to its evaluation: a mixed-method systematic review. J. Adv. Nurs. **67**, 1858–1875 (2011)
6. Paans, W., Sermeus, W., Nieweg, R.M., van der Schans, C.P.: Prevalence of accurate nursing documentation in patient records. J. Adv. Nurs. **66**, 2481–2489 (2010)
7. Tuinman, A., de Greef, M.H.G., Krijnen, W.P., Paans, W., Roodbol, P.F.: Accuracy of documentation in the nursing care plan in long-term institutional care. Geriatr. Nurs. **38**, 578–583 (2017)
8. Paans, W., Nieweg, R.M., van der Schans, C.P., Sermeus, W.: What factors influence the prevalence and accuracy of nursing diagnoses documentation in clinical practice? A systematic literature review. J. Clin. Nurs. **20**, 2386–2403 (2011)
9. Hayrinen, K., Lammintakanen, J., Saranto, K.: Evaluation of electronic nursing documentation–nursing process model and standardized terminologies as keys to visible and transparent nursing. Int. J. Med. Inform. **79**, 554–564 (2010)
10. D'Agostino, F., Vellone, E., Tontini, F., Zega, M., Alvaro, R.: Development of a computerized system using standard nursing language for creation of a nursing minimum data set. Professioni infermieristiche **65**, 103–109 (2012)
11. Zega, M., D'Agostino, F., Bowles, K.H., De Marinis, M.G., Rocco, G., Vellone, E., Alvaro, R.: Development and validation of a computerized assessment form to support nursing diagnosis. Int. J. Nurs. Knowl. **25**, 22–29 (2014)
12. Sanson, G., Alvaro, R., Cocchieri, A., Vellone, E., Welton, J., Maurici, M., Zega, M., D'Agostino, F.: Nursing diagnoses, interventions, and activities as described by a nursing minimum data set: a prospective study in an oncology hospital setting. Cancer Nurs., 13 March 2018, [Epub ahead of print]. https://doi.org/10.1097/NCC.0000000000000581
13. D'Agostino, F., Barbaranelli, C., Paans, W., Belsito, R., Juarez Vela, R., Alvaro, R., Vellone, E.: Psychometric evaluation of the D-Catch, an instrument to measure the accuracy of nursing documentation. Int. J. Nurs. Knowl. **28**, 145–152 (2017)
14. Paans, W., Sermeus, W., Nieweg, R.M., van der Schans, C.P.: D-Catch instrument: development and psychometric testing of a measurement instrument for nursing documentation in hospitals. J. Adv. Nurs. **66**, 1388–1400 (2010)
15. Dunn, O.J.: Multiple comparisons using rank sums. Technometrics **6**, 241–252 (1964)
16. D'Agostino, F., Zeffiro, V., Vellone, E., Ausili, D., Belsito, R., Leto, A., Alvaro, R.: Cross-mapping of nursing care terms recorded in italian hospitals into the standardized NNN terminology. Int. J. Nurs. Knowl., 12 Jan 2018, [Epub ahead of print]. https://doi.org/10.1111/2047-3095.12200

Effectiveness of E-Learning Training on Drug – Dosage Calculation Skills of Nursing Students: A Randomized Controlled Trial

Ferrari Manuela[1], Tonella Simone[1], Busca Erica[2](✉), Mercandelli Stefano[3], Vagliano Liliana[4], Aimaretti Gianluca[2], and Dal Molin Alberto[1,2]

[1] Nursing School, Biella Hospital, Biella, Italy
manuela.ferrari@infermieribiella.it
[2] Department of Translational Medicine, Università del Piemonte Orientale, Novara, Italy
erica.busca@uniupo.it
[3] Biella Hospital, Biella, Italy
[4] Department of Public Health and Pediatrics, School of Medicine, Università di Torino, Torino, Italy

Abstract. Drugs administration errors are often used as patient safety indicators in hospitals, being deemed a common and potentially risky event for individual safety. Poor mathematical skills might contribute to medication calculation errors. E-learning is a new training tool that takes advantage of new technology. It constitutes an important development in training and teaching in general.

The purpose of this study was to evaluate the effectiveness of an e-learning teaching course using on-line platform compared to traditional teaching.

Between 1 April 2016 and 22 April 2016, 198 second-year nursing students were randomized: 98 were allocated to the e-learning group (intervention group) and 100 in the control group received lectures with a professor. Students in both groups completed a questionnaire prior (T_0) and after the course (T_1).

The findings of this study indicated that students, in both groups, boosted their calculation abilities between T_0 and T_1 test (22.4 vs 25.6 $p < 0.001$; 20.3 vs 23.5 $p < 0.001$). However, there was no statistically significant difference between the number of correct responses on both tests or the number of individuals showing improvement for either training course.

Improvements in drug-dosage calculation skills have been observed in e-learning group, although the intervention was no more efficient than traditional lecture. We believe that more studies should be carried out in the future in order to understand better the phenomenon.

Keywords: Drug calculation skills · E-learning · Nursing education

T. Di Mascio et al. (Eds.): MIS4TEL 2018, AISC 804, pp. 253–260, 2019.
https://doi.org/10.1007/978-3-319-98872-6_30

1 Background

One of the most controversial aspects of medical care is the possibility of involuntary causing harm that could result in disabilities. For this reason, the percentage of iatrogenic damage has become an important and significant marker of quality of care [1]. Nowadays, therapeutic errors remain the most common type of nursing/medical errors registered worldwide [2]. It has been estimated that therapy error represents 12%–20% of total errors within the healthcare environment [3]. Although errors occur at every stage of the medication preparation and distribution process, one third of those that harm patients are attributed to the administration phase [4].

Drug administration is one of the most important nursing responsibilities [5]. Efficient and safe drug administration requires not only knowledge of essential information, which are patient data, diagnosis, age, and pathological condition; but, above all, theoretical and clinical, both pharmacological and mathematical, knowledge of dosages [6, 7]. Poor medication calculation skills can lead to erroneous dosages administration, which in turn can damage the patient or put his life in danger [8]. The Nursing Midwifery Council established the obligation for every English nurse to prove, before admission, suitable math skills that will be enhanced and reevaluated throughout her career [9]. In to 2002, Brown's study pointed out that many students are admitted to university with very poor basic math skills, such as calculations of fractions, decimals, and percentages [10]. In Italy, during nursing student education at university, one of the abilities, which is part of practicals, is drug dose calculation. This ability should have already been acquired during secondary school since it is part of the ministerial program, but many studies stated this skill is scarce [11].

In the last few years, great importance has been given to the use of new technology in teaching environment, especially given the effectiveness of e-learning and its capability to offer autonomous and personalized learning [12–14]. It constitutes an important development in terms of global training and teaching, since it allows for a teaching methodology characterized by flexibility in type of education, respect for learning pace, personalized forms of support, and assessment times. Economic and organizational aspects are considered key elements for e-learning sustainability. However, it is important not to place these components above others, in order to guarantee appreciable quality levels [15].

Recent studies investigated the effects of Web-based teaching during nursing education, specifically on the development of math skills and drug calculations [16–19]. More attention has been given to identifying appropriate teaching/learning strategies that facilitate the development of calculation skills, but there isn't a clear conclusion about the best teaching method. Few studies compared e-learning course with traditional classroom lecture, generally it was an integrative program in addition to standard course.

2 Aim

The aim of this study was to evaluate the effectiveness of an e-learning training course by using an on-line platform compared to traditional training on drug-dosage calculation skills.

3 Methods

3.1 Study Design

A randomized control trial in which second-year students have been randomized to an e-learning training group versus a face-to-face course by a random number generator (www.random.org).

3.2 Setting and Participants

All second – year nursing students at a university of northern Italy (academic year 2015–2016) were recruited to participate in the study after obtaining their signed informed consent form. Students who had already failed the course in a previous year were excluded.

3.3 Intervention Group

Students in the intervention group received an e-learning course. It was developed using a custom-made online platform. The course included general exercises for dosage calculation and case-based exercises. Students received feedback for every response given, whether correct or incorrect. The e-learning course was web-based and available to undergraduate students by an access key for 5 days, at multiple locations.

3.4 Control Group

Students in the control group received a four hour-lecture with a professor. The exercises provided were pertinent to the e-learning course. After the lecture, hand-outs of the PowerPoint presentation were given to the students.

3.5 Data Collection Tool

Calculation skills were evaluated for all students at their baseline (T_0) from 1 May 2016 to 15 May 2016 and after the course (T_1) from 7 June 2016 to 14 June 2016. The instrument, originally used in a pediatric environment [11], was adapted to an adult context. Six experts evaluated the content validity: three nurses, a pharmacist, a chemist, and a physician. The final tool contained 33 questions on calculation abilities (equivalences, percentages, proportions) and case-based exercises on medication calculation. Students received one point for every correct answer. Uncompleted exercises or wrong answers were appointed a score of zero. The maximum time to complete the test was 1 h.

Social demographic data (gender, age, high school diploma) were collected at the baseline (T_0).

3.6 Outcome

The outcome was drug-dose calculation skills, considered not only as basic knowledge of arithmetic but also as an ensemble of abilities such as problem conceptualization and mental representation of a solution [20].

3.7 Data Analysis

A descriptive statistic was performed: mean and standard deviation for quantitative variables and absolute and relative frequencies for qualitative variables (gender, secondary education).

Comparison between age, in the group was performed by t-test for unpaired data, while comparison between gender and secondary education by Chi-squared test.

Comparison between mean scores at T_0 and T_1 was carried out, for each group, with t-student test for paired data

Comparisons between the means of scoring increments (from T_0 to T_1), in the two groups, were performed by the t-test for unpaired data.

The percentage of individual improvements (from T_0 to T_1) within the groups were analyzed with the Chi-squared test without Yates's correction.

The significance level was fixed at 5%. All statistical analyses were performed using STATA Vers.13.

3.8 Sample Size

It was necessary to include 80 students in each group. Sample size was estimated assuming a standardized effect size (improvement in drug – dosage calculation skills between the two groups divided by the common standard deviation) of 0.45 (alpha: 0.05; Beta: 0.20) [21, 22].

3.9 Ethical Consideration

Ethical approval was accorded by the review committee of Azienda Ospedaliera Universitaria Maggiore delle Carità di Novara (approval n. 317/CE – 1st April 2016). Students received oral and written information about the study. Participants were asked to fill in and sign an informed consent form before recruitment.

4 Results

In this study, 198 students were enrolled. Ninety-eight of them were allocated in the intervention group (e-learning training) and one hundred in the control group (traditional training). Fourteen students did not complete the study (4 in the intervention group and 10 in the control group) due to their absence during testing. One hundred and eighty - four students were included in the analysis (see Table 1).

Table 1. Baseline characteristics of students

	Intervention group (n = 94)	Control group (N = 90)	p
Female. n (%)	71 (75.5%)	65 (75.2%)	ns
Age. mean (sd)	24 (4.8)	23 (3.34)	ns
High school diploma:			ns
Lyceum. n (%)	59 (62.76%)	64 (71.1%)	
Technical Institute. n (%)	27 (28.72%)	21 (23.3%)	
Vocational school. n (%)	7 (7.44%)	5 (5,55%)	

We observed a statistically significant increase in drug – dosage calculation skills in both groups. Students in the intervention group had a total mean score of 22.4/33 (sd 7.6) in the T_0 test and an average of 25.6 points (sd 6.1) in the T_1 test, with an average improvement of 3.2 points (sd 5) ($p < 0.001$). Control group students had a total mean score of 20.3/33 (sd 7.8) in the T_0 test and an average of 23.5 points (sd 8.3) in the T_1 test, with an average improvement of 3.2 points (sd 6) ($p < 0.001$). There was no statistically significant difference between the number of individuals showing improvement for either training course or the number of correct responses on both tests (see Table 2).

Table 2. T_0–T_1 improvement.

	Intervention group (n = 94)	Control group (N = 90)	p
Students who improved their skills between T_0–T_1. n (%)	68 (72.3%)	59 (65.5%)	0.32†
Scores improvement between T_0–T_1. mean (sd)	3.245 (4.97)	3.2 (6.02)	0.94‡

† Chi-squared test without Yate's correction
‡ T-Test for Unpaired data

5 Discussion

Medication calculation courses have a positive effect on drug-dosage calculation skills. Our study was not able to demonstrate any superiority of e-learning education program efficacy compared to a traditional one. Both methods resulted in improved medication calculations after the course. These results are in line with the study performed by Simonses et al. which proved how the calculation abilities of 183 nurses, coming from different task groups, improved after training [23]. Also, Van Lancker compared the effectiveness of an e-learning course with a traditional course among a group of nursing students in their last year of university. The results showed that both methods had increased the students' knowledge but, after three months, their medication calculation skills declined. In the group subjected to the face-to-face method, the decrease was not significant like in intervention group [18].

McMullan et al. on the other hand, reported better outcomes in favor of the e-learning program [16]. A possible explanation for this discrepancy is the high drop-out rate of students. In addition, in the control group of McMullan study, the provision of hand-outs involves less contact with a teacher.

One other study used a non-randomized design to evaluate the effect of an e-learning program integrated in a regular pediatric course. The students had access to the Pediatric Medication Safety program using an on-line platform, which could be utilized any time during the entire semester. Students in the e-learning group had significantly higher scores on medication calculation, however the intervention and control groups had significant baseline differences [17].

The study results suggest that implementation of an e-learning course in a teaching environment provides students with sufficiently realistic scenarios, allowing them to safely conceptualized and choose and appropriate solution [24]. In fact, the university should evaluate opportunities offered by new technology and include alternative and efficient teaching methodologies in university paths offered to students, which can also reduce educational costs in line with ever more pressing cost containment. All of this can be guaranteed, contributing considerably to the expansion of knowledge, thanks to the employment of 2.0 technologies and to the production of "flexible" educational raw materials, which develop potential to integrate raw material with sources [14].

This study had several limitations. First, fourteen students were lost to follow up, it could compromise the results. Second, the students voluntarily participated, and the e-learning program allowed the students self-directed learning in accordance with their needs. The amount of time students spent studying could not be precisely measured. Third, we only evaluated the effects of the intervention immediately after the course.

The traditional pedagogical model, which is teacher-centered, acknowledges authority in professors and written work. The knowledge process is rapid, but this kind of approach reveals its weakness as being mainly based on memorization rather than on thinking and problem solving and, hence, on the professor's ability to lead the student to solve the problem. In a modern learning environment, which is student-centered, the teacher's role continues to become more that of a facilitator, a group learning leader. The benefits of such a method are interaction, active problem-solving aptitude, and instant feedback from students. Students' weak areas can be easily bridged by focused teaching support. Moreover, evaluation can be carried out both objectively and by self-evaluation methods [13]. Technology provides new and interesting opportunities for learning experiences. An e-learning program allows students to access their course materials at any time or place and to study at their own pace. This kind of learning also provides chances to select alternative learning paths based on prior knowledge and experience. From this point of view, the evaluation process is primarily a self-evaluation process as well.

6 Conclusion

In the end, even if were not able to confirm that e-learning education is more efficient than a traditional education, we can instead infer that both modalities increased students' calculation abilities, highlighting the possibility that the two teaching methods are both efficient. We believe that equivalence or non-inferiority studies should be carried out in future in order to prove this data and to determine which are the best methods offered from the e-learning platforms.

Acknowledgement. We wish to thank: Abelli Gianfranco, Busca Gabriella, Fanton Monica, Ferrante Francesco, Pronsati Stefano, Spagarino Ermanno, for the content validity evaluation of the collection tool. Also, we would like to thank: Faggiano Fabrizio, Bambaci Marilena, Barresi Federica, Bertozzi Alessandra, Bidone Sara, Borsari Virna, Chilin Giovanni, Ferrauto Laura, Moscatiello Mimma, Scapparone Paola, Suardi Barbara, Tibaldi Antonietta, Zavattaro Irene, for the support given in the conduction of this trial and the statistician Silvano Andorno. In addition, we would like to express our gratitude to Julian Michelle.

References

1. Thomas, E.J., et al.: Incidence and types of adverse events and negligent care in Utah and Colorado. Med. Care **38**, 261–271 (2000)
2. Weeks, K.W., Sabin, M., Pontin, D., Woolley, N.: Safety in numbers: an introduction to the nurse education in practice series. Nurse Educ. Pract. **13**(2), e4–e10 (2013)
3. Guchelaar, H.J., Colen, H.B.B., Kalmeijer, M.D., Hudson, P.T.W., Teepe-Twiss, I.M.: Medication errors hospital pharmacist perspective. Drugs **65**, 1735–1746 (2005)
4. Cloete, L.: Reducing medication administration errors in nursing practice. Nurs. Stand. **29**, 50–59 (2015)
5. Armitage, G., Knapman, H.: Adverse events in drug administration: a literature review. J. Nurs. Manag. **11**, 130–140 (2003)
6. Sung, Y.H., Kwon, I.G., Ryu, E.: Blended learning on medication administration for new nurses: integration of e-learning and face-to-face instruction in the classroom. Nurse Educ. Today **28**, 943–952 (2008)
7. Koohestani, H., Baghcheghi, N.: Comparing the effects of two educational methods of intravenous drug rate calculations on rapid and sustained learning of nursing students: formula method and dimensional analysis method. Nurse Educ. Pract. **10**, 233–237 (2010)
8. Bagnasco, A., et al.: Mathematical calculation skills required for drug administration in undergraduate nursing students to ensure patient safety: a descriptive study: drug calculation skills in nursing students. Nurse Educ. Pract. **16**, 33–39 (2016)
9. Nursing and Midwifery Council.: Standards for pre-registration nursing education. Nurs. Midwifery Counc. 1–152 (2010)
10. Brown, D.L.: Does 1 + 1 still equal 2? A study of the mathematic competencies of associate degree nursing students. Nurse Educ. **27**, 132–135 (2002)
11. Bambaci, M., et al.: OC16 – Calculation skills and e-learning platform: study pre-post test on students of paediatric nursing in Italy. Nurs. Child. Young People **28**, 67 (2016)
12. Ekenze, S.O., Okafor, C.I., Ekenze, O.S., Nwosu, J.N., Ezepue, U.F.: The value of internet tools in undergraduate surgical education: perspective of medical students in a developing country. World J. Surg. **41**, 672–680 (2017)
13. Della Corte, F., La Mura, F., Petrino, R.: E-learning as educational tool in emergency and disaster medicine teaching. Minerva Anestesiol. **71**, 181–195 (2005)
14. Pinto, A., Brunese, L., Pinto, F., Acampora, C., Romano, L.: E-learning and education in radiology. Eur. J. Radiol. **78**, 368–371 (2011)
15. Calvani, A., Bonaiuti, G., Fini, A.: Lifelong learning: what role for e-learning 2.0? J. e-Learn. Knowl. Soc. **4**, 179–187 (2009)

16. McMullan, M., Jones, R., Lea, S.: The effect of an interactive e-drug calculations package on nursing students' drug calculation ability and self-efficacy. Int. J. Med. Inform. **80**, 421–430 (2011)
17. Lee, T.Y., Lin, F.Y.: The effectiveness of an e-learning program on pediatric medication safety for undergraduate students: a pretest-post-test intervention study. Nurse Educ. Today **33**, 378–383 (2013)
18. Van Lancker, A., et al.: The effectiveness of an e-learning course on medication calculation in nursing students: a clustered quasi-experimental study. J. Adv. Nurs. **72**, 2054–2064 (2016)
19. Aydin, A.K., Dinç, L.: Effects of web-based instruction on nursing students' arithmetical and drug dosage calculation skills. CIN - Comput. Inform. Nurs. **35**, 262–269 (2017)
20. Wright, K.: Resources to help solve drug calculation problems. Br. J. Nurs. **18**(878–80), 882–883 (2009)
21. Machin, D., Campbell, M.J., Tan, S.B., Tan, S.H.: Sample Size Tables for Clinical Studies, pp. 18–20. Wiley, Hoboken (2009)
22. Cohen, J.: Statistical Power Analysis for the Behavioral Sciences, 2nd edn, p. 567. Academic Press, Cambridge (1988)
23. Simonsen, B.O., Daehlin, G.K., Johansson, I., Farup, P.G.: Improvement of drug dose calculations by classroom teaching or e-learning: a randomised controlled trial in nurses. BMJ Open **4**, 6–25 (2014)
24. Wright, K.: Supporting the development of calculating skills in nurses. Br. J. Nurs. **18**, 399–402 (2009)

Efficacy of High-Fidelity Patient Simulation in Nursing Education: Research Protocol of 'S4NP' Randomized Controlled Trial

Angelo Dante(✉), Carmen La Cerra, Valeria Caponnetto, Ilaria Franconi, Elona Gaxhja, Cristina Petrucci, and Loreto Lancia

Department of Health, Life and Environmental Sciences, University of L'Aquila, Edificio Delta 6 – Via San Salvatore, 67100 Coppito, L'Aquila, Italy
angelo.dante@univaq.it,
{carmen.lacerra,valeria.caponnetto,ilaria.franconi,
elona.gaxhja}@graduate.univaq.it,
{cristina.petrucci,loreto.lancia}@cc.univaq.it

Abstract. High-Fidelity Patient Simulation is defined as a replicated clinical experience in a controlled learning environment that closely represents reality. This learning method creates the opportunity for students to gain specific and fundamental clinical skills in a controlled, safe, and forgiving environment. Even though the available evidence shows the impact of High-Fidelity Patient Simulation in improving nursing students' learning objectives, several reasons do not allow confirming its effectiveness. Many studies show weakness in validity and reliability due to the low quality of study design, inadequate sample size and/or convenience sample recruitment. Furthermore, most studies appear uneven in research methods and unstable in outcome measurements. Finally, little is known about learning outcome retention. Therefore, high-level studies are required to strongly confirm the impact of High-Fidelity Patient Simulation on students' learning outcomes, especially on long-term retention, and on patients' advantages. This trial project, called S4NP (Simulation for Nursing Practice), was designed to confirm if the nursing students' exposure to complementary training based on High-Fidelity Patient Simulation could produce long-term beneficial effects on learning outcomes compared to the traditional training program.

Keywords: High Fidelity Simulation Training · Nursing students
Learning outcomes · Clinical trial

1 Introduction

High-Fidelity Patient Simulation (HFPS) is defined as a replicated clinical experience in a controlled learning environment that closely represents reality [1, 2]. Generally, HFPS sessions are realized with a computerised full-body manikin that can be manipulated to reproduce various physiological parameters and to provide a realistic response to students' actions [3]. HFPS creates the opportunity for students to gain specific and fundamental clinical skills [4, 5] in a controlled, safe, and forgiving

T. Di Mascio et al. (Eds.): MIS4TEL 2018, AISC 804, pp. 261–268, 2019.
https://doi.org/10.1007/978-3-319-98872-6_31

environment free from the risk of harm both for students [6–8] and patients, to which best-quality care in safety conditions must be guaranteed [9]. In fact, the involvement of students in an appropriate learning environment along with the improvement of their knowledge and performances are shared responsibilities of healthcare professionals and educators. As much as healthcare professionals and educators struggle to create appropriate learning environments, the risk of harm to patients by students in training cannot be completely eliminated [10]. Similarly, students in training can be exposed to several risks, such as percutaneous and mucocutaneous exposures to biological materials potentially infected with blood-borne viruses [11, 12]. These risks usually decrease with the rise of clinical skills of students, which they could improve preliminarily and safely before beginning their clinical training [12, 13]. It is evident from the literature that HFPS is becoming an increasingly important aspect of learning in undergraduate nursing programs [9, 14]. Numerous available evidence shows the impact of HFPS on nursing students in terms of achievement of learning objectives [2, 15]. Particularly, adding HFPS to traditional teaching methods, such as classroom lectures and clinical training placements, could result in an effective educational strategy to simulate the 'real-world' clinical settings and enhance the application of theory to practice [16]. However, despite this encouraging evidence, at least, two reasons do not allow confirming the effectiveness of the HFPS simulation in improving nursing students' learning outcomes. Firstly, many studies show weakness in validity and reliability due to the low quality of study design, inadequate sample size and/or convenience sample recruitment [2, 15]. Secondly, most studies appear uneven in research methods and unstable in outcome measurements. Furthermore, little is known about learning outcome retention [2], and the effect of HFPS training on patient outcomes should also be confirmed [17, 18]. For these reasons, high-level studies are required to confirm the impact of HFPS on students' learning outcomes, especially on long-term retention and on patients' outcomes [19–22].

1.1 Aims and Study Hypothesis

Taking these requirements into account, this trial project called S4NP (Simulation for Nursing Practice) was designed to confirm if the nursing student's exposure to complementary training based on HFPS could produce long-term beneficial effects on learning outcomes, compared to the traditional training program. Therefore, the hypothesis of this study is that the second-year nursing students enrolled to receive HFPS show differences in learning outcome acquisition and retention compared to students enrolled to receive traditional training programs based on classroom lecture, clinical placement and static manikin.

2 Methods

2.1 Study Design and Population

The S4NP study will be conducted through a 24-month stratified single-blind, randomized, and controlled waiting-list trial involving nursing students attending an Italian Nursing Degree Program (NDP).

2.2 Setting

This study will be conducted into the NDP of the University of L'Aquila, located in central Italy. In Italy, the NDP lasts 3 years and is based on 180 credits (5400 h), including theoretical courses (96 credits), clinical placements (60 credits), and other activities (24 credits). Students obtain credits by successfully passing theoretical and clinical exams, 5 (on average) and one per year, respectively. Students attend their clinical training in hospitals and social or health facilities under the supervision of trained tutors. At the University of L'Aquila, an Advanced Clinical Simulation Laboratory, called 'Nursing Lab', is going to be equipped as a realistic replication of a hospital intensive care unit room. The patient under various clinical conditions can be simulated by a wireless-controlled Gaumard® simulator 'HAL® S1000', supported by an advanced audio-video recording system that allows both to perform high-level debriefing and to broadcast real-time connections with classrooms. Providing students attending the classroom with 'tele-assisted' lectures represents an innovative, strong point of this teaching system. Nursing Lab training is based on reproducible algorithms designed on the basis of real clinical scenarios and managed by a qualified instructor that operates according to pre-established training objectives. Finally, many part-task trainers allow students to train on specific clinical skills (e.g. injection, urinary catheterization, blood sampling, etc.).

2.3 Inclusion and Exclusion Criteria

Participants will be included in the study if: (1) they are enrolled in the second NDP year for the first time; (2) they have passed the exam of 'Clinical Nursing'; (3) they give their informed consent.

Participants will be excluded from the study if: (1) they were transferred from another degree course or from another university; (2) they are enrolled in an ERASMUS Project; (3) they are workers or volunteers in healthcare facilities.

2.4 Sample Size

Before the S4NP research project was developed, a systematic review of the learning effects of HFPS was conducted by the same team of researchers, and its results were used to estimate preliminary sample sizes for each identified learning outcome. Therefore, to document any difference in end-point scores with a 5% two-tailed alpha and 80% power, 78 students will be required under 1:1 parallel allocation (i.e. 39 students per group). The estimate calculation was performed by G* Power software 3.1.9.2. considering a medium effect size ($d = 0.65$).

2.5 Groups, Intervention, and Follow-up

According to the study design, after having obtained informed consent, participants will be randomized into two groups: Intervention Group (IG) and Control Group (CG).

Intervention Group
In association with traditional training, students in the IG will be exposed to 3 HFPS different scenarios consistent with second-year NDP learning objectives: (1) acute respiratory failure (ARF), (2) heart attack (HA), and (3) heart failure (HF). One week before the HFPS sessions, IG students will receive a two-hour lecture for each of these three clinical conditions. The 60-min HFPS session will start with an initial briefing to introduce students to the characteristics of this teaching and learning method and will finish with a structured and video-assisted debriefing aiming to explore and understand the relationships among events, actions, feelings, performance and outcomes of the simulation. To improve learning outcomes, after having reflected critically on their own weaknesses, students will also be invited to repeat once more the section under the supervision of a tutor. Overall each IG student will be exposed to six HFPS sessions throughout the second year (Fig. 1). Three IG students will be engaged per single session.

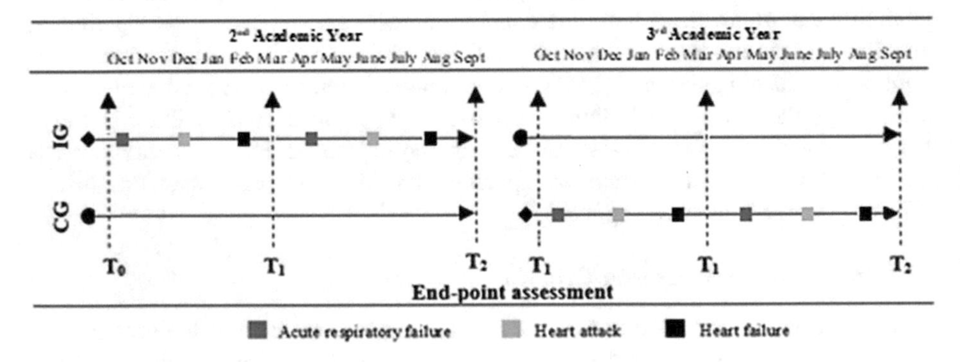

Fig. 1. HFPS sessions and end-point assessment

Control
Control group students will be exposed only to the traditional training based on classroom lecture, clinical placement, and static manikin.

Follow-up
According to the waiting list design, control group students will receive the intervention during the 3rd academic year. For each academic year, end-point measurements will be carried out as follows: T_0 (baseline), T_1 (6 months), and T_2 (12 months) (Fig. 1).

2.6 End-Points and Instruments

In this study, primary end-points include self-reported (self-efficacy, self-confidence, satisfaction) and objective outcomes (performance, knowledge, ability in nursing care) measured immediately after the HFPS session and longitudinally in order to detect the retention over the time. As regards self-reported outcomes, self-efficacy is how students think, feel, motivate themselves, and perceive their clinical practice [22], while self-confidence is how students believe in their own abilities when performing nursing care [23], whereas satisfaction corresponds to the feeling of pleasure that students experience in their learning environment. Regarding the objective outcomes, performance explains how students use their clinical skills [24, 25], knowledge represents the level of the theoretical bases about caring [26], and ability in nursing practice means the ability to achieve objectives in real nursing care settings [25]. Each endpoint will be assessed through ad hoc instruments, questionnaires reported in the literature, and considering current guidelines. Socio-demographic data will be collected with a pre-tested data collection form.

2.7 Randomization

To eliminate confounding effects produced by variables associated with nursing students' academic success and performances [27, 28], a computer-generated stratified randomization will be carried out based on gender, type of upper-secondary school attended, and average number of university credits achieved. Participants will be randomized following a 1:1 allocation distribution (IG:CG). An external investigator will perform the randomization procedure. To guarantee the allocation concealment, participants will be associated with a sequential number and their names will not be revealed to the external investigator, neither will the group to which they will be allocated.

2.8 Blinding

Since it is not possible to blind the intervention, the allocation will be revealed to the principal investigator and students only immediately before the study commences. Otherwise, investigators assessing the end-points and performing data analysis will be blinded.

2.9 Considerations on Guarantees for Students

According to the regulation of the course [29], simulation can be an integral part of the clinical training. Therefore, the participation to HFPS sessions does not compromise the regular training path of the students enrolled in the experimental group. Students who will consider dropping out of the experimental group will be guaranteed to continue their studies without any negative repercussions. Furthermore, to guarantee the same teaching and learning opportunities to the students, a waiting list trial design will be adopted [30, 31]. Therefore, initial intervention group students will receive additional training during the second academic year while initial control group students will receive the intervention after a waiting period, i.e. during the third academic year.

2.10 Data Management and Statistical Analysis

Data will be blindly processed and analysed by an investigator adopting an intention-to-treat approach (ITT). ITT analysis is a pragmatic approach that compares nursing students on the basis of the groups to which they were originally randomly assigned. The ITT approach is recommended to provide an unbiased estimate of the effect of intervention assignment on the outcome [32, 33]. A per-protocol analysis will also be performed as a secondary analysis to evaluate whether the trends observed in intent-to-treat results are robust across student subgroups [34]. The homogeneity of demographic characteristics of the two groups at baseline (T_0) will be explored preliminarily. Descriptive and bivariate statistics will be used to describe participants' characteristics and any possible group differences measured at the end of the first and second years of study. Data entry and processing will be done using the Statistical Package for Social Science (SPSS) version 19.0 software (IBM Corp., Armonk, NY, USA). The statistical level of significance accepted as valid will be two-tailed p-values ≤ 0.05.

2.11 Ethics

Ethical approval for this trial will be required from the Ethics Committee of the University of L'Aquila. Permission to reach students will be asked from the degree course board. Students will be informed of the purpose, requirements, duration of the study and that their participation in the study is voluntary. Students will also be informed that they have the right to drop out from the study at any time without having to give any justification. According to national law [35], confidentiality will be maintained throughout the data collection and analysis.

Informed consent and personal data intervention acceptance will be required through templates approved by the ethical board. These templates are based on national laws [35] and the ethical principles of the Helsinki Declaration.

2.12 Expected Outcomes of the Study

The study will provide scientific evidence on the effectiveness of HFPS in improving undergraduate nursing students' learning outcomes. Furthermore, this research will make data available on long-term effects of HFPS, apart from produce information about nursing students' clinical performances.

References

1. Lewis, R., Strachan, A., Smith, M.M.: Is high fidelity simulation the most effective method for the development of non-technical skills in nursing? A review of the current evidence. Open Nurs. J. **6**, 82–89 (2012)
2. Cant, R.P., Cooper, S.J.: Simulation-based learning in nurse education: systematic review. J. Adv. Nurs. **66**(1), 3–15 (2010)
3. Helyer, R., Dickens, P.: Progress in the utilization of high-fidelity simulation in basic science education. J. Adv. Nurs. **40**(2), 143–144 (2016)

4. Cannon-Diehl, M.R.: Simulation in healthcare and nursing: state of the science. Crit. Care Nurs. Q. **32**(2), 128–136 (2009)
5. Decker, S., Sportsman, S., Puetz, L., Billings, L.: The evolution of simulation and its contribution to competency. J. Contin. Educ. Nurs. **39**(2), 74–80 (2008)
6. Petrucci, C., La Cerra, C., Caponnetto, V., Franconi, I., Gaxhja, E., Rubbi, I., Lancia, L.: Literature-based analysis of the potentials and the limitations of using simulation in nursing education. In: Methodologies and Intelligent Systems for Technology Enhanced Learning, pp. 57–64. Springer, Cham (2017)
7. Pittman, O.A.: The use of simulation with advanced practice nursing students. J. Am. Assoc. Nurse Pract. **24**(9), 516–520 (2012)
8. Yuan, H.B., Williams, B.A., Fang, J.B., Ye, Q.H.: A systematic review of selected evidence on improving knowledge and skills through high-fidelity simulation. Nurse Educ. Today **32**(3), 294–298 (2012)
9. Vincent, M.A., Sheriff, S., Mellott, S.: The efficacy of high-fidelity simulation on psychomotor clinical performance improvement of undergraduate nursing students. Comput. Inform. Nurs. **33**(2), 78–84 (2015)
10. Wolf, Z.R., Hicks, R., Serembus, J.F.: Characteristics of medication errors made by students during the administration phase: a descriptive study. J. Prof. Nurs. **22**(1), 39–51 (2006)
11. Dante, A., Natolini, M., Graceffa, G., Zanini, A., Palese, A.: The effects of mandatory preclinical education on exposure to injuries as reported by Italian nursing students: a 15-year case–control, multicentre study. J. Clin. Nurs. **23**(5–6), 900–904 (2014)
12. Petrucci, C., Alvaro, R., Cicolini, G., Cerone, M.P., Lancia, L.: Percutaneous and mucocutaneous exposures in nursing students: an Italian observational study. J. Nurs. Scholarsh. **41**(4), 337–343 (2009)
13. Cicolini, G., Di Labio, L., Lancia, L.: Prevalence of biological exposure among nursing students: an observational study. Prof. Inf. **61**(4), 217–222 (2008)
14. Roberts, D., Greene, L.: The theatre of high-fidelity simulation education. Nurse Educ. Today **31**(7), 694–698 (2011)
15. Warren, J.N., Luctkar-Flude, M., Godfrey, C., Lukewich, J.: A systematic review of the effectiveness of simulation-based education on satisfaction and learning outcomes in nurse practitioner programs. Nurse Educ. Today **46**, 99–108 (2016)
16. Byrd, J.F., Pampaloni, F., Wilson, L.: Hybrid simulation. In: Human Simulation for Nursing and Health Professions, pp. 267–271. Springer (2012)
17. Haut, C., Fey, M.K., Akintade, B., Klepper, M.: Using high-fidelity simulation to teach acute care pediatric nurse practitioner students. J. Nurse Pract. **10**(10), e87–e91 (2014)
18. Scherer, Y.K., Bruce, S.A., Runkawatt, V.: A comparison of clinical simulation and case study presentation on nurse practitioner students' knowledge and confidence in managing a cardiac event. Int. J. Nurs. Educ. Scholarsh. **4**(1) (2007)
19. Kardong-Edgren, S.: Striving for higher levels of evaluation in simulation. Clin. Simul. Nurs. **6**(6), e203–e204 (2010)
20. Rutherford-Hemming, T.: Learning in simulated environments: effect on learning transfer and clinical skill acquisition in nurse practitioner students. J. Nurs. Educ. **51**(7), 403–406 (2012)
21. Shin, S., Park, J.-H., Kim, J.-H.: Effectiveness of patient simulation in nursing education: meta-analysis. Nurse Educ. Today **35**(1), 176–182 (2015)
22. Zulkosky, K.: Self-efficacy: a concept analysis. Nurs. Forum **44**(2), 93–102 (2009)
23. Perry, P.: Concept analysis: confidence/self-confidence. Nurs. Forum **46**(4), 218–230 (2011)
24. Robb, Y., Dietert, C.: Measurement of clinical performance of nurses: a literature review. Nurse Educ. Today **22**(4), 293–300 (2002)

25. Garside, J.R., Nhemachena, J.Z.: A concept analysis of competence and its transition in nursing. Nurse Educ. Today **33**(5), 541–545 (2013)
26. Hunt, D.P.: The concept of knowledge and how to measure it. J. Intellect. Cap. **4**(1), 100–113 (2003)
27. Lancia, L., Petrucci, C., Giorgi, F., Dante, A., Cifone, M.G.: Academic success or failure in nursing students: results of a retrospective observational study. Nurse Educ. Today **33**(12), 1501–1505 (2013)
28. Dante, A., Petrucci, C., Lancia, L.: European nursing students' academic success or failure: a post-Bologna Declaration systematic review. Nurse Educ. Today **33**(1), 46–52 (2013)
29. Department of Life, Health, and Environmental Sciences: Didactic regulation of Bachelor Degree in Nursing Science in academic year 2017/2018
30. Ronaldson, S., Adamson, J., Dyson, L., Torgerson, D.: Waiting list randomized controlled trial within a case-finding design: methodological considerations. J. Eval. Clin. Pract. **20**(5), 601–605 (2014)
31. Cunningham, J.A., Kypri, K., McCambridge, J.: Exploratory randomized controlled trial evaluating the impact of a waiting list control design. BMC Med. Res. Methodol. **13**(1), 150 (2013)
32. Shrier, I., Steele, R.J., Verhagen, E., Herbert, R., Riddell, C.A., Kaufman, J.S.: Beyond intention to treat: what is the right question? Clin. Trials **11**(1), 28–37 (2014)
33. Sedgwick, P.: Intention to treat analysis versus per protocol analysis of trial data. BMJ **350**, 681 (2015)
34. Wassertheil-Smoller, S., Kim, M.Y.: Statistical analysis of clinical trials. Semin. Nucl. Med. **40**(5), 357–363 (2010)
35. Gazzetta Ufficiale no. 174 of 29 July 2003 - Supplement no. 123, (2003) Personal Data Protection Code - Legislative Decree no. 196 of 30 June 2003

High-Fidelity Patient Simulation in Critical Care Area: A Methodological Overview

Carmen La Cerra(✉), Angelo Dante, Valeria Caponnetto, Ilaria Franconi, Elona Gaxhja, Cristina Petrucci, and Loreto Lancia

Department of Health, Life, and Environmental Sciences, University of L'Aquila, Edificio Delta 6 – via San Salvatore, 67100 L'Aquila, Coppito, Italy
{carmen.lacerra, valeria.caponnetto, ilaria.franconi, elona.gaxhja}@graduate.univaq.it, angelo.dante@univaq.it, {cristina.petrucci, loreto.lancia}@cc.univaq.it

Abstract. High-Fidelity Patient Simulation (HFPS) allows the nursing students' immersion in a learning environment marked by complexity, dynamism, reality, and safety, connecting theory to practice. The increased use of the HFPS was also sustained by the beneficial impact that HFPS seems to produce on nursing students' learning outcomes. However, available research is still conflicting, and several methodological issues do not allow achieving best-quality evidence. To analyse the impact of the HFPS on nursing students' learning outcomes in critical care area, a methodological overview of the studies included in a systematic review was conducted. Heterogeneity about the geographic area, methods, simulation techniques, and control approaches was pointed out. However, studies based on cardiocirculatory scenarios showed the highest standardization of methods and assessment tools. The development and application of shared guidelines derived from adequately-sampled randomized controlled trials are expected to demonstrate the high-fidelity simulation to be a valuable adjunct to the traditional training program.

Keywords: Simulation · Nursing · Education · Nursing students Systematic review · Methodological overview

1 Introduction

In the nursing field, simulation-based devices have markedly advanced up to current high-fidelity patient simulators with human traits that can mimic vital parameters resembling reality [1] and that can respond to physical and pharmacological interventions based on pre-established algorithms or on-the-fly inputs. High-fidelity patient simulation (HFPS) allows the nursing students' immersion in a learning environment marked by complexity, dynamism, reality, and safety, being the bridge between theory and practice [2, 3]. Clinical training can expose patients to adverse events [4] and students to the risk of occupational accidents due to exposure to biological material [5, 6]. However, nursing students must be guaranteed their clinical skill learning, especially in critical care area where rapid and effective interventions are often required [7, 8]. Considering that the

T. Di Mascio et al. (Eds.): MIS4TEL 2018, AISC 804, pp. 269–274, 2019.
https://doi.org/10.1007/978-3-319-98872-6_32

safety of patients and students increases along with the rise in students' clinical skills [5, 6] and that the reshaping of hospitals is resulting on a shortage of training opportunities [1, 5, 6, 9–11], a tendency for HFPS utilization in both undergraduate and post-graduate education has shown to be increased [11, 12]. This positive trend was also sustained by the beneficial impact that HFPS seems to produce on nursing students' learning outcomes [13–15]. However, available research is still conflicting, and several methodological issues do not allow achieving best-quality evidence [14–18]. For this reason, to analyse the impact of HFPS on nursing students' learning selected outcomes in critical care area, a systematic review was conducted. In this paper, a methodological overview of the studies included in this systematic review is reported.

2 Methods

A systematic review was conducted based on the Cochrane Handbook for Systematic Reviews of Interventions [19] and its reporting was checked against the PRISMA checklist [20]. Before implementing the search strategy, a pilot search was performed to obtain keywords and MeSH headings that could fulfil the research objectives. PubMed, Scopus, CINAHL with Full Text, Wiley, and Web of Science were searched. To perform an exhaustive search, reference and citation lists were checked for other relevant references. EndNote Web was used for the reference management of selected articles. To be included, the studies had to meet the following criteria: (a) be observational, quasi-experimental, or experimental; (b) involve academic nursing students; (c) utilize HFPS scenarios to facilitate the learning process; (d) contain data on learning outcomes (i.e. performance, knowledge, self-confidence, self-efficacy, or satisfaction); (e) be available in full-text version; (f) be scenario-based on critical care conditions; (g) be published in English, French, Spanish or Italian language; (h) have control groups not tested on HFPS before the intervention. Three raters screened the titles and abstracts and, for each eligible study, they reviewed the full-text for its relevance based on the inclusion criteria. The consistency of raters' judgments was checked estimating the Krippendorff's alpha coefficient [21]. The included studies were screened for their methodological quality through the Quality Appraisal Checklist for Quantitative Intervention Studies designed by the National Institute for Health and Care Excellence (NICE) [22]. Disagreements between raters were solved by discussion. The study design was checked with 'List of study design features' [19]. Using a coding protocol, data on year of publication, journal, design, country, sample size, mean age, course attended, brand and model of HFPS, control group intervention, scenarios, time of exposure to HFPS scenario, measurement tool of learning outcomes, and modalities of measurements were extracted independently by the authors.

3 Results

The search yielded 2603 studies from the databases directly and 1857 references from the reference and citation searching. After removing duplicates, three raters examined 2130 abstracts and, consequently, 492 eligible full texts independently. A total of 57 studies conducted from 2000 to 2017 were included in this review (Fig. 1).

Good internal validity was reported for all included studies, but, when evaluated by the NICE checklist, most of the studies revealed low generalizability and only one-third showed good external validity. Lack of data on the participants' features such as gender, race, age, educational background, and previous HFPS experiences was pointed out in more than 50% of the studies. Also, a lack of HFPS information, such as equipment, duration of scenario, briefing and debriefing modalities, as well as information about the tools used to evaluate the learning outcomes, were detected. Retrieved studies had been mostly conducted in North America (60.27%), while only 9.59% were conducted in Europe (Table 1). All included studies accounted for 4281 participants ranging from 11 to 352 (median = 65). Enrolled students were mainly undergraduates (86.84%) with an overall mean age of 25.59 (SD 5.30). Most of the studies were quasi-experimental (64.91%) while only 7.02% were randomized controlled trials. Quasi-randomization was undertaken in 45.61% of studies, while in 47.37% no randomization was performed. Controlled studies accounted for the 68.42% of all studies. The most chosen study design was the pre-post approach. HFPS sessions were based mainly on cardio-circulatory (55.00%) and respiratory (37.00%) scenarios, and Laerdal® and METI™ products were the most frequently used simulators (48.65% and 16.21%, respectively). Although several methods were used in the treated groups, the control groups were mainly exposed to lectures (29.63%) or no intervention (22.22%). Anyway, all students comprising those enrolled into the non-intervention groups attended the traditional clinical training.

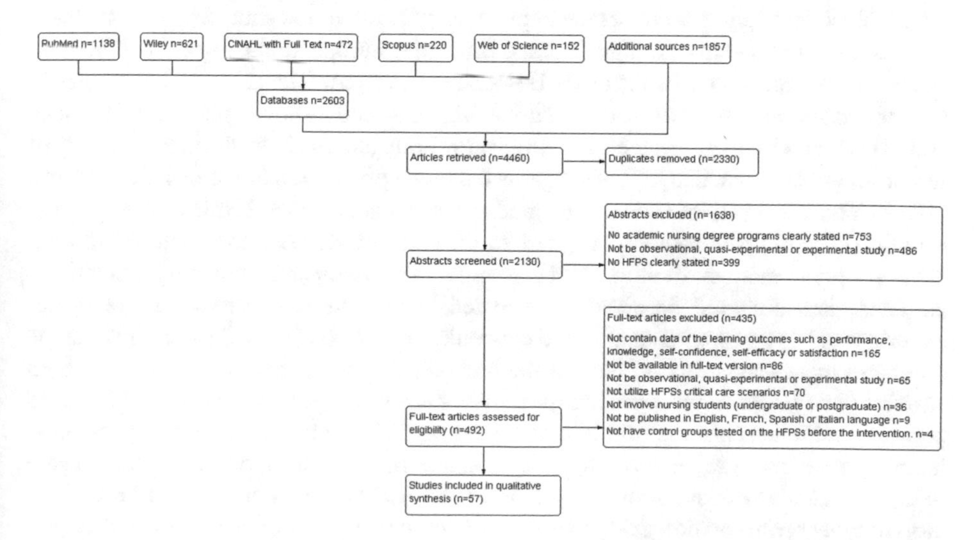

Fig. 1. Flowchart of study search and selection process

Table 1. Studies, students, and simulation characteristics

COUNTRY	
North America	**60.27%**
Asia	**23.29%**
Europe	**9.59%**
Oceania	**6.85%**

STUDENTS	
Undergraduate	**86.84%**
Bachelor	85.53%
Licensed Practical Nurse	1.32%
Diploma	5.26%
Postgraduate	**13.16%**
Advanced Practice Nurse	5.26%
Associate	2.63%

YEAR OF COURSE	
First	**24.56%**
Second	**17.54%**
Third	**21.05%**
Fourth	**31.58%**
Unspecified	**5.26%**

DESIGN	
Randomized Controlled Trial	**7.02%**
Quasi-experimental	**64.91%**
Quasi-Randomized Controlled Trial	45.61%
Non-Randomized Controlled Trial	8.77%
Controlled before-after	10.53%
Observational	**28.07%**

DESIGN	
Concealed randomization	**7.02%**
Quasi-randomization	**45.61%**
No randomization	**47.37%**
One-group	28.07%
Two-group	19.30%

DESIGN	
Controlled	**68.42%**
Pre-post	40.35%
Post-test only	28.07%
Uncontrolled	**31.58%**
Pre-post	31.58%

SIMULATORS	
Laerdal®	**48.65%**
SimMan®	37.84%
Resusci Anne® with Skillmeter	8.11%
SimBaby™	1.35%
Resusci Anne® with iStan®	1.35%
METI™	**16.21%**
Unspecified	9.46%
PediaSIM®	4.05%
BabySIM®	2.70%
Other	**2.70%**
Med Sim-Eagle	1.35%
Gaumard Noelle®	1.35%
Unspecified	**32.43%**

SCENARIOS	
Cardio circulatory	**55.00%**
Respiratory	**37.00%**
Others	**8.00%**

OTHER CONTROL TEACHING METHODS	
Lecture	29.63%
No intervention	22.22%
Low-fidelity simulation	9.26%
Problem-based learning	7.41%
Manikin with VitalSim®	5.56%
Web-based learning	5.56%
Audio-video watching/listening	5.56%
Standardized patient	3.70%
ACTAR® manikin	3.70%
Static half-torso manikin	1.85%
Laerdal vSim®	1.85%
Role-playing	1.85%
Laerdal Family Trainers®	1.85%

4 Discussion

This paper reports a methodological overview of the studies included in an ongoing research project based on a systematic review of the effects of HFPS on academic nursing students' learning outcomes. Low external validity and incompleteness of data reporting characterized a relevant part of the reviewed studies. Since unreported data could be potential confounders on the learning outcome analysis and the low external validity could affect the application of results in practice, more attention should be paid on these two factors to draw stronger conclusions. Since the European nursing education standards are homogeneous, thanks to the Bologna Process umbrella [23, 24], but studies in this field came mainly from North America, a greater body of evidence from European countries is needed to guarantee the ecological validity and facilitate the transferability of research results in practice [25]. As regards the target population of the included studies, little evidence is available about post-graduate nursing students. Furthermore, the distribution of participants among the years of the course does not allow identifying what the appropriate moment to utilize HPFS to obtain the best results on learning outcomes is. These issues should be better investigated in the future. Observational or quasi-experimental design and/or small-sized convenience samples [13–15] determined on the one hand more available data but, on the other side, lower-quality evidence. The high variability of HFPS approaches across studies makes comparative analyses difficult to perform and represents a barrier to understanding their effects on the nursing students' learning out-comes [26], even though the most implemented scenarios in the critical care field, i.e. 'cardiac arrest', showed common traits, as they were usually based on cardiopulmonary resuscitation guidelines and their outcome measurement tools were in part developed by the American Heart Association. Furthermore, since a high variability was detected also about measurement methods of self-perceived outcomes, standardizing these measurement tools it could be useful to make results more comparable. Heterogeneity also emerged in the control groups which had received different interventions, e.g. low-fidelity simulation, lecture, audio-listening, problem-based learning or no intervention, further making the results difficult to compare. For this reason, it would be desirable that control groups were exposed only to the traditional training program. As highlighted by other authors [26], to reach a consistent reproducibility in developing and implementing simulation-based educational strategies, shared guide-lines are required, and standardized protocols and algorithms are advisable as an integrative part of the learning process. This would also allow conducting a replication of studies currently unavailable.

5 Conclusion

HFPS studies were heterogeneous as regard geographic area, methods, simulation techniques, and control approaches. Most studies have been conducted in the field of critical care and based on cardiocirculatory scenarios, for which the highest standardization of methods and assessment tools was detected. Highlighting the presence of several limitations in the current evidence, this methodological review should stimulate the conduction of best-quality studies in order to add more reliable data to the literature. Specifically, the development of shared guidelines derived from randomized controlled studies performed with adequate sample sizes is expected to recognize the high-fidelity simulation as a valuable adjunct to the traditional training program.

References

1. Petrucci, C., La Cerra, C., Caponnetto, V., Franconi, I., Gaxhja, E., Rubbi, I., Lancia, L.: Literature-based analysis of the potentials and the limitations of using simulation in nursing education. In: Methodologies and Intelligent Systems for Technology Enhanced Learning, Cham 2017, pp. 57–64. Springer International Publishing (2017)
2. Al-Elq, A.H.: Simulation-based medical teaching and learning. J. Fam. Community Med. **17**(1), 35 (2010)
3. Weller, J.M.: Simulation in undergraduate medical education: bridging the gap be-tween theory and practice. Med. Educ. **38**(1), 32–38 (2004)
4. Wolf, Z.R., Hicks, R., Serembus, J.F.: Characteristics of medication errors made by students during the administration phase: a descriptive study. J. Prof. Nurs. **22**(1), 39–51 (2006)
5. Cicolini, G., Di Labio, L., Lancia, L.: Prevalence of biological exposure among nursing students: an observational study. Prof. Inferm. **61**(4), 217–222 (2008)
6. Petrucci, C., Alvaro, R., Cicolini, G., Cerone, M.P., Lancia, L.: Percutaneous and mucocutaneous exposures in nursing students: an Italian observational study. J. Nurs. Scholarsh. **41**(4), 337–343 (2009). https://doi.org/10.1111/j.1547-5069.2009.01301.x
7. Department of Health: Didactic regulation of Bachelor degree in Nursing Science 2017/2018 (2017)
8. Kneebone, R., Nestel, D., Vincent, C., Darzi, A.: Complexity, risk and simulation in learning procedural skills. Med. Educ. **41**(8), 808–814 (2007)

9. Gordon, J.A., Wilkerson, W.M., Shaffer, D.W., Armstrong, E.G.: "Practicing" medicine without risk: students' and educators' responses to high-fidelity patient simulation. Acad. Med. **76**(5), 469–472 (2001)
10. Jeffries, P.R.: A framework for designing, implementing, and evaluating: simulations used as teaching strategies in nursing. Nurs. Educ. Perspect. **26**(2), 96–103 (2005)
11. Kardong-Edgren, S., Adamson, K.A., Fitzgerald, C.: A review of currently published evaluation instruments for human patient simulation. Clin. Simul. Nurs. **6**(1), e25–e35 (2010)
12. Nehring, W.M.: US boards of nursing and the use of high-fidelity patient simulators in nursing education. J. Prof. Nurs. **24**(2), 109–117 (2008)
13. Cant, R.P., Cooper, S.J.: Simulation-based learning in nurse education: systematic review. J. Adv. Nurse **66**(1), 3–15 (2010)
14. Dante, A., Petrucci, C., Lancia, L.: European nursing students' academic success or failure: a post-Bologna Declaration systematic review. Nurse Educ. Today **33**(1), 46–52 (2013)
15. Warren, J.N., Luctkar-Flude, M., Godfrey, C., Lukewich, J.: A systematic review of the effectiveness of simulation-based education on satisfaction and learning outcomes in nurse practitioner programs. Nurse Educ. Today **46**, 99–108 (2016)
16. Haut, C., Fey, M.K., Akintade, B., Klepper, M.: Using high-fidelity simulation to teach acute care pediatric nurse practitioner students. J. Nurse Pract. **10**(10), e87–e91 (2014)
17. Newell, D.J.: Intention-to-treat analysis: implications for quantitative and qualitative research. Int. J. Epidemiol. **21**(5), 837–841 (1992)
18. Scherer, Y.K., Bruce, S.A., Runkawatt, V.: A comparison of clinical simulation and case study presentation on nurse practitioner students' knowledge and confidence in managing a cardiac event. Int. J. Nurs. Educ. Scholarsh. **4**(1), 1–14 (2007)
19. Higgins, J.P., Green, S.: Cochrane Handbook for Systematic Reviews of Interventions, vol. 4. Wiley, Hoboken (2011)
20. Moher, D., Liberati, A., Tetzlaff, J., Altman, D.G., Group, P.: Preferred reporting items for systematic reviews and meta-analyses: the PRISMA statement. PLoS Med. **6**(7), e1000097 (2009)
21. Krippendorff, K.: Agreement and information in the reliability of coding. Commun. Methods Meas. **5**(2), 93–112 (2011)
22. Jackson, R., Ameratunga, S., Broad, J., Connor, J., Lethaby, A., Robb, G., Wells, S., Glasziou, P., Heneghan, C.: The GATE frame: critical appraisal with pictures. ACP J. Club **144**(2), A8–A11 (2006)
23. Davies, R.: The Bologna process: the quiet revolution in nursing higher education. Nurse Educ. Today **28**(8), 935–942 (2008)
24. Millberg, L.G., Berg, L., Lindström, I., Petzäll, K., Öhlén, J.: Tensions related to implementation of postgraduate degree projects in specialist nursing education. Nurse Educ. Today **31**(3), 283–288 (2011)
25. Polit, D.F., Beck, C.T.: Generalization in quantitative and qualitative research: myths and strategies. Int. J. Nurs. Stud. **47**(11), 1451–1458 (2010)
26. Cant, R.P., Cooper, S.J.: The value of simulation-based learning in pre-licensure nurse education: a state-of-the-art review and meta-analysis. Nurse Educ. Pract. **27**, 45–62 (2017)

The Use of a Dedicated Platform to Evaluate Health-Professions University Courses

Giovanni Galeoto[1(✉)], Raffaella Rumiati[2], Morena Sabella[2], and Julita Sansoni[1]

[1] Department of Public Health, "Sapienza" University of Rome, Rome, Italy
{giovanni.galeoto,julita.sansoni}@uniroma1.it
[2] National Agency for the Evaluation of Universities and Research Institutes, Rome, Italy
{raffaella.rumiati,morena.sabella}@anvur.it

Abstract. The aim of the current study is to discuss a national platform for evaluating nursing education in Italy by means of a progress test and to compare digital versus paper administration of the test. In 2016, the agency updated the research design, including the domains, the methodological approach, and the tests for both Transversal Competencies (TECO-T) and Disciplinary Competencies (TECO-D). The TECO project aims to construct indicators that reflect the skills developed from the first through the third year of the university degree. For the digital study, 8516 students at 19 Italian university universities were recruited; 5975 students of degree courses in nursing took the electronic TECO, and 4326 used the paper format. Asked to evaluate their satisfaction in completing the TECO, the students found it simple, clear, and understandable, but reported difficulty in answering questions due to a lack of practicality in the paper test. The project encourages the development of shared core disciplinary contents and their compatibility with the Dublin Descriptors; allows the development of disciplinary tests (TECO-D) whose results can be used for self-assessment and inter- and intra-university comparisons; and ensures centralized management of the collection of data.

Keywords: Progress test · Platform · Health professions

1 Introduction

Progress testing is a longitudinal testing approach devised by the University of Missouri-Kansas City School of Medicine and the University of Limburg in Maastricht [1–3]. It is a test of the complete domain of knowledge considered as a core requirement for a medical student on completion of his or her undergraduate program.

The main advantage of the progress test is that it breaks the link between learning and revision [3]. It is widely believed that what is asked in examinations drives what students learn [4, 5].

On this principle, in Italy in 2012, the National Agency for the Evaluation of Universities and Research Institutes (ANVUR) started a project with the aim of testing and monitoring the learning outcomes of Italian undergraduate students (the TEst of

T. Di Mascio et al. (Eds.): MIS4TEL 2018, AISC 804, pp. 275–284, 2019.
https://doi.org/10.1007/978-3-319-98872-6_33

COmpetencies, or TECO). The idea behind this project was to develop an important tool for monitoring the quality of the education process. TECO results are part of the Italian Quality Assurance system in the quality-of-education process. In 2016, the agency updated the research design, including the domains, the methodological approach, and the tests for both transversal competencies (TECO-T) and disciplinary competencies (TECO-D). The disciplinary competencies are now closely linked to the specific educational contents of the students' program, and as such they can be compared only to those from similar programs. The survey is coordinated by ANVUR with coordinators chosen by the disciplinary groups. The agency provides an electronic platform (CINECA) to the over 8500 students enrolled in the health-professions programs. Meanwhile, with the aim of measuring and comparing the usefulness and versatility of the CINECA platform for the administration of the TECO-T and TECO-D questionnaires, a parallel administration of the same tests in paper form was offered to 4326 nursing students of the Sapienza University of Rome.

The aim of the current study is to discuss a national platform for evaluating nursing education in Italy by means of a progress test and to compare digital versus paper administration of the test. A secondary objective is to evaluate the differences in satisfaction between paper delivery and electronic administration from the point of view of students.

2 Materials and Methods

2.1 Population

The sample was recruited from October 2017 through January 2018 at 19 Italian universities on a voluntary basis of participation. In order for the students to be included in the study, they had to meet the following inclusion criteria:

(1) Enrolled at one of the 19 universities participating in the experimental administration;
(2) Enrolled in a three-year degree program in nursing;
(3) Enrolled in the first, second, or third year, or were graduating students of the November–December or March–April sessions of nursing courses offered in the Italian language.

Exclusion Criteria

(1) Students not enrolled at one of the 27 universities participating in the experimental administration;
(2) Students not enrolled in a three-year degree program in nursing; or
(3) Students enrolled in nursing courses offered in English.

Electronic and/or Paper TECO: In 2016, the agency redefined the research design, including the domains, the methodological approach, and the tests for both Transversal Competencies (TECO-T) and Disciplinary Competencies (TECO-D). The TECO project aims to construct indicators that reflect the skills developed from the first through the third year of the university degree. The generic/transversal competencies defined by ANVUR are Literacy and Numeracy. The research hypothesis is that these competencies draw on general formative education and are taught throughout the term of higher education, assuring comparable information between institutions and program studies. The disciplinary competencies, by contrast, are closely linked to the specific educational contents of the students' program and consequently can be compared only with those from similar programs.

Times of Administration. The session of the TECO-T is 75 min in total: 35 min for the Literacy module and 40 for the Numeracy module. After the TECO-T, the students answer the TECO-D questions in a session lasting 90 min. Enrollment by students at the Universitaly site takes approximately 10–15 min.

Prerequisites. For successful testing, students must be instructed to bring an identity document, their tax code, and Universitaly credentials (if already registered).

TECO-Cineca Platform. A CINECA institutional platform was created for the administration of the TECO on skills. The platform includes three persons involved in the process of creation/administration of the test: University contact (the person who prepares the sessions for the administration of the test), Classroom tutor (the person who opens and closes the administration of the test) and the student (who compiles and sends the test data).

2.2 Activity of the University Representative

Reserved Area. The university contact can access his or her reserved area through the following link: https://verificheonline.cineca.it/teco/login_ateneo.html, inserting the credentials received via email from CINECA (Fig. 1).

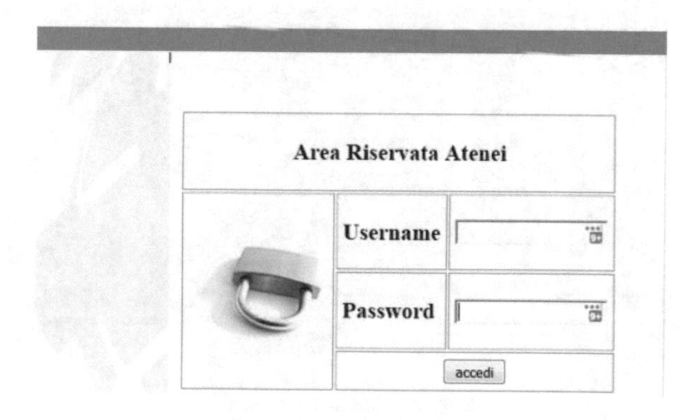

Fig. 1. Reserved area

Session Creation. After accessing the reserved area, the university contact can generate the classroom sessions. Sessions can be created by clicking on “Manage Test Session” (Fig. 2).

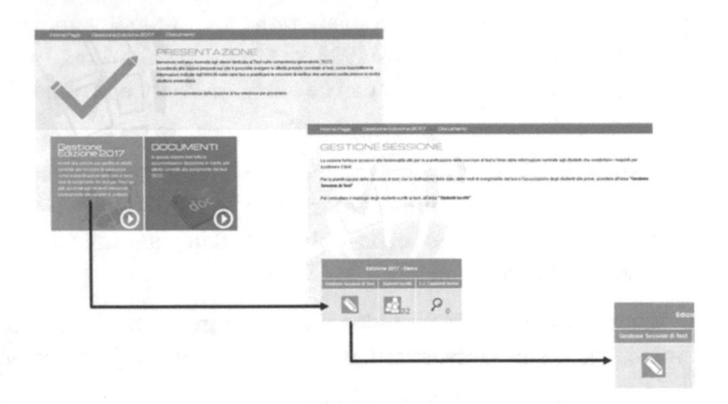

Fig. 2. Session creation

Definition of Meeting Space and Supervision (Tutors). The university contact must complete all of the fields on the “Insert New Session” screen. The number of available seats is defined by the same university contact during the session-creation phase (Fig. 3).

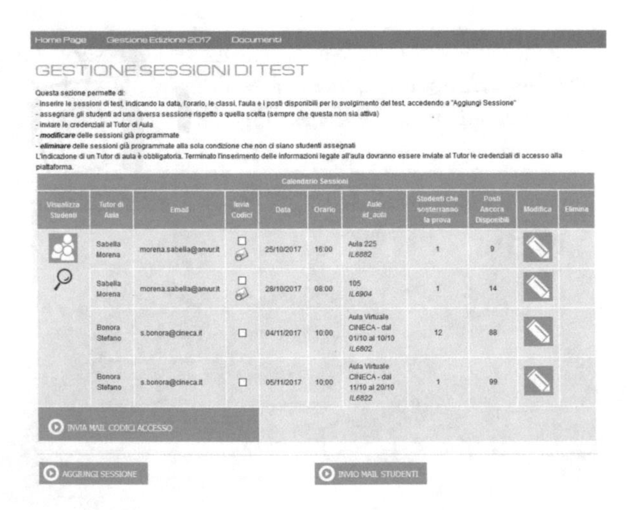

Fig. 3. Sessions of test

2.3 Activity of Supervising Tutor

Reserved Area. The supervising tutor can access his or her private area through the following link: https://verificheonline.cineca.it/teco/login_commissione_aula.html, inserting the credentials received from the university contact.

Student List. On the "Student List" page, all students who have been selected are present; the "Status" column makes it possible to monitor the progress of student tests (Fig. 4).

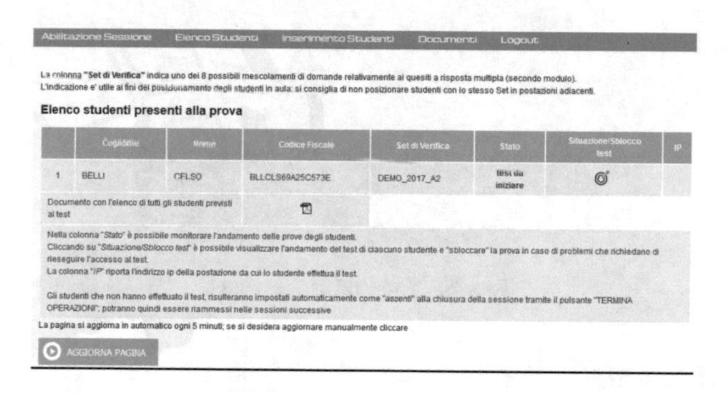

Fig. 4. Page of student list

Generation of the Tests. When the system has not associated the test to a student, the classroom tutor must generate the test. By accessing the "Session Enable" menu page, the classroom tutor can define an access key (test password) for the test, enable the session, and communicate the password to the classroom. The system will request confirmation of the operation: "Do you want to proceed with activation of the session for all students?" (who have been registered for the session) (Fig. 5).

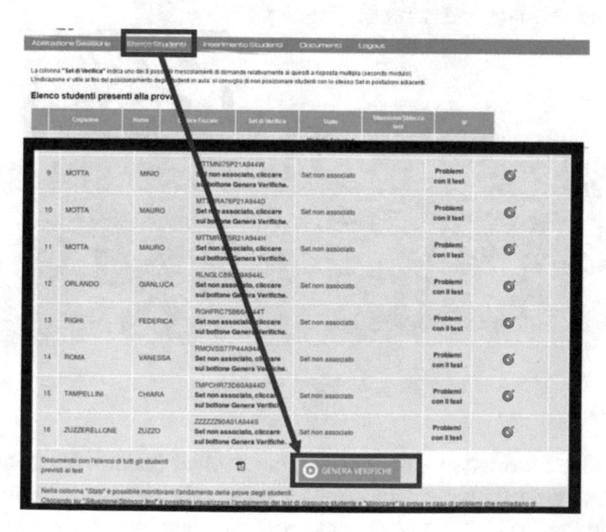

Fig. 5. Page of generation of the test

End of the Test Session. When all students have completed the test (including students who have completed the test after using the test release), click "End Task" to confirm the conclusion of the test session. The session can no longer be enabled, and absent students can be re-admitted in a new session (Fig. 6).

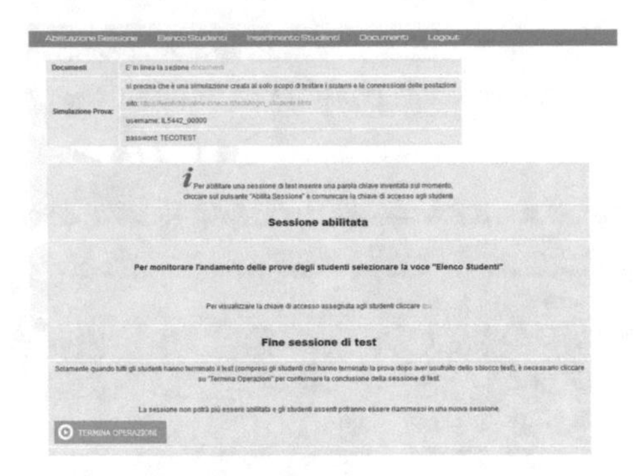

Fig. 6. Page of session test

2.4 Student Activity

Students must access the area: https://verificheonline.cineca.it/teco/login_studente.html
.

Students Not Yet Registered on Universitaly. An essential requirement for the test is registration on the Universitaly website. If the student is not registered, he or she can do it directly by clicking on the blue "Subscribe to the test" box on the home page. Through the "Subscribe" button, students not registered on the Universitaly portal will first fill out the regular registration form for Universitaly (first screen) and then the one to register for the test (second screen) (Fig. 7).

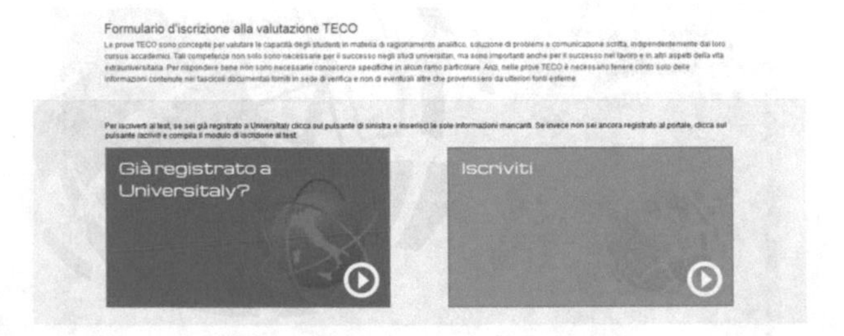

Fig. 7. Page of student

Test Access. Once enrolled at Universitaly, the student (in the classroom) will have to access the private area: https://verificheonline.cineca.it/teco/login_studente.html, click on "Subscribe to the Test" (blue box), and access the test by entering its fiscal code and the password defined by the class tutor. The buttons are present in each phase of the test. The "Summary" page contains: the number of missing answers; a list of answers provided with an indication in red of the missing answers; an explanation of how to return to the last page consulted; and the delivery button to end the current phase. All of the answers are clickable so that the student can access and answer the questions directly or just verify the answer. The "Delivery" button terminates the current phase in a definitive and irreversible way. A confirmation message warns the student of the impossibility of being able to re-enter the current phase: "Attention: By delivering, it will no longer be possible to make changes to this form. Do you want to confirm delivery?"

2.5 Satisfaction Questionnaire

A questionnaire was constructed to investigate through nine questions the students' satisfaction with the administration of the TECO (Appendix A).

2.6 Statistics Analysis

A descriptive analysis of the student cohorts was carried out, and a t student for independent samples was performed for the evaluation of the differences in satisfaction with the questionnaire. The software SPSS Statistics V22.0 was used.

3 Results

For the digital study, 8516 students at 19 Italian university universities were recruited; 5975 students of degree courses in nursing took the electronic TECO, and 4326 used the paper format. Four hundred eighty-one sessions were opened and 146 classroom tutors were used (Table 1).

Table 1. Number of student, session and tutor

University	Sessions	Tutor	Nursing electronic TECO	Nursing paper TECO
1	28	13	248	
2	12	5	257	
3	22	11	266	
4	8	3	160	
5	23	6	285	
6	19	15	415	
7	6	3	118	
8	16	5	136	
9	25	11	102	
10	15	4	317	
11	28	11	166	

(continued)

Table 1. (*continued*)

University	Sessions	Tutor	Nursing electronic TECO	Nursing paper TECO
12	27	27		4326
13	2	1	80	
14	8	2	206	
15	43	13	796	
16	12	3	179	
17	28	6	437	
18	7	1	121	
19	107	9	1686	
Total	481	146	5975	4326

Data from the qualitative assessments of tests administered through the IT platform are not yet available.

From the evaluation of the students using the paper questionnaire, it emerged that there are no statistically significant differences between the two cohorts of students as regards the simplicity, clarity, and comprehensibility of the contents.

On the other hand, the difference between Question 4 and Question 5 is statistically significant, where the students replied that the electronic TECO is satisfactory with regard to the TECO style and graphics ($p < 0.05$). As regards the time of compilation, students who responded with the paper TECO are not congruous for 88%.

4 Discussion

The progress test generates a great deal of information, and this can be used in a number of ways. Because every student takes the same test, meaningful comparisons can be made between students within the same cohort and between cohorts. When plotted as a line chart [1, 6], this gives a visual aid that is easy to understand for the wide variety of people who support the test, the students, and the medical school as a whole. The role of feedback and the importance it plays in medical schools is not in doubt [7]. Progress tests generate a variety of information that can be used [8], and the digital form is indicated as the best.

Due to the number of participants and the lack of digital rooms, it was decided to present TECO in paper format with supervision of students by tutors. The students completed the three questionnaires, and the process was very demanding in terms of commitment, presence, and timing of administration.

From the evaluation of the questionnaire, it emerged that 96% of the students consider TECO very effective to monitor the preparation of students in the degree courses in nursing; 78% consider TECO very useful for the evaluation of degree courses; and 86% find it very useful for the evaluation of preparation. As for the satisfaction of completing the TECO, the students find it simple, clear, and understandable. However, it is difficult to answer the questions due to the lack of practicality of the paper test.

5 Conclusion

The project encourages the development of shared core disciplinary contents and their compatibility with the Dublin Descriptors; allows the development of disciplinary tests (TECO-D) whose results can be used for self-assessment and inter- and intra-university comparisons; and ensures centralized management of the collection of data.

A Appendix

1. Do you think it is easy to answer the questions of the paper/electronic TECO?
   □Yes □In part □No
2. Are the contents of the paper/electronic TECO clear?
   □Yes □In part □No
3. Did you understand the contents of the paper/electronic TECO?
   □Yes □In part □No
4. Did you like the way the paper/electronic TECO was graphically designed?
   □Yes □In part □No
5. Are you satisfied with the style of the paper/electronic TECO?
   □Yes □In part □No
6. In your opinion, is the time sufficient to compile the paper/electronic TECO?
   □Yes □In part □No
7. In your opinion, how useful is the TECO for monitoring the preparation of students in the nursing-degree program?
   □Extremely effective □Very effective □Moderately effective □Slightly effective □Not effective
8. In your opinion, how useful is the TECO for evaluating the degree courses in the nursing-degree program?
   □Extremely effective □Very effective □Moderately effective □Slightly effective □Not effective
9. In your opinion, how useful is the TECO for evaluation of your degree-course preparation?
   □Extremely effective □Very effective □Moderately effective □Slightly effective □Not effective

References

1. Van der Vleuten, C.P.M., Verwijnen, G.M., Wijnen, H.F.W.: Fifteen years of experience with progress testing in a problem-based learning curriculum. Med. Teacher **18**, 103–109 (1996)
2. Van Heeson, P.A.W., Verwijnen, G.M.: Does problem-based learning provide other knowledge? In: Bender, W., Hiemstra, R.J., Scherpbier, A.J.J.A., Zwierstra, R.P. (eds.) Teaching and Assessing Clinical Competence. Boek Werk, Groningen, The Netherlands (1990)
3. Arnold, L., Willoughby, T.L.: The quarterly profile examination. Acad. Med. **65**, 515–516 (1990)

4. Newble, D.I., Jaeger, K.: The effect of assessments and examinations on the learning of medical students. Med. Educ. **17**, 165–171 (1983)
5. Frederiksen, N.: The real test bias. influences of testing on teaching and learning. Am. Psychol. **39**, 193–202 (1984)
6. Freeman, A.C., Ricketts, C.: Choosing and designing knowledge assessments: experience at a new medical school. Med. Teach. **32**(7), 578–581 (2010)
7. Archer, J.: State of the science in health professional education: effective feedback. Med. Educ. **44**, 101–108 (2010)
8. Coombes, L., Ricketts, C., Freeman, A., Stratford, J.: Beyond assessment: feedback for individuals and institutions based on the progress test. Med. Teach. **32**(6), 486–490 (2010). https://doi.org/10.3109/0142159X.2010.485652

Systems Thinking, Complex Adaptive Systems and Health: An Overview on New Perspectives for Nursing Education

I. Notarnicola(✉), A. Stievano, A. Pulimeno, and G. Rocco

Centre of Excellence for Nursing Scholarship, OPI Rome, Italy
ippo66@live.com

Abstract. This article describes a new concept of health, originally developed and published in 2014. This model offers a new understanding of health, disease and healing that may be very useful for patient care. In fact, the Meikirch model leads to a new comprehension of health as a complex adaptive system. Systems thinking and complex adaptive systems share a number of components, namely: emergence, self-organization, and hierarchies of interacting systems. Systems thinking, especially with simulation models, facilitates understanding of health as a complex phenomenon. Therefore, the simulation model is becoming an excellent translator of complex problems in easily understandable results. Systemic thinking is a process that can influence cause and effect and prompts solutions to multifaceted tribulations. In conclusion, this paper describes the principles of complementarity and of differences between scientific approaches for systemic thinking and traditional thinking and suggests that it is time for research approaches that fosters a non-mechanistic thinking.

Keywords: Systems thinking · Complex adaptive systems · Health Systems modeling · Nursing

1 Introduction

Health is defined by the World Health Organization (WHO) as a *"state of complete physical, mental and social well-being and not just the absence of disease"*; therefore, it is considered a fundamental right of every human being. This concept plays an important role in every State because it determines what a health system must deliver equally to all individuals. As such, it is of fundamental importance to identify and try to modify those factors that negatively affect health, while, at the same time, supporting the positive ones [1]. Health can be considered a constant resource, which allows every human being to lead a profitable life at economic, social and individual levels. The WHO concept of health is very important and always stirs reflections and discussions in the scientific environment. However, the concept of health continues to be accepted, particularly in the biomedical sciences, as a disease-free condition or state [2].

T. Di Mascio et al. (Eds.): MIS4TEL 2018, AISC 804, pp. 285–292, 2019.
https://doi.org/10.1007/978-3-319-98872-6_34

2 Background

We can consider human beings as an open system. A system that cannot be considered closed, but a complex system with the ability to interact with the external environment in a dynamic and continuous way. The relationships that each individual has in society are fundamental for his/her social life. We do not have to have a vision of the human being as an individual, but as a system of relationships or rather a complex adaptive system (CAS) [3]. Therefore, the quantity and quality of information that each individual is able to receive, communicate and, then, process is important. A human being can be considered a multilevel agent that operates in different environments incessantly and is capable of participating in its own overall and collective development. We can consider our health as a manifestation and result of a myriad of complex interrelationships in structure and function, both within and throughout many levels of organization—from molecules and cells to systems functioning throughout the body and their interfaces with whole ecosystems [4]. The traditional, *"reductionist"* view of health care has unquestionably succeeded and led to great progress and remarkable insights, but it may be time for a more integrated approach to integrate traditional nursing. According to some researchers, the science of complexity is simply the next stage in understanding how systems work [5]. In fact, nursing is the tradition of studying systems, seen as connections and interactions within a systems paradigm, and must continue in a modern vision [6]. Therefore, it is important to clarify the meaning of the well-being of each individual in a systemic vision. According to Bircher & Hahn, "*Health is a state of well-being emerging from the conductive interactions between the potentials of individuals, the needs of life and social and environmental determinants. Health outcomes throughout the course of life when the potential of individuals - and social and environmental determinants - are sufficient to respond satisfactorily to the demands of life. The demands of life can be physiological, psychosocial or environmental, and vary between individual and context, but in any case, unsatisfactory answers lead to disease.*" [7].

In recent years, the science of complexity has been gradually entering nursing field. This becomes relevant as an explanation of health and disease as different states of a CAS. An understanding of the state of health care requires a complexity science approach, introducing a new dimension to patient care and assistance [8]. According to some Netherlands researchers [9], a systemic approach, such as systemic thinking, incorporates interactions between relevant factors by providing additional information for planning and evaluating health promotion [9].

Precisely for this reason, systemic thinking has been defined as: a process applied to individuals, teams and organizations to influence the cause and the effect where solutions to complex problems are realized through a collaborative effort based on personal abilities compared to the improvement of components and to the complex. The main attributes that characterize systemic thought are: the dynamic system, the holistic perspective, the identification of the model and the transformation [10].

Systemic thinking, therefore, is the ability to recognize, understand and synthesize interactions and interdependencies in a system. This method includes the ability to recognize patterns and repetitions in interactions and an understanding of how actions and components can reinforce or neutralize each other [11]. In other words, systemic thinking relates the environment of a person to his/her way of acting within a system. In nursing delivery, this implies the nurse understands the assessment of how the components of a complex health care system affect the care of an individual patient [11]. Systemic thinking is an essential attitude for nurses. Even if, this ability is important, there is little knowledge of this thought process, so we need to work towards ensuring that it plays a key role in improving patient care in increasingly complex health organizations. To educate nurses in systemic thinking, it is important to improve their awareness of this topic. Nurses should also be encouraged to see interdependencies between people, processes and services, and to see problems that have occurred as part of a chain of events of a larger system, rather than as independent events. Learning is a dynamic process; as nurses discover new knowledge and build new insights, the development of systemic thinking will help to improve their decision-making processes. For example, in an academic and health setting, nurses can use systemic thinking, both to improve decision-making skills and to improve clinical practice by increasing essential skills.

Systemic thinking, through simulation models, can facilitate the understanding of complex health policy problems. In fact, simulation models educate health professionals to improve their skills, serving as tools capable of translating complex scientific evidence into easily understood results [12]. In this sense, a system that could help clarify what health is, in other words, to propose a valid concept of health to apply to the care of people and public health has been developed by Bircher and Kuruvilla [13], with their Meikirch Model.

3 Aim

The purpose of this article is to summarize the relevant characteristics of the Meikirch Model and to show, in detail, how to apply this Model to better understand patient care. The Meikirch Model is a new definition of health that has all the characteristics of a CAS.

4 Data Source

A bibliographic search, without time limits, was undertaken to retrieve articles published using the following databases: Cumulative Index for Nursing and Allied Health Literature (CINAHL©), PubMed© and Google Scholar©. Once the articles were identified, with the analytical support of ENDNOTE X8 (Thomson Reuters©, New York), duplicates were excluded and articles of interest were selected. Only articles in English were retrieved and included. Furthermore, both the title and the abstract had to contain the following keywords: Meikirch model, health, system thinking, complex adaptive system, nursing. The aim was to identify the evidence in the literature that outlined the concept of health and how this was examined through systemic thinking. Initially, 811 articles were found; after the removal of duplicates, 646 articles were left. The articles found used various research methodologies. The material available was interesting for the various topics analyzed, above all, those on systemic thinking according to the concept of health. This led to the identification of 278 articles; these articles were then analyzed, focusing on those in which systemic thought and the concept of health were in the title and in the abstract or those which contained contents related to this concept. This led to an analysis and discussion of 10 articles. These were read and re-read analyzed carefully through a content based process (See Fig. 1).

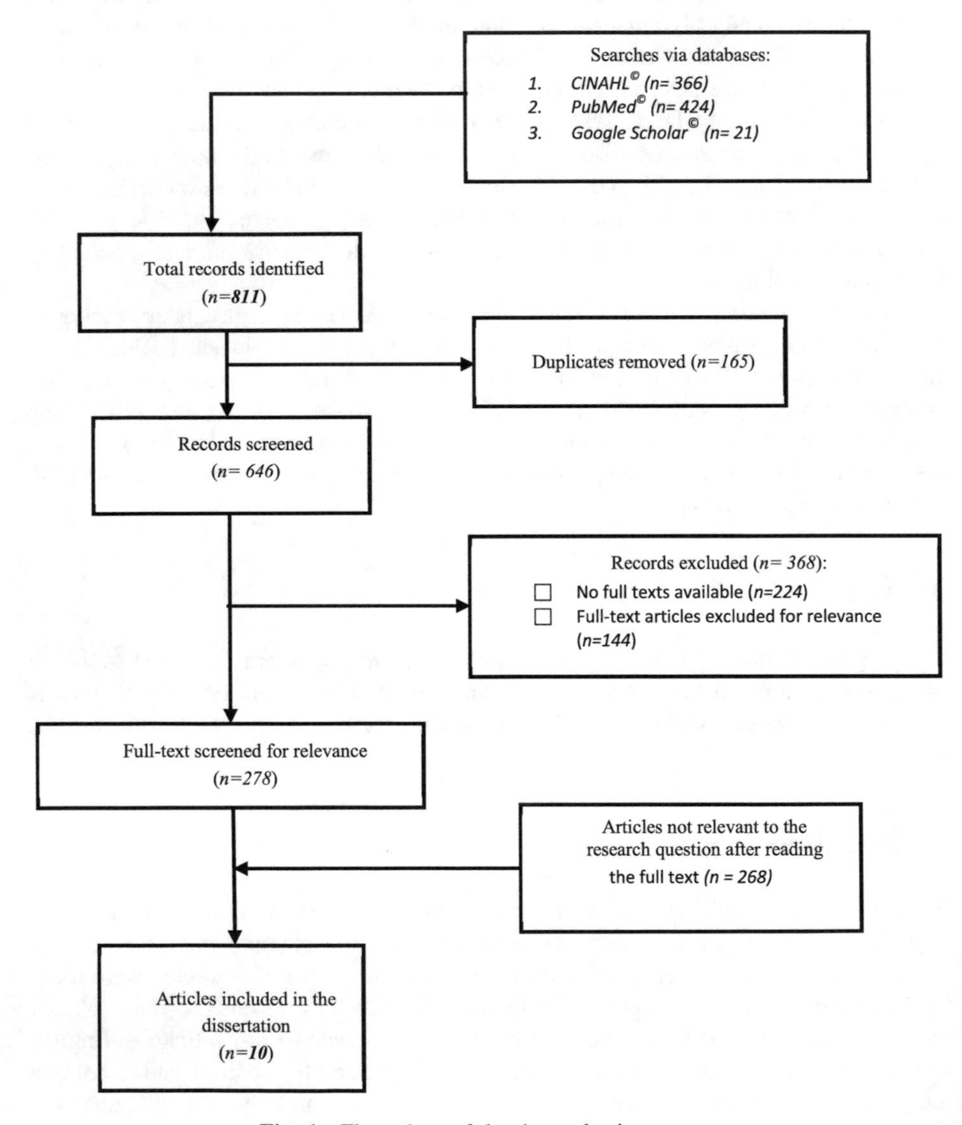

Fig. 1. Flow chart of the data selection process

5 Results and Discussion

The concept of health, when explained as a CAS, offers new perspectives, particularly to nurses and nursing staff. We agree with Bircher & Kuruvilla that the Meikirch Model could lead to significant improvements in the world of health [13]. This discussion seeks to briefly clarify the Meikirch Model, describing its parts and what possible consequences it can have on the individual and his/her health, as well as its possible interactions with nursing.

The Meikirch Model is based on five components—*demands of life* (LD*); biologically given potential* (BGP); *personally acquired potential* (PAP*); social determinants* of health (SD); and, finally, *environmental determinants* of health (ED)—and 10 complex interactions (See Fig. 2). This framework allows us to define health and diseases as a CAS. Therefore, according to the Meikirch Model, health can be defined as *"a state of dynamic well-being emerging from the interactions conducted between the potentials of an individual, the needs of life and the social and environmental determinants,"* [13] while the following can explain disease: The results of health in the course of life when the potentials of an individual and the social and environmental determinants are sufficient to respond satisfactorily to the needs of life, these may be physiological, psychosocial or environmental and they vary between individuals and contexts, but in any case, the unsatisfactory answers lead to the disease [13].

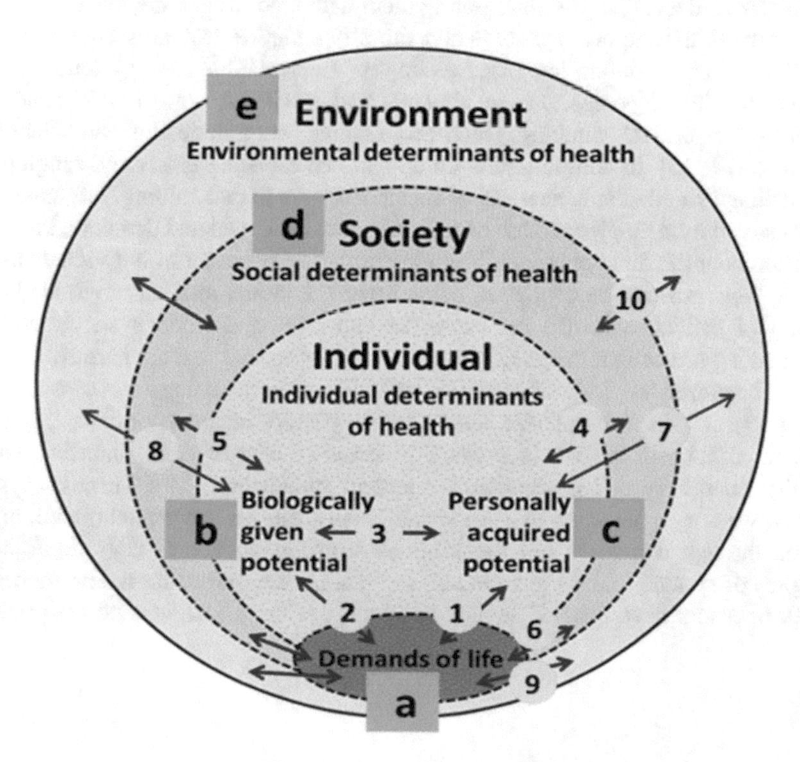

Fig. 2. The Meikirch model

Figure 2 also shows five components, while interactions are exposed as double-edged arrows from 1 to 10.

Every human being potentially seeks to satisfy his/her life's demands. In part, this is due to the biologically given potential, congenital in every person, which has been acquired to a certain extent during a person way of life, and the personally acquired potential. Equity and equality, social concerns, working conditions, autonomy and social participation interact strongly with the needs of life and the potential of the individual, constituting the social determinants of health. We can consider these are the main determinants of health [13]. While the living and working conditions of every human being constitute the environmental determinants of health, sometimes these can have global significance, such as natural resources, climate change and population growth [13]. In nursing, it is especially important to understand the health status of patients in order to improve their essential competence and clinical practice. In this, the Meikirch Model takes on particular importance, as it considers the patient a CAS, with all the properties of a system. All of this is significant because the relationships between CAS and other concepts provide the basis for new perspectives of nursing theory development. In fact, this discussion provides us with further elements to outline the CAS concept within nursing, developing different theoretical perspectives of the discipline itself, which can be used in different clinical and academic fields [3]. For the most part, nurses analyzing CASs in particular, have the opportunity to redesign clinical and academic practice, integrating them with theories and models used by other health professionals, in order to solve and simplify complex problems, such as the state of health. At an academic level, for example, in simulation labs, systemic thinking increases the initial perceptions of students and teachers, as well as expands their capacities for critical thinking, inter-professional communication and laboratory operations [14, 15]. In addition, new computational tools are emerging, ranging from computational simulation to analysis of social networks to data mining. For example, in computational modeling, a number of tools are interconnected and developed to answer questions about the functioning of CASs, including the behavior of individuals in those systems, which cannot be addressed using other traditional research methods [16]. In the care of patients who do not have the capacity to develop systemic thinking, understanding of health remains fragmented, creating factors that are harmful to patient safety. Therefore, the lack of systemic thinking creates a negative care treatment environment in which the patient is seen as the passive recipient of care; this creates additional deficits in the whole system, in terms of cost of care, financial burdens, mortality, morbidity and decline of the patient satisfaction [10]. Through systemic thinking, we can organize, model, guide and, finally, build a conceptual model, offering a vision through which we can focus on ideas and relationships [17]. Applying the principles of systemic thinking and CAS in nursing care and using a new model (See Fig. 3), opens up new ideas of thought and study, offering insights and new points of view.

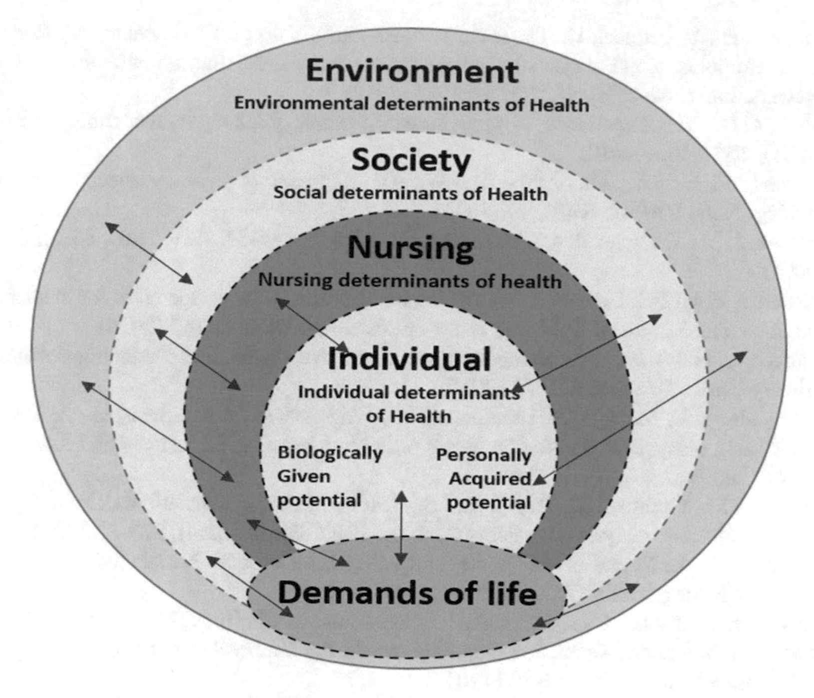

Fig. 3. The Meikirch model with Nursing

6 Conclusion

The Meikirch Model is a theoretical framework based on scientific evidence. Compared to other definitions of health, the model seeks to understand the concept of health in a rational way by offering innovative opportunities [13]. It further tries to combine both the concept of health and illness. The theoretical framework the model provides is in line with the theory and practice of evidence-based medicine and nursing. In addition, the care is centered on the person, offering the opportunity for each individual to self-motivate and, therefore, to improve their health [18]. Until today, traditional nursing has not dealt with health care in a rational manner, but instead has used an intuitive process, which has not provided new evidence to better comprehend the concept of health. In the care of patients who do not have the capacity to develop systemic thinking, the understanding of health remains fragmented, creating factors that are harmful to patient safety.

References

1. WHO. The world health report 2000: health systems: improving performance: World Health Organization (2000)
2. Simmons, S.J.: Health: a concept analysis. Int. J. Nurs. Stud. **26**(2), 155–161 (1989)

3. Notarnicola, I., Petrucci, C., De Jesus Barbosa, M.R., Giorgi, F., Stievano, A., Rocco, G., et al.: Complex adaptive systems and their relevance for nursing: an evolutionary concept analysis. Int. J. Nurs. Pract. **23**(3), e12522 (2017)
4. West, G.B.: The importance of quantitative systemic thinking in medicine. Lancet **379** (9825), 1551–1559 (2012)
5. Fawcett, J., Barrett, E.A., Wright, B.W.: Development of a new conceptual model of nursing. Nurs. Outlook **56**(2), 49 (2008)
6. Holden, L.M.: Complex adaptive systems: concept analysis. J. Adv. Nurs. **52**(6), 651–657 (2005)
7. Bircher J, Hahn EG. Understanding the nature of health: New perspectives for medicine and public health. Improved wellbeing at lower costs. F1000Res **5**, 167 (2016)
8. Bircher, J., Hahn, E.G.: Applying a complex adaptive system's understanding of health to primary care. F1000Res **5**, 1672 (2016)
9. Naaldenberg, J., Vaandrager, L., Koelen, M., Wagemakers, A.M., Saan, H., de Hoog, K.: Elaborating on systems thinking in health promotion practice. Global Health Promot. **16**(1), 39–47 (2009)
10. Stalter, A.M., Phillips, J.M., Ruggiero, J.S., Scardaville, D.L., Merriam, D., Dolansky, M.A., et al.: A concept analysis of systems thinking. Nurs. Forum. **52**(4), 323–330 (2017)
11. Dolansky, M.A., Moore, S.M.: Quality and Safety Education for Nurses (QSEN): the key is systems thinking. Online J. Issues Nurs. **18**(3), 1 (2013)
12. Powell, K.E., Kibbe, D.L., Ferencik, R., Soderquist, C., Phillips, M.A., Vall, E.A., et al.: Systems thinking and simulation modeling to inform childhood obesity policy and practice. Public Health Rep. **132**, 33S–38S (2017)
13. Bircher, J., Kuruvilla, S.: Defining health by addressing individual, social, and environmental determinants: new opportunities for health care and public health. J. Public Health Policy **35**(3), 363–386 (2014)
14. Gill, T.: Development and implementation of a simulated laboratory information system to enhance student critical thinking skills and laboratory operations. Clin. Lab. Sci. **30**(2), 88–89 (2017)
15. Petrucci, C., La Cerra, C., Caponnetto, V., Franconi, I., Gaxhja, E., Rubbi, I., et al. (eds.) Literature-based analysis of the potentials and the limitations of using simulation in nursing education. In: International Conference in Methodologies and Intelligent Systems for Technology Enhanced Learning. Springer (2017)
16. Clancy, T.R., Effken, J.A., Pesut, D.: Applications of complex systems theory in nursing education, research, and practice. Nurs Outlook. **56**(5), 248–256.e3 (2008)
17. Chaffee MW, McNeill MM. A model of nursing as a complex adaptive system. Nursing Outlook **55**(5), 232–241.e3 (2007)
18. Bircher, J., Hahn, E.G.: Will the Meikirch model, a new framework for health, induce a paradigm shift in healthcare? Cureus. **9**(3), e1081 (2017)

Use of High Fidelity Simulation: A Two-Year Training Project Experience for Third Year Students in Nursing Course Degree of Reggio Emilia

Daniela Mecugni[1,2](✉), Giulia Curia[1,3], Alessandra Pisciotta[2], Giovanna Amaducci[1], and Tutors of the Nursing Course Degree

[1] Nursing Degree Course, University of Modena and Reggio Emilia, Modena, Italy
daniela.mecugni@unimore.it
[2] Department of Surgery, Medicine, Dentistry and Morphological Sciences, University of Modena and Reggio Emilia, Modena, Italy
[3] Department of Biomedical, Metabolic and Neural Science, University of Modena and Reggio Emilia, Modena, Italy

Abstract. Nursing students at the end of their studies are supposed to own skills that allow them, in a short time, to act effectively and safely. Therefore, it is of primary importance that during the training period they have the opportunity, under protected conditions, to practice the management of scenarios realistically representative of the clinical setting.

High-fidelity simulation workshops, aimed to provide adequate skills in the management of vital criticality, represent innovative and exciting learning tools, thanks to teaching method and forefront technology involved. The laboratory activities included allow the students to completely descend in simulated scenarios either by the use of computerized interactive cases and by computerized manikins. In our study both these types of simulation laboratories were proposed to 3rd year Nursing students attending the academic years 2015/2016 and 2016/2017. The survey administered to both groups at the end of the workshop revealed a highly positive feedback towards this innovative teaching approach that allowed the students to understand the correlation between pseudorealistic simulations and theoretical notions learned during lectures.

The experience built through this two-year study allowed us to lay the bases for further studies on this topic.

Keywords: Simulation · High fidelity · Baccalaureate nursing education Vital criticality

1 Introduction

1.1 Background

The complexity of current clinical settings requires that Nursing students, at the end of their studies, acquire skills that make them able, in a short time, to act effectively and

T. Di Mascio et al. (Eds.): MIS4TEL 2018, AISC 804, pp. 293–301, 2019.
https://doi.org/10.1007/978-3-319-98872-6_35

safely [1]. As early as during the training, students are supposed to have the opportunity, in a protected context, to experience themselves with the management of scenarios adequately representative of clinical contexts. Today more than ever, limiting knowledge to just transmission of theoretical notions means limiting professional education, depriving it from the relational, negotiating, and emotional features typical of the relationship between nurses and patients. Simulation is a teaching strategy that allows students to experiment with this complexity, by simultaneously and effectively acting the different types of professional skills, such as technical-gestural, communicative-relational, organizational-management expertise and diagnostic thinking [2].

Indeed, through the reproduction of real, dynamic, complex and unpredictable situations, simulated scenarios require the students to completely descend into the specific situation and to show their skills in decision-making, problem solving, definition of priorities, and contextually in managing emotions and participating in team work [3]. Combining active teaching methods with traditional approaches makes it possible to apply the contents of the theoretical lessons to concrete problems to be solved. This allows the students, as future professionals, to make decisions in a protected setting where mistakes are a source of learning as well [4].

Two types of simulations are reported in the literature: Low and High Fidelity. Traditional simulation in form of static manikins, partial task-trainers and role-playing is defined as Low Fidelity because it does not reproduce authentic scenarios and allows the acquisition of skills and knowledge related only to specific areas, such as hygiene, elimination needs (urinary catheters), nutrition (nasogastric tube), mobilization, peripheral vein cannulation, oxygenation therapy, injection and wound care.

Electronic patient is the most complete type of simulation and can be either manikin-based or virtual reality-based, with sophisticated computer control that can provide various physiological parameter outputs replicating several clinical environments. High-fidelity simulation technique mimics essential aspects of a clinical situation in the three major dimensions of fidelity: physical, psychological and conceptual [3, 5]. In a high-fidelity simulation, the use of a software allows the student to manage ongoing care scenarios, modifying the behavior of the simulator and of the actors involved. The student can interact with such virtual or simulated reality and has the possibility to direct its evolution, based on the decisions made. These processes depend on the capability to simultaneously put into practice different skills of professional knowledge which consist in "know-how" (knowledge and application of specific theoretical content), "know how to do" (effectively acting technical-gestural skills), "knowing how to be" (ability to self-control emotions and work in a team), "knowing how to become" (being able to self-evaluate and learning from own mistakes) [6].

1.2 Context of the Study

In Nursing Degree Course, laboratory activities are planned with the intent to emphasize the experience and the use of active teaching methodologies [7]. During the laboratory activity, the tutor - a professional nurse with advanced skills in clinical and pedagogical settings - sets up the learning through the experience made. Indeed, according to Kolb, experience is to be intended as the moment in which the learner is drawn into the practice (experiential learning). Experience creates learning through a continuous and mutual exchange of inputs and feedbacks on perceptions and reactions to the experience itself, in order to redefine inadequate attitudes and enhance constructive behaviors.

In the 3rd Year of the degree course, the laboratory activities are aimed to allow the students to experience in the management of clinical, relational, and organizational complexities, in particular in vital criticality, providing a training path (workshop) that combines and integrates multiple phases in a systematic sequence: (1) a lecture phase (course in Nursing in Critical Area), with students being provided with specific disciplinary contents; (2) a laboratory activity where students individually interact in a virtual scenario on a computer platform, avoiding the emotional component; (3) an advanced laboratory activity of a critical event, with the use of a computerized manikin, requiring the students to manage the event, their own and others emotional experiences and the inter-professional integration.

This workshop aim to allow the students: (i) to apply the systematic method learned through the lecture phase, while evaluating the patient; (ii) to implement skills such as diagnostic thinking and decision making; (iii) to carry out technical-gestural interventions; (iv) to coordinate own actions with those of other professionals; (v) to remain lucid during the situation by controlling own emotions.

1.3 Aim of the Study

Our study describes (a) the laboratory experience with the use of high fidelity simulation proposed to the 3rd year students attending the Academic Years (AYs) 15/16 and 16/17, and (b) the results of the survey to measure their level of satisfaction about the laboratory experienced with high fidelity simulation.

2 Study Design

A descriptive observational study was conducted for 2 AYs (15/16 and 16/17). 156 students (83 in the AY 15/16 and 73 in the AY 16/17, respectively), experienced 2 laboratory activities with the use of high-fidelity simulation during the 2nd semester.

Inclusion criteria were students at the 3rd year of the Nursing Degree Course in Reggio Emilia. No students were excluded from the project; however, it could happen that a student was not present the day of the laboratory. In this case, in the result section, n indicates the students who were present the day of the laboratory and effectively experienced the high fidelity simulation activity.

At the beginning of each AY, all students were informed that their evaluation on the laboratory activities would have used for a research study that was approved by the President of the Degree Course. All students agreed to be part of the study.

The laboratory activity on computer platform was addressed to groups of 15 students each time. It involved the use of computerized interactive cases, which were presented in a repetitive way, for a total of 5 h. It was performed in a dedicated room equipped with 15 PC workstations. The software consisted in 8 modules, each of them containing 5 different possible scenarios for a total of 40 interactive cases. At the beginning of the workshop, the tutor explained the use of the software and showed a guided simulation through one of the potential scenarios. Then, each student was given the time to practice independently, for about 2 h, on an own chosen scenario. In the management of the scenario, the student was required to activate a clinical thinking in order to define the data to be collected, the actions to be done and how to perform them. Using the computer system, the real situation is rebuilt in three dimensions, with actors, devices and action timelines. The student is required to interact with this 3D reconstruction and modify it, based on the choices made. When the student has accomplished the management of the scenario, the software provides a detailed report showing the correctness and completeness of each decisions taken, with regard to data collected and carried out interventions. The student is not allowed to start a new scenario before a minimum of 72% of decision taken are correct in the formerly executed one. The first laboratory activity ends with a debriefing session, conducted by the tutor to stimulate critical thinking regarding the decisions taken, the difficulties encountered and the performance score. According to self-assessment and level of awareness about own areas of improvement, the students were given the opportunity to continue self-training with other scenarios.

The laboratory activity performed with the use of a computerized manikin was addressed to groups of 20 students each time, in a repetitive way, for a total of 5 h. The clinical situation of a patient with progressive respiratory failure was simulated. The scene took place in an in-hospital room, with actors playing roles as follows: the computerized manikin representing the patient, the tutor playing the medical doctor and two students acting the part of the nurses. The other students were observers of the simulation. Based on the parameters of the manikin – undergoing edits by a second tutor who acted as an observer during the simulation – the nursing students were required to collect data, attribute meaning to the data collected, and interact with each other or with the doctor, in a coordinated fashion, for the implementation of the interventions. At the end of the simulation, the tutor guided the students pondering on the experience just ended, through the following questions: how did you feel? what did you observe? what knowledge did you put into effect and what one should you have used? what difficulties did you encounter? if you could repeat the simulation, what would you have done differently?

The learning generated during the thinking moment consolidate the experience made, allowing students to identify the thoughts that have oriented their actions, the mistakes made, the corrective for future (live or simulated) experiences, and to prefigure the skills related to the professional role [8].

At the end of the workshop, a satisfaction survey was administered to the students, through 5 closed questions, using the Likert scale that ranges from ‘Not satisfied’ (Not at all/Little) to ‘Extremely satisfied’ (Very much/Completely), or the dichotomous template ‘Yes’, or ‘No’. The survey also contained an open question (“in case this workshop were proposed again, would you…”) which offered the students the opportunity to provide suggestions, personal notes, expectations, unfulfilled expectations.

3 Results

3.1 Laboratory Activity on Computer Platform

77/83 (92,8%) students in AY 15/16 and 72/72 (100%) students in AY 16/17 (Fig. 1) declared they were able to understand the correlation between the simulated experience and contents of the course of Nursing in Critical Area.

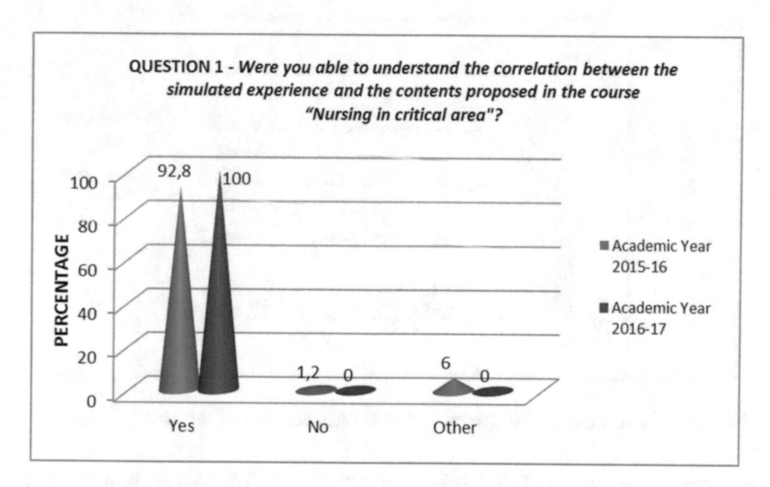

Fig. 1. Students' level of understanding about the correlation between the simulated experience and the contents proposed in the course "Nursing in Critical Area".

74/83 (88%) students in AY 15/16 and 65/72 (89%) students in AY 16/17 declared that the teaching method proposed favored "extremely" the learning (Fig. 2).

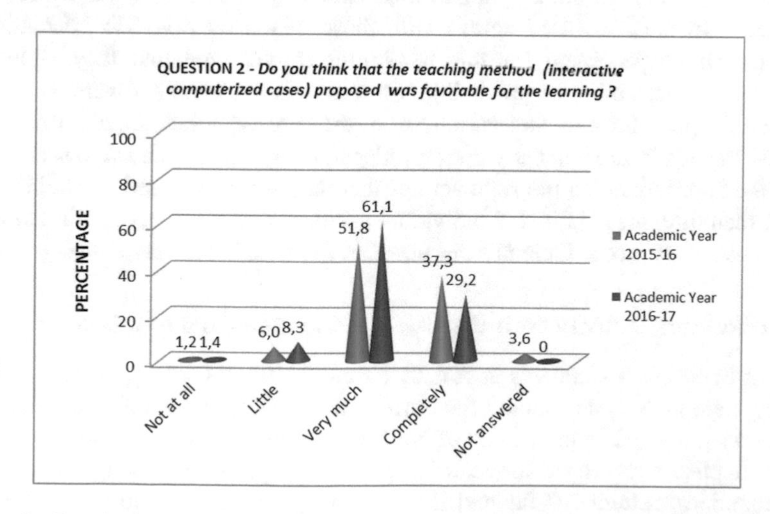

Fig. 2. Level of students' satisfaction with the use of interactive computerized cases

The contents of the proposed first laboratory activity were considered useful for the future professional role for 70/83 (84%) students in the AY 15/16 and for 68/72 (94%) students in the AY 16/17 (Fig. 3).

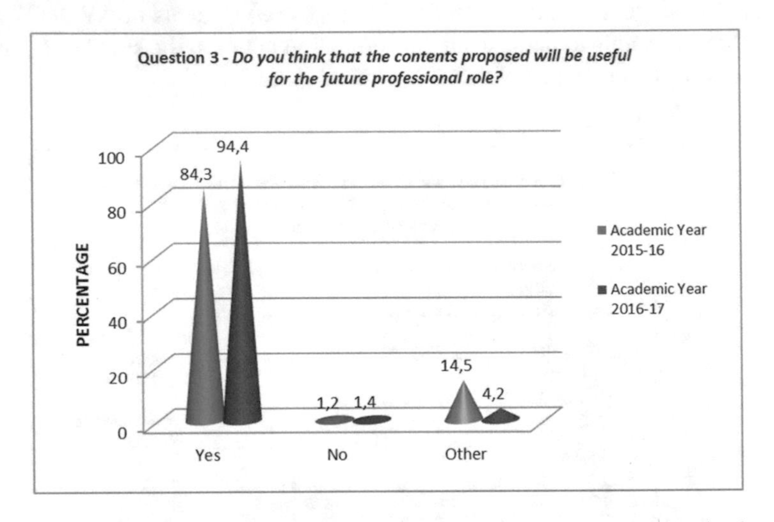

Fig. 3. Usefulness of the proposed contents for the future professional role

When openly asked to, the students of the AY 15/16 answered, in an almost univocal way, that a longer-time simulation and/or more sessions of the laboratory experience were needed as early as in the first semester. Moreover, they requested the possibility of practicing at home with the software, while one student highlighted the need to receive previously indications about specific contents required for each scenario. Another student reported that this teaching approach should not be limited to the vital criticality area, but other contents/areas should be included in the project.

Students in the AY 16/17 agreed with those of the previous AY, declaring that more time should be spent for this laboratory activity and that they would have appreciated the possibility of practicing at home with the same software. One student expressed the need for a greater variability of the cases. Another student reported that he would have had all students working independently but on the same scenario, and having the tutor indicating the right actions that they were supposed to put in act step-by-step; therefore, his idea is that individual exercise was poorly useful due to the fact that mistakes were not sufficiently discussed at the time they were made.

3.2 Laboratory Activity with the Use of a Computerized Manikin

A total of 81/83 (98%) students in AY 15/16 and 70/70 (100%) in AY 16/17 declared that they were able to understand the correlation between the simulated scenario and the contents proposed in the course of Nursing in Critical Area (Fig. 4). The teaching methods applied encouraged the learning process. In particular, the use of a computerized manikin "extremely" favored the learning process, according to 77/83 (92%) students in AY 15/16 and to 68/70 (96%) in AY 16/17 (Fig. 5).

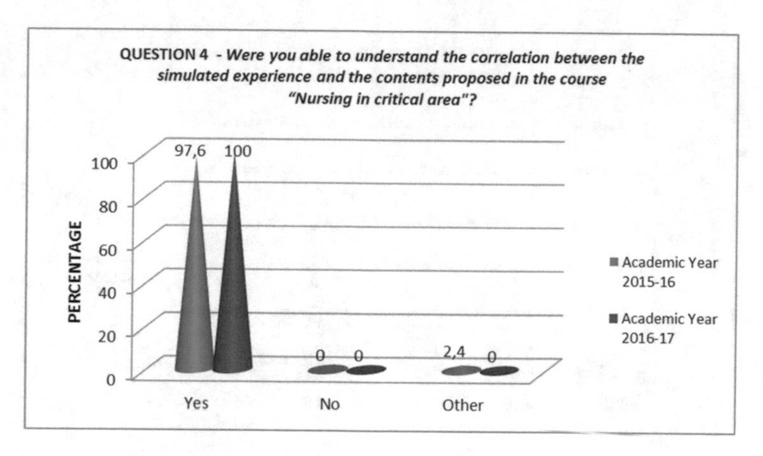

Fig. 4. Students' level of understanding about the correlation between the simulated experience and the contents proposed in the course "Nursing in Critical Area".

In addition, the use of role-playing "extremely" favored the learning process, according to 70/83 (84,3%) students in AY 15/16 and to 67/70 (96,7%) in AY 16/17 (data not shown). The contents of the proposed second laboratory activity were considered useful for the future professional figure by 80/83 (96,4%) students of the AY 15/16 and by 69/70 (98,6%) of the AY 16/17, respectively (data not shown).

When openly asked to, the students of the AY 15/16 answered, in an almost univocal way, that a longer-time simulation and/or more sessions of the laboratory experience would have been useful. Moreover, they expressed the need for more than one role-playing, thus allowing all the students to participate as actors, at least once. These observations were raised by the students of AY 16/17 as well. In addition, two students suggested to perform more laboratory experience and more simulations, including different areas of expertise and with progressively increasing difficulty level.

4 Discussion

Our results show how students recognize in simulations a teaching strategy that has favored (very/completely) their learning process. Students suggested extending this teaching methodic to learn skills as soon as in the first year of the nursing degree course. In the specific relation to interactive computerized cases, students wish to have the possibility to exercise on their own PC having the software and the different scenarios available. Several studies show indeed that both students' satisfaction and student's confidence in themselves increase if simulations are proposed repetitively and for several consecutive years [8].

The simulation is considered a learning strategy useful to prepare nurse students to their future professional role. This suggests that: (i) the competencies performed by the students during the 2 laboratory activities of high fidelity simulation are specific and fundamental for their professional identity, and (ii) the proposed scenarios are adequate representation of the reality allowing students to dip themselves in a realistic, although

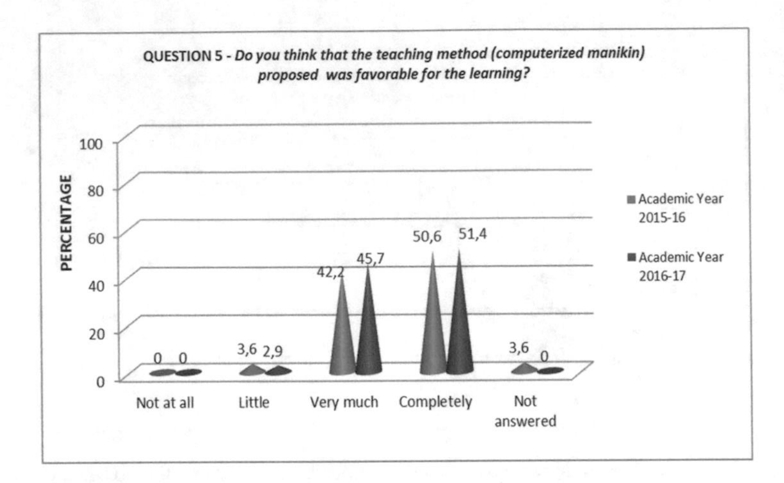

Fig. 5. Level of students' satisfaction with the use of computerized manikins

simulated, scenarios [8, 9]. Based on results obtained, we have proposed high fidelity laboratories to the 3rd year students of the current AY (17/18).

More robust study design will be necessary to analyze deeper and with more details the results of this, just descriptive, study. In particular, it would be interesting to better understand how simulation favors learning and whether experience learning correlates with better performance in clinical placement.

The limit of this study is related to the fact that a validated tool for this type of study does not exist, as already noted by Levett-Jones et al. [10]; therefore, data are collected using a non-validated implement.

5 Conclusions

The simulation workshops, aimed to the acquisition of skills in the management of vital criticality, represent an innovative training strategy thanks to the teaching methods and the high-level technology used. These simulation workshops intend to provide the future nursing professionals with appropriate skills, in order to cope to the complexity of care and organization-management of the clinical settings.

With the use of innovative and endearing teaching methods, the proposed laboratory activities allow the students to completely descend into the simulated scenarios (virtual reality in 3D and pseudo-reality under equipped settings with a computerized manikin). Moreover, such workshops allow the student "to act as if…", by feeling themselves stimulated to implement all the skills and behaviors that the situation requires, as well as to build in a safe setting their own learning through the mistakes.

References

1. Jeffries, P.R.: A framework for designing, implementing, and evaluating simulations used as teaching strategies in nursing. Nurse Educ. Perspect. **26**(2), 96–103 (2005)
2. Cioffi, J.: Clinical simulations: development and validation. Nurse Educ. Today **21**, 477–486 (2001)
3. Tosterud, R., Hedelin, B., Hall-Lord, M.L.: Nursing students' perceptions of high- and low-fidelity simulation used as learning methods. Nurse Educ. Pract. **13**(4), 262–270 (2013)
4. Sponton, A., Iadeluca, A.: La simulazione nell'infermieristica. Metodologie, tecniche e strategie per la didattica. 1st edn. Casa Editrice Ambrosiana, Milano (2014)
5. Petrucci, C., La Cerra, C., Caponnetto, V., Franconi, I., Gaxhja, E., Rubbi, I., Lancia, L.: Literature-based analysis of the potentials and the limitations of using simulation in nursing education. In: Vittorini, P., Gennari, R., Di Mascio, T., Rodríguez, S., De la Prieta, F., Ramos, C., Azambuja Silveira, R. (eds.) MIS4TEL 2017. AISC, vol. 617, pp. 57–64. Springer, Cham (2017). https://doi.org/10.1007/978-3-319-60819-8_7
6. Hyland, J.R., Hawkins, M.C.: High-fidelity human simulation in nursing education: a review of literature and guide for implementation. Teach. Learn. Nurs. **4**, 14–21 (2009)
7. Kolb, D.A.: Experiential Learning: Experience as the Source of Learning and Development, vol. 1. Prentice-Hall, Englewood Cliffs (1984)
8. Levett-Jones, T., Lapkin, S.: A systematic review of the effectiveness of simulation debriefing in health professional education. Nurse Educ. Today **34**(6), e58–e63 (2014)
9. Zapko, K.A., Ferranto, M.L.G., Blasiman, R., Shelestak, D.: Evaluating best educational practices, student satisfaction, and self-confidence in simulation: a descriptive study. Nurse Educ. Today **60**, 28–34 (2018)
10. Levett-Jones, T., McCoy, M., Lapkin, S., Noble, D., Hoffman, K., Dempsey, J., Arthur, C., Roche, J.: The development and psychometric testing of the satisfaction with simulation experience scale. Nurse Educ. Today **31**(7), 705–710 (2011)

Author Index

T. Di Mascio et al. (Eds.): MIS4TEL 2018, AISC 804, pp. 303–304, 2019.
https://doi.org/10.1007/978-3-319-98872-6

Printed in the United States
By Bookmasters